PreCalculus
Transformed

Extending Transformations to Explore PreCalculus

Chris Harrow and Nurfatimah Merchant

Venture Publishing
9 Bartlet Street, Suite 55
Andover, MA 01810

Preface

Pre-Calculus Transformed highlights the under-explored role of transformations in visualizing, interpreting, and understanding Pre-Calculus concepts. This text develops function composition as a broader description of transformations (algebraic application of one function to another transforms the original function) far beyond constant translations and dilations . Through this lens of transformations on numeric, algebraic, graphical and verbal representations of objects, students learn there are a myriad of ways to explore functions. Transformations allow entire families of functions to be understood through immutable distinguishing characteristics, and students discover that many problems which initially appear complicated can be more easily analyzed and deeply understood after the application of a transformation. Students learn to discern underlying patterns and to bring out links between otherwise seemingly disconnected ideas and function families. Finally, familiarity with the concept of transformations enhances the understanding of different forms of a function, and the ability to manipulate them to one's advantage: if you don't like the way a problem is posed, transformations allow you to change it. In short, *Pre-Calculus Transformed* provides an interconnected and dynamic Pre-Calculus experience with a wide range of problems from basic practice to those requiring creativity, insight and meaningful connections.

Pre-Calculus Transformed integrates computer algebra system (CAS) technology throughout, beginning with the very first sections, but it is certainly possible to grasp its concepts without a CAS. Routine CAS use enables students to recognize even larger patterns and make broader predictions about function behavior. A CAS also enables students to maintain their focus on critical characteristics of a problem without losing themselves in algebraic manipulations when those manipulations are not the point.

Chris Harrow   Nurfatimah Merchant

The Westminster Schools
Atlanta, GA

To Katie, Susanna, Rebecca, and Michelle - CH
To mom, dad, Irfan, Faiz, and Iman - NM

We're finally coming home.

Acknowledgements

We are thankful to:

- Jere Wells and The Westminster Schools for curriculum support and the innumerable professional development opportunities without which none of this would have been possible.

- Jerry Carnes, Landy Godbold, Henry Pollack, & Diamond Rahemtulla for inspiration, mentorship, and for those first few peeks outside the box.

- The hundreds of our students over the last four years whose boundaries we tried to push while they taught us that we hadn't yet pushed them far enough.

- Ray Klein, David Matloff, & Paul Foerster for deep reviews of materials, major revision suggestions, and for helping us strengthen many parts of the text.

- George Best, Lina Ellis, & Ilene Hamilton for great feedback on the early drafts.

Table of Contents

UNIT 0: ALGEBRA & CAS FUNDAMENTALS

To read without reflecting is like eating without digesting

– Edmund Burke

The goals of this unit are to

- Review some algebra skills.

- Introduce mathematical characters and shorthand writing notation.

- Introduce a wide range of Computer Algebra Systems (CAS) functions and techniques.

- Introduce the method of partial fraction decomposition.

Enduring Understandings

- CAS are a tool to facilitate deeper exploration and understanding of mathematics.

- Despite the power and multiple capabilities of a CAS, the student remains solely responsible for understanding mathematical relationships and for mastering basic algebraic algorithms.

This section assumes you have already been exposed to the following topics. As needed, refresh your basic familiarity with any of these topics on your own.

- Linear, quadratic, cubic, roots, absolute value, reciprocal, exponential, logarithmic, sine, cosine, and tangent basic function families and their graphs.

- The Natural, Whole, Integer, Rational, Real, and Complex number sets.

- Basic computations $(+, -, *, \div)$ on the number i and complex numbers.

- The concept of a function, domain, range, and function notation.

- Basic information about polynomials including graphing, factorization, expansion, division (synthetic and long).

- Arithmetic and geometric sequences and series and sigma notation.

- Factorials and Combinatorics

Other Notes:

- There are a wide variety of handheld and computer-based CAS available. The demonstrations and syntax provided in this text are all for a handheld TI-Nspire CAS.

- This section introduces **_many_** CAS techniques and syntax. You are *not* expected to master all of this immediately. The purpose of this unit is to introduce you to what a CAS is capable of doing and to serve as reference material as you progress through the course. We expect you to frequently return to these first pages for CAS ideas and syntax. For now, enjoy the power and flexibility of this tool for learning mathematics.

0.1: Mathematics Grammar & A Toolbox of Functions

You don't understand anything until you learn it more than one way.

– Marvin Minsky

Being proficient at rote skills is not the same thing as being educated. And training that develops skills, important as they may be, is a different thing from schooling in the art and the science of thinking.

– HH Aga Khan, IV

Mathematics is a language, and translating between languages is often difficult. Understanding what is asked and translating that information into the language of mathematics is often the most challenging part of any problem. As with any endeavor, the more actively you engage the material, the more you learn. For this text to be a useful reference for your continued mathematics learning, you must engage, read, ponder, and question.

As an initial step toward realizing Minsky's implied goal, this text offers definitions, explanations, and solutions from multiple perspectives. Its basic framework is the **Rule of Four**; mathematical ideas should be understood algebraically, graphically, numerically, and verbally. Much like translating between languages, it is sometimes difficult to translate between these four representations. Nevertheless, skilled mathematicians have a refined ability to recognize aspects of old, familiar problems in the midst of the new and seemingly unfamiliar. Converting a problem between forms often reveals elegant solutions and deeper understandings. At the very least, it allows an opportunity to select more efficient solution strategies from the available pathways. You are encouraged to absorb and evaluate the strengths and weaknesses of the multiple approaches presented throughout this work. A very basic example to illustrate this involves solving the following question presented verbally.

Example 1: Find the zeros of $f(x) = x^2 + 2x - 8$.

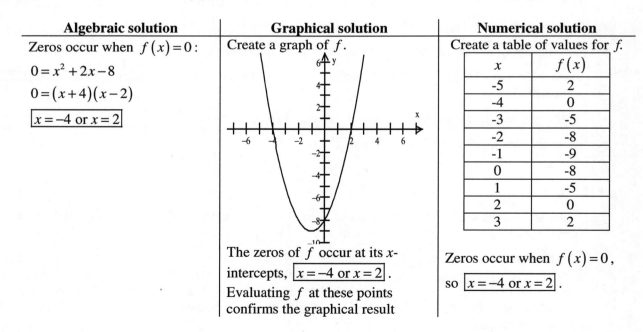

Algebraic solution	Graphical solution	Numerical solution
Zeros occur when $f(x) = 0$: $$0 = x^2 + 2x - 8$$ $$0 = (x+4)(x-2)$$ $\boxed{x = -4 \text{ or } x = 2}$	Create a graph of f.	Create a table of values for f.

x	$f(x)$
-5	2
-4	0
-3	-5
-2	-8
-1	-9
0	-8
1	-5
2	0
3	2

The zeros of f occur at its x-intercepts, $\boxed{x = -4 \text{ or } x = 2}$.
Evaluating f at these points confirms the graphical result

Zeros occur when $f(x) = 0$, so $\boxed{x = -4 \text{ or } x = 2}$.

Hopefully you knew the form of the graph in Example 1 before reading the solution. This text assumes previous mathematics studies exposed you to several function families and their parent graphs. Figure 0.1a summarizes the basic functions this text assumes you already understand.

Graphs of parent functions of basic function families studied in previous courses

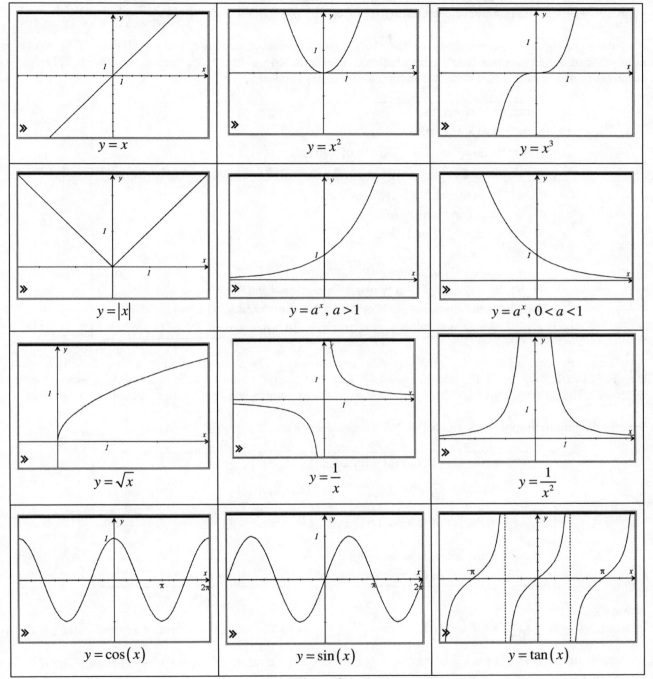

Figure 0.1a

Functions versus relations

All of the families represented in Figure 0.1a are functions. Mathematicians draw a critical distinction between relations and functions.

> A *relation* is any correspondence or pairing between objects.
>
> A relation that has only one output value for each input value is called a *function*.

Every family represented in Figure 0.1a is a function because there is precisely one output value of y for every input value of x in the domain of each family member.

As with any language, mathematics has its own grammar, sentence structure, and vocabulary. This text uses notation common in upper level mathematics. These include a few basic terms and several phrases so common in mathematics that there are accepted shorthand versions.

$\equiv$ means "defined as,"
$\subseteq$ $\equiv$ "is a subset of,"
$\cup$ $\equiv$ "union," as in the joining of two sets,
$\cap$ $\equiv$ "intersection," as in the overlap between two sets,
$\in$ $\equiv$ "is an element of" or "is in the set,"
$\notin$ $\equiv$ "is not an element of,"
$\ni$ $\equiv$ "so that" or "such that,"
$\forall$ $\equiv$ "for all,"
$\exists$ $\equiv$ "there is" or "there exists,"
iff $\equiv$ "if and only if" – a two-way conditional statement, and
QED is an abbreviation for Quod Erat Demonstrandum (Latin for "that
         which was to be demonstrated"). In other words: Proof Complete.

As a simple example of the use of "iff," consider the sentence: The number rolled on a six-sided die is odd ***iff*** the result is a 1, 3, or 5. Here, odd guarantees a 1, 3, or 5, **and** a result of 1, 3, or 5 would guarantee an odd outcome. The iff statement says that the existence of *either* conditional is sufficient to guarantee the other.

When referring to number sets, the following are commonly used.

$\mathbb{N}$ $\equiv$ the set of Natural numbers: $\{1,\ 2,\ 3,\ ...\}$.

$\mathbb{Z}$ $\equiv$ the set of Integers: $\{...,-3,-2,-1,0,1,2,3,\ ...\}$.

$\mathbb{Q}$ $\equiv$ the set of Rational numbers: $\left\{\dfrac{a}{b}\right\}$ $\forall a,b \in \mathbb{Z}$ $\ni b \neq 0$.

$\mathbb{R}$ $\equiv$ the set of Real numbers.
$\mathbb{C}$ $\equiv$ the set of Complex numbers.

Example 2:
Interpret the sentence "$\forall n \in \mathbb{Z} \geq 0$, let $x = 2n + 1$" in words and describe the values of x which satisfy it.

The sentence says x is the set of numbers that are one more than double every non-negative integer. This is one definition of the positive odd integers.

Problems for Section 0.1:

NOTES:

- Unless specifically mentioned in a problem or unit, assume all problems refer to $\mathbb{R}$ <u>only</u>.

- In the problem sets, a [NC] before any problem means "No Calculator" and indicates that it can and should be solved without using a calculator of any sort. Of course, it is always valid to check your work after solving.

Exercises

[NC] For each equation in questions 1-9, sketch a graph of the given function or relation without referring back to the reading for this section.

1. $y = x^2$

2. $y = x^5$

3. $y = |x|$

4. $y = \sqrt{x}$

5. $y = \sqrt[3]{x}$

6. $y = 2^x$

7. $y = \dfrac{1}{x}$

8. $y = \cos x$

9. $y = (\sin x)^0$

10. Consider the data set.

s	3	0	1	-1	6	2	7
t	-2	5	3	1	2	5	11

   A. Could t be a function of s? Explain. *NO, for a input there would be two*

   B. Could s be a function of t? Explain. *yes because for a given outputs input there is only one output* *switch*

11. What values of x are described by each statement?

   A. $x \in \mathbb{Z} \geq 0$

   B. $x \in \mathbb{Z} < 0$

   C. $x \in \mathbb{Q} \cup x \notin \mathbb{Z}$

   D. $\forall x \in \mathbb{R}, \sqrt{x} \in \mathbb{Z}$

   E. $x \in \mathbb{Q} > 0$, but $x \notin \mathbb{N}$

   F. $ax^2 + bx + c = 0$ iff $(b^2 - 4ac) \in \mathbb{R}$

Explorations

Identify each statement in questions 12-16 as true or false. Justify your conclusion.

12. $x \in \mathbb{R}$ iff $x \in \mathbb{N}$.

13. If $x = y^2$, then y is a function of x. *False*

14. If $x = y^3$, then y is a function of x.

15. {The set of all odd integers} $\cup$ {The set of all even integers} $= \mathbb{Z}$

16. $\exists$ two values of $x \ni x! = 1$.

For questions 17-22, state a possible equation for the graph of each function or relation. Assume all graphs are shown with identical the *x*- and *y*-axis scales.

17.

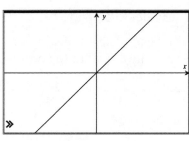

18.

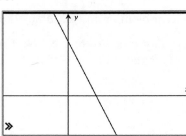

19.

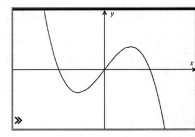

20.

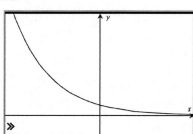

21.

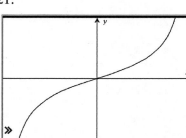

22.

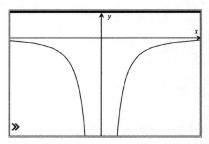

23. Write the digits 1-9, in order, on your paper. Without rearranging the numbers, insert exactly two subtraction signs and one addition sign to create an expression with four integers.

    A. What is the smallest possible value of this expression?

    B. What is the largest possible value of the expression?

    C. Where could you insert the operators so that the expression was equal to exactly 100?

24. [NC] If $14 < y^2 < 108$ and $y \in \mathbb{Z}$, then what is the result of subtracting the smallest possible value of y from its largest possible value?

25. Given $a^2 b < 0$ and $b^2 c < 0$. Which is larger, zero or the product of b and c? Explain.

26. A family of functions is defined by the equation $Ax + By + C = 0$ where $\{A, B, C\}$ forms an arithmetic sequence. What is special about this family of lines? Prove your claim.

27. Let f be a function $\forall x \notin \{0,1\}$ such that $f(x) + f\left(\dfrac{1}{1-x}\right) = x$. What is the value of $f(2)$?[1]

28. There are 10 books, all with different numbers of pages. The shortest book has 80 pages and the longest has 500 pages. If y is the average number of pages for all 10 books, determine the narrowest possible interval with integer endpoints containing all possible values of y.

---

[1] This problem was revised from the 21st University of South Carolina High School Mathematics Contest at
http://www.math.sc.edu/contest/2006-2007/exam/PDF/exam2006-2007.pdf .

0.2: CAS I: Basic Skills

The way to get good ideas is to get lots of ideas, and throw the bad ones away. – Linus Pauling

The examples in the next three sections are designed to introduce many computer algebra system (CAS) features and functions. Other functions will be introduced later, but these sections demonstrate almost all of the features you may want or need. While there are several CAS available, all of the examples in this text feature the TI-Nspire CAS[2] with keystroke directions for most commands provided in the footnotes.

Basic algebraic manipulations:
The primary difference between a CAS and other calculators is that a CAS can accept, manipulate, and output algebraic expressions. This difference can quickly be seen from a **Calculator Screen**[3] by entering $x+x+x+x+x+x$ and $x \cdot x \cdot x \cdot x \cdot x \cdot x$, the sum and product, respectively, of six xs (Figure 0.2a). This type of algebraic output is not possible on non-CAS machines without resorting to programming.

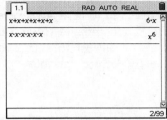

Figure 0.2a

A CAS also converts an exponent of $\frac{1}{2}$ into a square root and negative exponents into equivalent reciprocals (Figure 02.b).

While the Distributive Property of Multiplication over Addition states $4\left(x^2 + 7\sqrt{x} - 5\right) = 4x^2 + 28\sqrt{x} - 20$, the CAS did not automatically distribute. A CAS can be forced to distribute by using the **expand** command (Figure 0.2c).[4] [NOTE: While almost every command can be found in the **Catalog**, on the Nspire, it is also possible to use a function by typing its name directly using the alpha keys.]

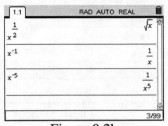

Figure 0.2b

The expand command also handles simple expressions like $(x-3)(7x-1)$ as well as more cumbersome expressions like $\left(2x^2 + 8x - 1\right)\left(5x^2 + 10x - 13\right)$ and $(x+y)^4$ (Figure 0.2d).

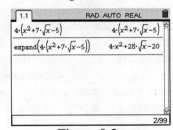

Figure 0.2c

The coefficients of the expansion of $(x+y)^4$ are from the fourth row of Pascal's Triangle. A CAS could compute those coefficients using combinations (Figure 0.2e).[5]

Factoring
The factored form of $x^2 + 6x - 16$ is $(x+8)(x-2)$. A nasty expression to factor by hand is $1692x^2 + 743x - 300$, but both expressions fall to the **Factor** command (Figure 0.2f).[6] Notice, however, that $x^2 - 7$ did not factor. The reason is that the Nspire's default factoring guidelines are to factor over $\mathbb{Z}$ only. To force the CAS to factor over $\mathbb{R}$, add a ",x" (or whatever variable is

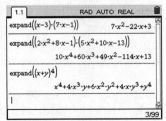

Figure 0.2d

---

[2] In most cases, the syntax for a TI-Nspire and a TI-89 is identical.
[3] A new calculation screen is created by pressing (⌂) → ①:**Calculator**.
[4] The **Expand** command can be found under (≡) or at at (menu) → ③:**Algebra** → ③:**Expand**.
[5] **nCr** is under (menu) → ⑤:**Probability** → ③:**Combinations** or by typing the command directly with the alpha keys.
[6] **Factor** is under (menu) → ③:**Algebra** → ②:**Factor**.

appropriate) to the end of the command inside the parentheses. The command cFactor forces factoring with complex numbers with integer coefficients while cFactor with the "*,x*" allows the broadest possible result: complex numbers with irrational coefficients (Figure 0.2g).[7]

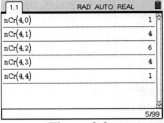

Figure 0.2e

Defining Functions

A CAS can also define and use functions. The top row of Figure 0.2h defines $f(x) = x^2 - x + 2$ and then evaluates using function notation. Notice that the CAS easily determines both numerical and algebraic answers.

NOTE: Once a function is defined, the variable name assigned to the function is reserved. There are several ways to change this, but the simplest is to use the **DelVar** command (Figure 0.2i).[8] If there are many items to be changed, it may be easiest to start over with a new document[9], effectively erasing all locally defined variables.

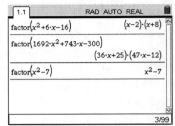

Figure 0.2f

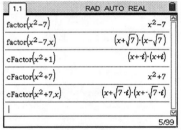

Figure 0.2g

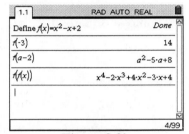

Figure 0.2h

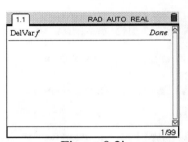

Figure 0.2i

Solving

Arguably the second most useful CAS command is **solve**.[10] The syntax for the basic solve command is **solve(equation,variable)** where **equation** is the mathematical sentence being solved and **variable** is the unknown quantity whose value or expression is desired. Figure 0.2j shows that the **solve** command can compute both numerical quantities (top line) and variable solutions (the two parts of the quadratic formula in the bottom line).

If any equation to be solved can be manipulated so that one side is zero, the **zeros** command[11] can be used as an alternative approach to solve. Figure 0.2k repeats the solutions of Figure 0.2j using zeros. Notice that the solutions to the zeros command are presented as lists (indicated by braces) while solve returns answers as equations separated by logical operators.

Since both commands produce the same algebraic answers, it may seem irrelevant which command is used. Depending on how an answer is to be used, though, sometimes one form is actually more utilitarian.

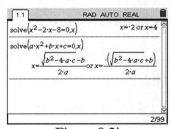

Figure 0.2j

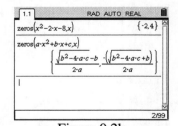

Figure 0.2k

---

[7] **cFactor** is under (menu) → (3):**Algebra** → (A):**Complex** → (2):**Factor**.

[8] **DelVar** is under (menu) → (1):**Actions** → (3):**Delete Variable**. Figure 0.2h shows the deletion of function *f* only, but multiple variables or functions could be deleted with a single DelVar command by separating the items by commas.

[9] A new document is opened at (⌂) → (6):**New Document.**

[10] **Solve** is under (menu) → (3):**Algebra** → (1):**Solve**.

[11] **Zeros** is under (menu) → (3):**Algebra** → (4):**Zeros**.

COMMON CAS SYNTAX ERRORS:
1. Because functions and variables can be named with multiple characters, the CAS does not assume any implied multiplication between alpha characters. For example, a, x, and ax are three different variable names on a CAS. If you want to compute $a \cdot x$, you must insert the multiplication symbol.
2. Also, users should be careful when multiplying a variable by a parenthetical expression. For example, while an acceptable factorization of $x^2 - 5x$ is $x(x-5)$, entering this latter expression in a CAS without a multiplication symbol between the x and the parentheses will force the CAS to think of the expression as some function x evaluated at the expression $x-5$.

Example 1:
Find all axis intercepts for the graph of
$y = x^4 + 2x^3 - 26x^2 - 22x + 165$.

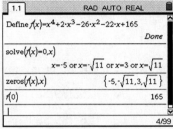
Figure 0.2l

Because the x-intercepts of a function's graph are equivalent to the zeros of the function, both the solve and the zeros command would work for this example. The y-intercept is found by evaluating the function at $x = 0$. Figure 0.2l shows the CAS results that give the x-intercepts at $x \in \{-5, 3, \pm\sqrt{11}\}$ and the y-intercept at $y = 165$. (Obviously, the y-intercept did not require the use of a calculator, but the point here is to demonstrate the power of a CAS's function notation.)

Many functions like the quartic in Example 1 have one input only, but a function can be more broadly understood as an operation that uses its input variable(s) to determine unique output values. The distance formula is a multivariable function. Figure 0.2m defines a function giving the generic distance between the points (a,b) and (c,d). This function can now be called upon with the coordinates of the desired points.

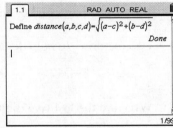
Figure 0.2m

Example 2:
A **locus** of points, (x, y), is the set of ordered pairs that satisfy a given relationship. Determine an equation for the locus of points equidistant from $(3,-4)$ and $(1,5)$.

Figure 0.2n uses the distance function to say that the distance from $(3,-4)$ to (x,y) is equivalent to the distance from $(1,5)$ to (x,y). The result is solved for y to produce the linear equation, $y = \frac{2}{9}x + \frac{1}{18}$.

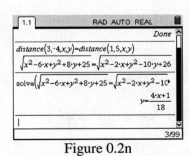
Figure 0.2n

Recall from geometry that the set of points equidistant from two points is a perpendicular bisector of the segment defined by the two points. That explains why the solution to Example 2 is linear.

Problems for Section 0.2:

NOTES:

- Several problems in this text require the use of a Computer Algebra System (CAS). Some early problems will specifically direct you to use these tools, but such directions become less frequent as the units develop. It is *your responsibility* to learn not only how to use such powerful tools for learning, but when it is appropriate to turn them on. **Use of a CAS is never an excuse to stop thinking.** In fact, you should discover that many CAS-required questions will require you to think more deeply about the problems while the computer handles the necessary algebra.

- Access to a CAS is assumed for the remaining problems in Unit 0.

Exercises

1. Apply a CAS factor command to each number. Which are prime?

   A. 126,289,587

   B. 126,289,588

   C. 126,289,589

   D. $\dfrac{29}{2769}$

2. Compute $111,111,111 * 111,111,111$. What is interesting about the result? Explain why it happens.

3. Factor $x - 5\sqrt{x} + 4 = 0$ on a CAS. Algebraically confirm the CAS results.

4. Use factoring to help determine a solution to each equation. (Do not use a solve command!)

   A. $2^{2x} - 5 \cdot 2^x + 4 = 0$

   B. $2^{2x} + 5 \cdot 2^x + 4 = 0$

5. What are the last five digits of 197^{197} ?

6. What is the 10th digit in the expansion of $100!$?

7. Compute $f(a-2)$ if $f(x) = 7x^4 + x - 5$. $7a^4 - 56a^3 + 168a^2 - 223a + 105$

8. Given $a \cdot x^2 + b \cdot x + c = 0$, solve for

   A. c

   B. a

   C. x

9. Simplify $(1 + 3i)^{13}$.

10. What is the largest coefficient in the expansion of $(2x + 3)^9$?

11. What is the *y*-coordinate of point $A = (4, y)$ if A is 5 units away from $(3, -1)$?

Harrow & Merchant © 2010

Explorations

Identify each statement in questions 12-16 as true or false. Justify your conclusion.

12. $\forall A \in \mathbb{R}$, $x^2 = A$ has two x-solutions.

13. The expansion of $\left(x^2 - A\right)^{12}$ has 12 terms.

14. zeros($f(x) - x^2 + 1, x$) and solve($x^2 - 1 = f(x), x$) produce the same CAS results for all functions, f.

15. $\forall a > 0$, let $D(a)$ be the distance between two points with coordinates $(a, 2a + 1)$ and $\left(a^2 + a, a + 1\right)$.
     For any values of b and c, if $b < c$, then $D(b) < D(c)$.

16. $y = 32x^9 + 16x^7 - 5x^4 + 2$ has no real-valued roots.

17. How many factors does 7,625,597,484,987 have? How many are prime?

18. The list of prime numbers begins 2, 3, 5, 7, ... What is the 25th number in this list?

19. Enter the equation $\sqrt{6} + \sqrt{2} = \sqrt{2 + \sqrt{3}}$ in your CAS. Prove the CAS answer algebraically.

20. Determine the least common multiple and greatest common divisor of 13542 and 9867.

21. What is the largest three-digit prime number whose last two digits are 99?

22. A, B, C, and D are prime numbers and their product is less than 8743. What is the maximum possible value of C?

23. How many zeros are at the end of the expansion of 200! ? (It is possible to answer this without actually counting the zeros.)

24. For $A, B, n \in \mathbb{Z}$, one of the terms in the expansion of $(Ax + By)^n$ was $27869184x^5y^3$. What are the values of A, B, and n?

25. How many digits long is the number 300! ?

26. What are the coordinates of the points on $y = x^2 - 1$ that are equidistant from both coordinate axes?

27. Determine an equation that does not contain any square root symbols for the set of all points one unit away from the origin.

28. Define a multivariable slope function on your CAS and use it to compute the slope of the line connecting the points $(3, -4)$ and $(1, 5)$. How does this answer partially confirm the equation found in Example 2?

0.3: CAS II: Graphing Skills

Mathematics is not a careful march down a well-cleared highway, but a journey into a strange wilderness, where the explorers often get lost. Rigour should be a signal to the historian that the maps have been made, and the real explorers have gone elsewhere.

— "Mathematics and History", *Mathematical Intelligencer*, v. 4, no. 4.

Example 1:

Graph the points $(3,-4)$ and $(1,5)$ and the locus of points equidistant from those points.

Example 2, Section 0.2 found the equation of the locus : $y = \frac{2}{9}x + \frac{1}{18}$.

To graph it, get a **Graphs & Geometry** (G&G) screen,[12] enter the equation in the command line at the bottom of the screen, and press **enter** (Figure 0.3a). The command line can be toggled on and off to alleviate clutter on the screen, if needed.[13]

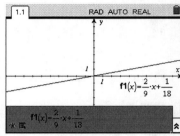
Figure 0.3a

Data for a scatter plot can be entered through a **Lists & Spreadsheet** (L&S) screen.[14] Figure 0.3b names the variables in the first row of an L&S screen with their corresponding coordinates listed below. The cursor wheel and the **tab** button are used to navigate around the window.

Then return to the graph window[15] and select a **Scatter Plot** screen.[16] The command line re-appears with empty spaces for the definitions of the x- and y-coordinates of the scatter plot. After selecting the list names defined in Figure 0.3b for each coordinate,[17] the scatter plot appears in the same window as the original graph (Figure 0.3d).

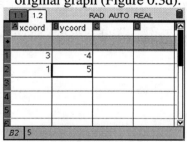
Figure 0.3b

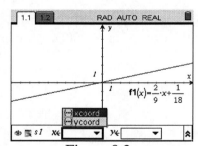
Figure 0.3c

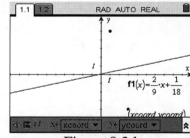
Figure 0.3d

---

[12] The G&G window is at (⌂) → (2):**Graphs & Geometry**.

[13] (ctrl) → (G) will toggle the graph command line on and off. To hide or show the equation label for a function's graph, press (menu) → (1):**Actions** → (3):**Hide/Show**. The cursor will appear in the graph window. Use the wheel to maneuver the cursor over the function label and click the mouse to gray out the label. When you press (esc), the label will be hidden. Repeat the process to show the label and to hide or show any other object in the graph window.

[14] The L&S window is at (⌂) → (3):**Lists&Spreadsheet**.

[15] (ctrl) → **left arrow on wheel** should do this. Alternatively, press (ctrl) → **up arrow**, navigate to the window you want, and press (click) or **enter**.

[16] Press (menu) → (3):**Graph Type** → (4):**Scatter Plot**.

[17] To select a list name, **click** on the field and use the **up** and **down arrows** to navigate among the variable names in the resulting drop-down menus. Press **enter** to select a variable and (tab) to move between the variable names in the menus. When both coordinates are defined, the scatter plot automatically appears.

Example 2:
Determine the coordinates of the points of intersection between $y = x^2 - 3$
and $y = -2x - 1$.

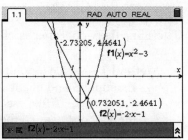

Figure 0.3e

Create a new G&G window or hide the graph and scatter plot from
Example 1.[18] Then make sure the graph type is set to Function and
graph the given equations. Adjust the viewing window as needed to
see the both intersection points and graph labels. The intersect tool[19]
automatically plots both intersection points and labels their coordinates.
The intersection   points are therefore approximately $(-2.732, 4.464)$
and $(0.732, -2.464)$ (Figure 0.3e).

Example 3:
Determine the axis intercepts of the graph of $y = x^2 - 3$.

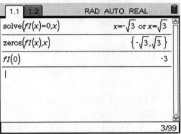
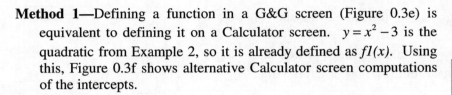
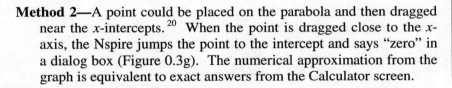

Figure 0.3f

Method 1—Defining a function in a G&G screen (Figure 0.3e) is
equivalent to defining it on a Calculator screen. $y = x^2 - 3$ is the
quadratic from Example 2, so it is already defined as *f1(x)*. Using
this, Figure 0.3f shows alternative Calculator screen computations
of the intercepts.

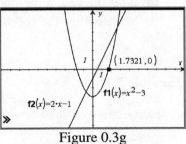

Figure 0.3g

Method 2—A point could be placed on the parabola and then dragged
near the *x*-intercepts.[20] When the point is dragged close to the *x*-
axis, the Nspire jumps the point to the intercept and says "zero" in
a dialog box (Figure 0.3g). The numerical approximation from the
graph is equivalent to exact answers from the Calculator screen.

NOTE: Dragging a point on a curve near a maximum or minimum point will also cause the Nspire to
jump the point to the extremum with the appearance of a corresponding dialog box.

Method 3—The intercepts could be found by computing the intersection(s) of each axis with the curve.
Just as two curves could be chosen for intersections (Example 2), either axis can be selected, and all
available intersection points will be displayed. You might need to turn coordinates on to see the
ordered pair.[21]

---

[18] There are three ways to do this.
- To hide an existing equation or scatter plot without deleting it, press (tab) until you are back on the graph
  command line. Press the **left arrow** until the eye at the far left of the command line is highlighted and press
  enter or (⌾) to toggle between showing and hiding the graph.
- Use the Hide/Show tool within the graph screen.
- Create a new problem within the document by pressing (ctrl) → (⌂) → (4):**Insert** → (1):**Problem**. In a new
  problem, all variable and graph definitions defined locally in the previous problem are erased.

[19] Intersection points between graphs are found by pressing (menu) → (6):**Points & Lines** → (3):**Intersection Point(s)**.
This puts you back on the graph screen. Maneuver the cursor to each graph and press (⌾). After the second function is
chosen, the points and their coordinates automatically appear.

[20] Press (menu) → (6):**Points & Lines** → (2):**Point On**. Then maneuver the mouse to the curve and press (⌾). The
coordinates automatically appear. Press (esc) to get out of Point On mode. Click and hold on the point to change the
open hand cursor to closed

[21] Te see the coordinates of a point that are hidden, press (menu) → (1):**Actions** → (7):**Coordinates and Equations** and
then click on the point whose coordinates are desired. The same tool gives equations of functions.

Example 4:

Determine the x-coordinates of the points on $y = x^2 - 3$ whose y-coordinate is 5.

> **Method 1**— Place a point on the parabola as was done in Method 2 of Example 3 and change the y-coordinate of the point to the desired value of 5. After escaping from Point On, move the mouse over the y-coordinate of the point until the word "text" appears on the screen (Figure 0.3h). Then double-click the mouse, delete the numerical values there, type a 5, and press **enter**. The point then jumps to the desired location, revealing an x-coordinate.

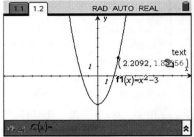

Figure 0.3h

> Here, there are two solutions. To get the other coordinates, recognize symmetry or drag the point somewhat close to the left-hand intersection and again change the y-coordinate to 5.

> **Method 2**— Use a solver to find when $f1(x) = 5$. This is equivalent to $f1(x) - 5 = 0$, allowing use of the zeros command (Figure 0.3i).

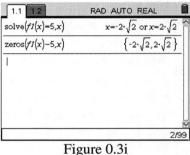

Figure 0.3i

> **Method 3**—Graph the line $y = 5$ and use the approach of Example 2 to determine the coordinates of the points of intersection.

Example 5:

Graph the curve defined by $\dfrac{x^2}{20} + \dfrac{(y-1)^2}{3} = 1$.

> One approach solves for y and graphs each resulting equation. This is reasonable, but cumbersome.

> A more sophisticated technique forces the zeros command as in Method 2 of Example 4. This works because the zeros command returns answers as lists which the Nspire can handle similar to its approach to scatter plots. NOTE: The key is using y as the parameter for the zeros command (Figure 0.3j).

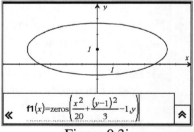

Figure 0.3j

Even vs. Odd Functions

When it exists, it is sometimes convenient to employ the symmetry of a function when solving problems. Two very common symmetries are those of even and odd functions.

> A function f is an **even function** iff $\forall x \in$ the domain of f, $f(-x) = f(x)$.
>
> A function g is an **odd function** iff $\forall x \in$ the domain of g, $g(-x) = -g(x)$.

Translating the algebra, an even function gives the same output for positive and negative values of the same input. Graphically, that is equivalent to a function whose graph is symmetric to the y-axis.

Similarly, an odd function has an opposite output when its input is opposite. This has two different sounding, but mathematically equivalent interpretations:
- The graph of an odd function is point-symmetric with respect to the origin, *and*
- The graph of an odd function has rotational symmetry about the origin.

Historically, these names were used to connect the symmetries of non-polynomials to those of monomials. Because monomials with even exponents (i.e. $\forall n \in \mathbb{N}$, $y = x^{2n}$) have y-axis symmetry, all other functions with similar symmetry are also called "even." Similarly, the graphical symmetry of monomials with odd exponents (i.e. $\forall n \in \mathbb{N}$, $y = x^{2n+1}$) is the basis for "odd" functions.

Example 6:
Identify each of the following functions as even, odd, or neither.
$$y_1 = x^2 + 2 \qquad y_2 = x^3 + x \qquad y_3 = \sqrt{x+1}$$

Conjectures: Figure 0.3k shows graphs of the three functions. The graph of y_1 appears symmetric to the y-axis and is likely even. The graph of y_2 looks symmetric about the origin and is likely odd. Since the graph of y_3 has neither symmetry, it is neither even nor odd.

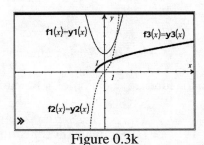

Figure 0.3k

The symmetry of these functions could be investigated on a CAS using Boolean algebra. After defining $y_1(x)$, $y_2(x)$, and $y_3(x)$ Figure 0.3l shows that y_1 is even, and y_2 is odd. That y_3 did not return a Boolean result for either test suggests that it is neither.

Proof: The CAS and the graphs suggest the answers, but proof of these properties is an algebraic endeavor. The simplest case is y_3. Because this function is defined for $x > 1$, but not for $x < -1$, it satisfies neither the even nor the odd function definitions. $y_1(-x) = (-x)^2 + 2 = y_1(x)$, so y_1 is an even function. The fact that $y_2(-x) = (-x)^3 + (-x) = -(x^3 + x) = -y_2(x)$ establishes y_2 as an odd function.   *QED*

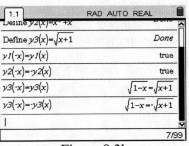

Figure 0.3l

Problems for Section 0.3:

Exercises

1. What are the coordinates of the point(s) where the graphs of $y = |x|$ and $y = x^3 - 2x$ intersect?

Identify each function in questions 2-5 as even, odd, or neither. Verify your claim algebraically and confirm graphically or with a CAS.

2. $y = |x|$

3. $y = x^4 - x^2 + 173$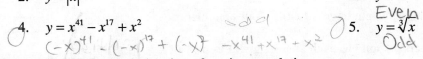

4. $y = x^{41} - x^{17} + x^2$

5. $y = \sqrt[3]{x}$

For questions 6-8, graph the given function or relation.

6. $x^2 + y^2 = 1$ 7. $x = y^2 + y$ 8. $x^2 - 8x - 4y^2 + 12 = 0$

9. If the graph of $y = 2x^2 - ax + 3$ has exactly one x-intercept, what is the value of a?

Explorations

Identify each statement in questions 10-14 as true or false. Justify your conclusion.

10. The minimum point on the graph of $y = x^4 - 3x^2 + 7x$ is at approximately $(-1.609, -12.327)$.

11. The graphs of $(x-3)^2 + y^2 = 9$ and $x^2 - 8x - 4y^2 + 12 = 0$ intersect exactly three times.

12. The sum of an even and an odd function is an odd function.

13. The product of any even and any odd function is an odd function.

14. The composition of two odd functions is an even function.

15. If the graphs of $f(x) = ax^4 - ax + 13$ and $g(x) = 5x^2 + 2x - 10b$ intersect at $(2, -7)$, what are the values of a and b? What are the coordinates of the other point of intersection?

16. Find an equation of a function that is both even and odd. How many functions are both even and odd? Why?

17. Does the graph of every **continuous**[22] odd function contain the origin? Explain.

18. Does the graph of every continuous even function contain the origin? Explain.

For questions 19-23, sketch a graph of the **conic section**[23] defined by the given equation.

19. $x^2 - 8x - 4y^2 + 12 = 0$

20. $x^2 - 8x + 2xy - 4y^2 + 12 = 0$

21. $x^2 - 8x + 5xy - 4y^2 + 12 = 0$

22. $x^2 - 8x + 4y^2 + 12 = 0$

23. $x^2 - 8x - 3xy + 4y^2 + 12 = 0$

24. Based on your graphs for questions 19-23, what are the apparent effects of adding an xy term to the equation of a conic section?

25. Given the system of equations $\forall\, h \in [0, \infty)$, $\begin{cases} x^2 - y^2 = 1 \\ (x-h)^2 + y^2 = 1 \end{cases}$.

    A. How many solutions could the system have? Justify.

    B. State the value(s) of h that correspond to each answer given in part A.

26. Given the system of equations $\forall\, h \in \mathbb{R}$ $\begin{cases} x^2 = y^2 \\ (x-h)^2 + y^2 = 1 \end{cases}$.

    A. How many solutions could the system have? Justify your conclusion.

    B. State the value(s) of h that correspond to each answer given in part A.

---

[22] Loosely defined, a continuous function is a function whose entire graph can be drawn without ever lifting your pencil.
[23] Conics are the family of curves whose members include parabolas, hyperbolas, circles, and ellipses.

0.4: CAS III: Advanced Skills

In matters of style, swim with the current; in matters of principle, stand like a rock.
                                                            – Thomas Jefferson

Advanced Solving
Example 1:

Solve the system defined by $\begin{cases} y = 2x - 1 \\ x + y = -7 \end{cases}$.

The CAS **solve** command also works on systems of equations.[24] Notice that while the syntax in Figure 0.4b asked for the value of x only, values for both coordinates were returned.

> **CAS Strategy:** When solving systems, a CAS sometimes returns different solution forms depending on which variables are solved for. If the output from a **solve** command is not ideal, sometimes a different form can be determined by solving for a different variable.

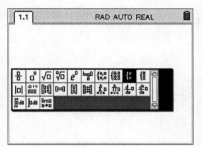

Figure 0.4a

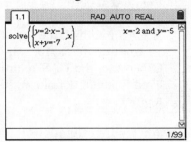

Figure 0.4b

The previous sections defined and used functions from both graph and calculator windows at individual points or algebraic expressions. Sometimes it is convenient to evaluate multiple values simultaneously. **Lists** and spreadsheets are the two primary ways to do this. Lists are defined on an Nspire using braces.[25]

Example 2:

Evaluate $f(x) = x^2 + x$ at $x = -1, 0, \sqrt{3}, a + 1, i$.

Values can be substituted using the | (**"such that"**) character/key. This command tells the CAS to evaluate the expression or command to the left of | using the information to the right of the vertical line. Figure 0.4c substitutes the given values directly into the expression for f and into a definition. A list was used to hold all input values and all output values were returned in list form. An advantage of this approach is expediency. Its disadvantage is that it is more difficult to connect corresponding input (independent) and output (dependent) values.

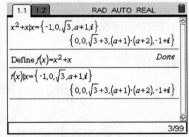

Figure 0.4c

Figure 0.4d resolves the connection issue in a Lists & Spreadsheets screen. Notice that both columns are named in the top row and that the output values are computed in function notation with f evaluated at the name of the dependent variable column. Another advantage of the tabular approach is that the columns are now dynamically linked. Changing any input values or adding new values to the list will get corresponding instantaneous updates in the output column.

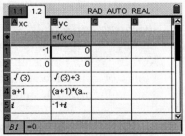

Figure 0.4d

---

[24] After starting the solve command, press (ctrl) → (palette) and choose the system palette highlighted in Figure 0.4a. Figure 0.4b shows the remainder of the syntax and the solution to given problem. To solve higher order systems, select the system palette to the right of the item highlighted in Figure 0.4a.
[25] (ctrl) → ({})

Example 3:
Determine an equation of the quadratic function whose graph contains the points $(-2, 27)$, $(1, 9)$, and $(3, 7)$.

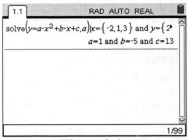

Figure 0.4e

A system of three equations similar to Example 1 could be used, but implementing the list tool from Example 2 allows another option. Figure 0.4e shows the solution of the generic quadratic equation, $y = ax^2 + bx + c$, with x-coordinate list $\{-2, 1, 3\}$ and corresponding y-coordinates, $\{27, 9, 7\}$. Note the "and" between the two list definitions after the "such that" command, |.

Example 4:
Use Boolean algebra[26] on a CAS to investigate whether the function $g(x) = \cos(x) + x^2$ is even, odd, or neither.

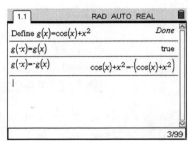

Figure 0.4f

Figure 0.4f defines g and uses the definitions of even and odd functions.

A CAS returns "true" or "false" when an entered equation or inequality is true or false for all possible values of the entered variables. Otherwise, the CAS returns an equation. The CAS said "true" for the even function definition, so g is an even function. Obviously, this is ***not*** a proof.

Parameter Investigations
Just as it convenient to be able to investigate the value of a function at several points simultaneously, there are many times when it is helpful to see the graphical effect of varying a parameter on a function. The list approach can do this for fixed values, but many CAS also have the ability to vary a parameter using a slider as shown in the following example.

Example 5:

Use a CAS to determine the effect of varying the parameter a in the relation $\left(\dfrac{x}{a}\right)^2 + y^2 = 1$.

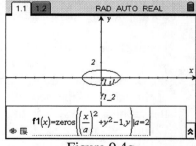

Figure 0.4g

Because this relation is not a function, an easy approach uses the zeros command. Figures 0.4g and 0.4h show the results of substituting values for a after the **such that** character to obtain different graphs. Negative values of a produce the same results as $|a|$. Varying a appears to horizontally stretch the ellipse.

NOTE: When not using the zeros command, you could enter multiple values of a parameter in a list after the **such that** command as was done in Figures 0.4c and 0.4e.

Alternatively, you could insert a **slider** on the Graphs & Geometry (G&G) screen[27] and use that to control the parameter (Figure 0.4i).

---

[26] Boolean algebra is the logical algebra of true and false statements. It is named for George Boole who developed its theory in his 1854 book, *An Investigation of the Laws of Thought*.
[27] From a G&G screen, press (menu) → ①:**Actions** → Ⓐ:**Insert Slider** and change the variable name. To change the settings on a slider, click on it once to select the slider, then press (ctrl) → (menu) → ①:**Settings…** (Figure 0.4j).

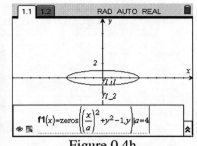
Figure 0.4h

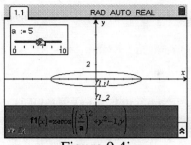
Figure 0.4i

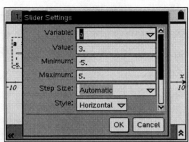
Figure 0.4j

Summations

A CAS can evaluate sums of series. For the series $1^2 + 2^2 + 3^2 + 4^2 + 5^2$, each term could be represented by the formula n^2 for $n \in \{1,2,3,4,5\}$. In sigma notation, this is equivalent to $\sum_{n=1}^{5} n^2$. Figure 0.4k shows the CAS computation of this sum.[28]

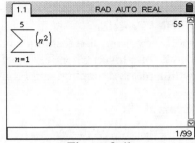
Figure 0.4k

Polynomial Division
Example 6:
Compute the quotient and remainder when $x^3 - 8x^2 + 12x + 7$ is divided by $x + 2$.

The equivalent long and synthetic versions of this division are shown in Figure 0.4l. Two CAS commands can produce this result, but they sometimes present the answer differently. On a TI-Nspire, both commands produce the same result for this example (Figure 0.4m).[29]

$$
\begin{array}{r}
x^2 - 10x + 32 \\
x+2 \overline{) x^3 - 8x^2 + 12x + 7} \\
\underline{x^3 + 2x^2} \\
-10x^2 + 12x + 7 \\
\underline{-10x^2 - 20x} \\
32x + 7 \\
\underline{32x + 64} \\
-57 = R
\end{array}
$$

$$
\begin{array}{r|rrrr}
-2 & 1 & -8 & 12 & 7 \\
   &   & -2 & 20 & -64 \\
\hline
   & 1 & -10 & 32 & \underline{-57}
\end{array}
$$

Figure 0.4l: Long and synthetic polynomial division

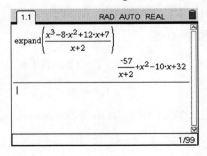

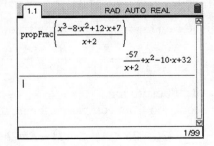

Figure 0.4m

---

[28] The summation template is in the operations palette at (ctrl) → (⬛x□) or at (menu) → ④:**Calculus** → ④:**Sum**

[29] The **expand** command is under (menu) → ③:**Algebra**, and **propFrac** is at (menu) → ③:**Algebra** → ⑦:**Fraction Tools** → ①:**Proper Fraction**. Fractions can be entered in Pretty Print form within each command by pressing (ctrl) (□÷□) after the command.

Example 7:

Use a CAS to compute the quotient and remainder when $x^3 - 8x^2 + 12x + 7$ is divided by $x^2 - 4$.

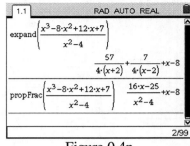

Figure 0.4n

With both the expand and propFrac commands, the quotient is clearly $x - 8$ (Figure 0.4n), but only the propFrac command clearly shows the remainder to be $16x - 25$.[30] The reason the two commands gave different answers is a mathematical procedure called **partial fractions**.

Partial Fractions

Partial fractions is a technique used to separate a more complicated fraction into a sum of simpler fractions. It can be employed whenever the denominator of a given fraction is factorable. Example 8 shows the technique with a rational number, and Example 9 explains the algebraic result of the Expand command in Figure 0.4n.

Example 8:

$\dfrac{11}{14}$ is the sum of two fractions with denominators less than 14. Find these fractions.

Because the only prime number factorization of the denominator is $14 = 7 \cdot 2$, the only possible denominators for the desired fractions are 7 and 2. So, $\dfrac{11}{14} = \dfrac{A}{7} + \dfrac{B}{2}$ for some values of A and B. Finding a common denominator allows a comparison of the numerators.

$$\frac{11}{14} = \frac{A}{7} + \frac{B}{2} = \frac{2A + 7B}{14} \quad \Rightarrow \quad 11 = 2A + 7B$$

Trial-and-error provides the unknown numerators: $A = 2$ and $B = 1$, so $\dfrac{11}{14} = \dfrac{2}{7} + \dfrac{1}{2}$.

Example 9:

$\dfrac{16x - 25}{x^2 - 4}$ can be written as the sum of two fractions with a linear denominators. Find them.

The partial fractions approach for variable fractions is identical to that for constant fractions. Factor the denominator to $(x + 2)(x - 2)$ to see that the only possible denominators for the unknown fractions are $(x + 2)$ and $(x - 2)$. Thus, $\dfrac{16x - 25}{x^2 - 4} = \dfrac{A}{x + 2} + \dfrac{B}{x - 2}$ for some values of A and B. A common denominator on the right again allows a comparison of the numerators.

$$\frac{16x - 25}{x^2 - 4} = \frac{A}{x + 2} + \frac{B}{x - 2} = \frac{A(x - 2) + B(x + 2)}{x^2 - 4} \quad \Rightarrow \quad 16x - 25 = A(x - 2) + B(x + 2)$$

Method 1: Substitution

The equation from comparing the numerators is true for all values of x, so pick convenient values.

$\underline{x = -2}$ $-32 - 25 = A(-2 - 2) + B(-2 + 2)$ $\Rightarrow$ $-57 = -4 \cdot A + 0 \cdot B$ $\Rightarrow$ $A = \dfrac{57}{4}$

$\underline{x = 2}$ $32 - 25 = A(2 - 2) + B(2 + 2)$ $\Rightarrow$ $7 = 0 \cdot A + 4 \cdot B$ $\Rightarrow$ $B = \dfrac{7}{4}$

---

[30] The remainder also could be found by recombining the terms from the expand command by using the **comDenom** command at (menu) → ③:**Algebra** → ⑦:**Fraction Tools** → ④:**Common Denominator** as in Figure 0.4n.

Method 2: Compare corresponding terms.

The coefficient of x is $16 = A + B$, and the constant term is $-25 = -2A + 2B$. This is just a system of equations that can be easily solved (Figure 0.4o) to give the unknown constants: $A = \dfrac{57}{4}$ and $B = \dfrac{7}{4}$. Both methods provide the same result. Therefore, $\dfrac{16x - 25}{x^2 - 4} = \dfrac{57/4}{x+2} + \dfrac{7/4}{x-2}$, that is equivalent to the CAS results from the Expand command in Example 7 (Figure 0.4n). There are algebraic techniques for more complicated forms in denominators (see problem 4, Section 0.5), but it is enough for this text to be able to split (decompose) distinct linear, constant factors.

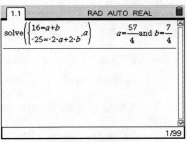

Figure 0.4o

Problems for Section 0.4:

Exercises

1. Solve the system $\begin{cases} 2x - 7y + z = 12\pi \\ x + 4y + z = 0 \\ x - y - 3z = -7\pi \end{cases}$.

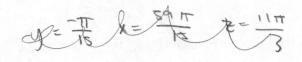

2. What happens when the command, **factor** $\left(x^2 - \{1,2,3,4\} \right)$, is used in a CAS? Be careful to use the precise syntax shown.

3. [NC] Divide each given rational expression. *After you have finished*, check your work on a CAS.

    A. $\dfrac{x^2 + 3x - 7}{x + 5}$ B. $\dfrac{x^4 - 8x - 7}{x - 1}$ C. $\dfrac{8x^3 - 5x^2 + 4x + 1}{2x^2 + x + 3}$

4. Decompose $\dfrac{3x^2 - 338x + 460}{(x+6)(7x-2)(10x+1)}$ using the method of partial fractions.

5. Compute each sum. Are there any surprises?

    A. $\displaystyle\sum_{n=1}^{20} \dfrac{1}{x}$ B. $\displaystyle\sum_{x=1}^{\infty} \dfrac{1}{2^x}$ C. $\displaystyle\sum_{A=0}^{\infty} \dfrac{1}{A!}$

    D. $\displaystyle\sum_{n=1}^{\infty} \dfrac{1}{n^2}$ E. $\displaystyle\sum_{n=0}^{\infty} \dfrac{x^n}{n!}$ F. $\displaystyle\sum_{n=0}^{\infty} \dfrac{1}{2^n \cdot n!}$

    G. $\dfrac{1}{2} + \dfrac{1}{2^2} + \dfrac{1}{2^3} + \ldots + \dfrac{1}{2^{30}}$ H. $\displaystyle\sum_{x=1}^{10} \dfrac{1}{(1+i)^x}$ where $i = \sqrt{-1}$

Explorations

Identify each statement in questions 6-10 as true or false. Justify your conclusion.

6. If $y = f(x)$ and $y = d(x)$ are random quadratic functions and $y = r(x)$ is linear, then there is a quartic function, g, which has quotient f and remainder r when g is divided by d.

7. There is a quadratic function which contains the points $(0,3)$, $(2,12)$, and $(-3,-9.5)$.

8. $\forall a \in \mathbb{R}$, varying the value of a in the graph of $x^2 - 8x - axy + 4y^2 + 12 = 0$ has the effect of rotating the conic section around the origin.

9. The CAS commands Propfrac and Expand always give the same *quotient* for polynomial division.

10. The CAS commands Propfrac and Expand always give the same *output* for polynomial division.

11. What are the coordinates of the point on the y-axis which is exactly 13 units from $y = 2x - 7$?

12. [NC] What is the exact value of the product $\left(1 - \dfrac{1}{2^2}\right)\left(1 - \dfrac{1}{3^2}\right)\left(1 - \dfrac{1}{4^2}\right)\cdots\left(1 - \dfrac{1}{1000^2}\right)$? [31]

13. Determine a CAS procedure to compute $\left(1 - \dfrac{1}{2^2}\right)\left(1 - \dfrac{1}{3^2}\right)\left(1 - \dfrac{1}{4^2}\right)\cdots\left(1 - \dfrac{1}{1000^2}\right)$.

14. Write $\dfrac{197}{4513} + \dfrac{4513}{197}$ as
    A. an improper fraction, and
    B. as a mixed number involving a proper fraction.

15. Decompose the following rational numbers using the method of partial fractions as shown in Example 3.
    A. $\dfrac{67}{70}$
    B. $\dfrac{35}{26}$

16. Decompose $\dfrac{3734}{1155}$ using the method of partial fractions as shown in Example 3.

17. Decompose the following rational expressions using the method of partial fractions.

    A. $\dfrac{-10x - 18}{x^2 - 9x}$
    B. $\dfrac{-11x^2 + x + 64}{x^3 + x^2 - 9x - 9}$

18. $\forall x \in \mathbb{Z} > 0$, let
    * $f(x) = 1$ if x is odd and $f(x) = 0$ if x is even, and
    * $g(x) = 1$ if x is even and $g(x) = 0$ if x is odd.

    Identify each statement as always, sometimes, or never true. Justify your conclusions.
    A. $f(ab) = f(a)f(b)$
    B. $g(ab) = g(a)g(b)$
    C. $f(g(x)) = f(x)$
    D. $g(f(x)) = g(x)$
    E. $f^{-1}(x)$ does not exist
    F. $f(x) \neq 1 - g(x)$

---

[31] Problem 16 (revised) from the 21st University of South Carolina High School Mathematics Contest at http://www.math.sc.edu/contest/2006-2007/exam/PDF/exam2006-2007.pdf

0.5: Further Explorations & Projects

Mathematics is not a spectator sport.

– Anonymous

1. **The Free Throw Problem**

   A basketball player has been keeping up with her career statistics. Before the beginning of her senior season, she was hitting less than 80% of her free throws. At the end of her senior season, she was hitting more than 80% of her free throws. **Must** there be a time during her senior season when she was hitting exactly 80% of her free throws? Prove your claim.

2. **Consecutive Integers Problem**

   The product of any four consecutive integers is always an output value for a polynomial of degree smaller than four. Determine that function and prove that it contains the product of *any* four consecutive integers.

3. **[Challenge]**[32] Evaluate $\displaystyle\sum_{n=1}^{2007} \frac{5^{2008}}{25^n + 5^{2008}}$.

4. **[Challenge]** Research or discover a technique for partial fraction decomposition when there are repeated linear factors in the denominator such as $\dfrac{12x^3 + 86x^2 + 219x + 183}{x^4 + 3x^3 - 19x^2 - 87x - 90}$.

---

[32] Problem S126, *Math Horizons*, April 2008 (Vol XV, Issue 4), page 32.

UNIT 1: POWERS, POLYNOMIALS, & DESCRIBING FUNCTIONS

Humans are allergic to change. They love to say, 'We've always done it this way.'
I try to fight that. That's why I have a clock on my wall that runs counter-clockwise.
– Grace Hopper, computer scientist and educator

Enduring Understandings

- The Power Function family connects many functions that appear different from each other.

- It is important to understand the behavior of polynomial functions and to be able to make connections among their different representations – algebraic, graphical, numeric, and verbal.

- Limit notation describes the behavior of a function as it approaches a point, independent of whether the function exists at that point.

- Piecewise functions describe function behavior that may not be able to be captured in a single function pattern throughout its domain.

This section assumes you understand the following concepts and procedures.

- Domain and range of a function

- Basic information about polynomials including details about their graphs, zeros, degree, etc.

- Basic constant horizontal and vertical translations and dilations of graphs

- Function composition and its notation ($f \circ g$)

- Interval notation

- Geometric relationship between a curve and a tangent line to that curve

- The relationships between the slopes of perpendicular and parallel lines

- Completing the square

- Special triangle values of sine and cosine

1.1: Power Functions

To change a culture requires more than new laws, it requires new insights.
      – Margaret Gallago, Sandra Hollingsworth, & David Whitnack

A new family

The common feature of the function family containing $y = x$, $y = x^5$, $y = \sqrt{x}$, $y = \sqrt[3]{x}$, and $y = \dfrac{1}{x}$ is that each can be written in the form, $y = ax^b$ where $a, b \in \mathbb{R}$ are constants. This is the **power functions** family. When exploring new mathematical ideas, it often helps to look at different cases. The constant a defines the vertical stretch of power functions, so a straightforward way to begin exploring the power functions is to let $a = 1$ (for now) and explore the possibilities for the values of b.

Case 1: $b \in \mathbb{Z} \ge 0$

This subset of power functions is referred to as the parent monomials. The graph for $b = 0$ is the horizontal line, $y = 1$. For $b \in \mathbb{N}$, graphs of $y = x^b$ take three basic forms; the simplest of each are $y = x$, $y = x^2$ and $y = x^3$ (Figures 1.1a-1.1c).

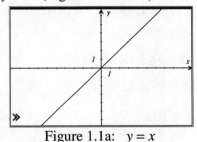
Figure 1.1a: $y = x$

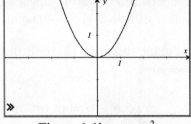
Figure 1.1b: $y = x^2$

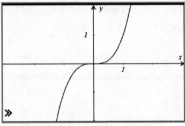
Figure 1.1c: $y = x^3$

Case 2: $b \in \mathbb{Z} < 0$

The most familiar examples of these are the reciprocal function, $y = x^{-1} = \dfrac{1}{x}$ (Figure 1.1d) and the reciprocal square function $y = x^{-2} = \dfrac{1}{x^2}$ (Figure 1.1e). Graphs of $y = x^b$ for negative, odd integer values of b look like Figure 1.1d, while negative, even values look like Figure 1.1e. Again, there is a convenient pattern, and the general shape depends on the whether b is even or odd. These patterns stem from the definition of even and odd functions discussed at the end of Section 0.3.

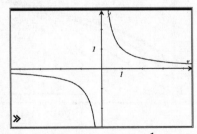
Figure 1.1d: $y = \dfrac{1}{x}$

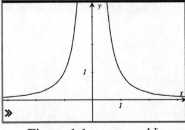
Figure 1.1e: $y = a \cdot b^x$

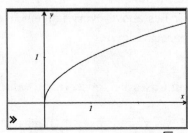
Figure 1.1f: $y = x^{1/2} = \sqrt{x}$

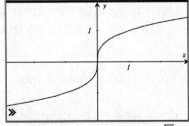
Figure 1.1g: $y = x^{1/3} = \sqrt[3]{x}$

Case 3: $b \in \mathbb{Q} \cup b \notin \mathbb{Z}$

For now, assume b is in its simplest form and its numerator is 1. An example for b with an even denominator is the square root function (Figure 1.1f). For values of b with odd denominators, the cube root function

(Figure 1.1g) is an example. As before, these power functions can be classified into odd and even examples based on the oddness or evenness of the denominator of b.

There is a big difference in the domains of the members of this subset. Recall that even roots have domain $x \in [0,\infty)$ while odd roots have domain $x \in \mathbb{R}$ (note the differences in the domains in Figures 1.1f-1.1g). This is because even roots of negative numbers do not produce real-valued solutions.

The **end behavior** of a function is a description of the graphical behavior of the function as the value of x approaches negative infinity ($x \to -\infty$)[33] and as it approaches infinity ($x \to \infty$). For power functions with $a = 1$ and $b > 0$, the right-side end behavior is "as $x \to \infty$, $y \to \infty$." The left side is more complicated depending on domain. For $y = \sqrt{x}$, the left side is "as $x \to 0$, $y \to 0$," and for $y = x^3$ (Figure 1.1c), the left side end behavior is "as $x \to -\infty$, $y \to -\infty$."

Generalizing Power Functions

These three cases dealt only with rational values of b where the numerator was one and ignored all irrational values of b. It is easiest to handle these last two cases, without proof, by generalizing the overall pattern.

First, all power functions are defined for $x \in \mathbb{R} > 0$, but a power function is defined for $x \in \mathbb{R} < 0$ iff its exponent can be written as a rational number with an odd denominator.

Notice several things about the combined graph (Figure 1.1h) of the results of Figures 1.1a-1.1g.

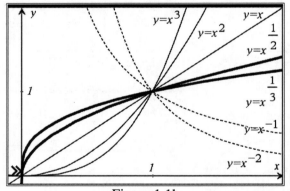

- Every graph contains the point $(1,1)$.

- Graphs for $b > 0$ contain $(0,0)$.

- For $x \in (1,\infty)$, as b decreases, so does the right hand side of the graph, and the reverse occurs for $x \in (0,1)$. Point $(1,1)$ can be seen as a fulcrum; pushing down on one side lifts the other side.

- Power functions for $b \in (0,1)$ are the only power functions with downward **concavity**.

Figure 1.1h

The term concavity describes a function's curvature. If every possible tangent line to a piece of a graph lies above the curve, that piece of the graph is called **concave down**. Likewise, part of a graph is called **concave up** when the graph lies above the tangent line to any part of the curve. For example, $y = x^2$ is always concave up because the parent quadratic lies above every tangent line to the curve. For $x > 0$, only power functions with $a > 0$ and $b \in (0,1)$ are concave down.

To graph a power function not covered in Cases 1-3, observe the value of b. For $y = \sqrt[3]{x^2}$, $b = \frac{2}{3} \in \left(\frac{1}{2},1\right)$, so its graph must be between the graphs of $y = x^1$ and $y = x^{1/2}$ (Figure 1.1i). For $y = x^{\pi/2}$, $b = \frac{\pi}{2} \approx 1.57 \in (1,2)$ and its graph is between those of $y = x^1$ and $y = x^2$ (Figure 1.1j).

---

[33] The notation $a \to b$ is read "a approaches b."

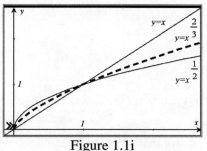

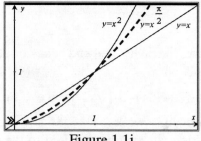

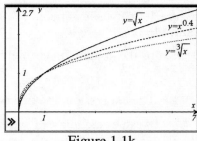

Figure 1.1i   Figure 1.1j   Figure 1.1k

Interpreting $y = x^{0.4}$

While the Quadrant I portion of $y = x^{0.4}$ lies between the graphs of $y = \sqrt{x}$ and $y = \sqrt[3]{x}$ (Figure 1.1k), what happens for $x < 0$? For domain reasons, $y = x^{4/10}$ does not exist for $x < 0$, but $y = x^{2/5}$ does. Some graphers produce Figure 1.1l for $y = x^{0.4}$, but others provide the right-hand branch only (Figure 1.1m). Ultimately, the answer comes down to the grapher's interpretation of the exponent, 0.4. Is it $\dfrac{2}{5}$, $\dfrac{4}{10}$, or some other equivalent value? On the surface, there seems to be no difference in these as they all reduce to the same decimal equivalent.

Assume $f(x) = x^{0.4} = x^{2/5}$. By exponent laws, $f(x) = \left(x^2\right)^{1/5} = \left(x^{1/5}\right)^2$. In both cases, the inner expressions (x^2 and $x^{1/5}$) have domain $x \in \mathbb{R}$ and therefore, both exist for $x < 0$. Further, both representations square at some point, so both functions are ultimately non-negative, creating a graph above the x-axis in Quadrant II for $x < 0$ (Figure 1.1l).

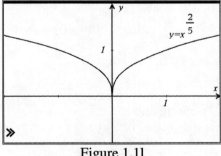

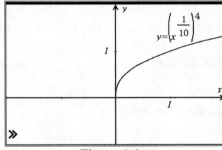

Figure 1.1l   Figure 1.1m

What if $f(x) = x^{0.4} = x^{4/10}$? If this is interpreted as $f_1(x) = \left(x^4\right)^{1/10}$, Figure 1.1l again appears. But, $f_2(x) = \left(x^{1/10}\right)^4$ gives only the right hand side (Figure 1.1m).[34] For the first time, an even root was the first function applied to the x-values, eliminating $x < 0$ from the domain. For $x > 0$, the various representations of f are identical, but the interpretation of the exponent determines its left side.

So what can be safely concluded about $y = x^{0.4}$? Graphers which default to Figure 1.1l most likely assume the simplest form of the decimal exponent, 2/5, under all interpretations. A more conservative approach recognizes that $b = 0.4$ has several interpretations, and its graph should include only the portion of the graph guaranteed under *all* such interpretations (Figure 1.1m). Different sources will expect different answers to this dilemma, but all have merit.

---

[34] If your grapher continues to provide Figure 1.1l, even when $y = x^{0.4}$ is entered as $y = \left(x^{1/10}\right)^4$, then its programming may be permitting complex numbers in intermediate computational steps. Whether this is permissible in real-valued functions is debatable.

Example 1:

Determine the domain, range, symmetry, concavity, and end behavior of $f(x) = x^{6/5}$. Graph $y = f(x)$.

- The expression $x^{6/5}$ can be viewed either as $\left(\sqrt[5]{x}\right)^6$ or $\sqrt[5]{x^6}$. Either way, it involves an odd root, so all x-values are allowed and its domain is $x \in \mathbb{R}$.

- At some point in the composition, all values are raised to the 6th power making them non-negative. Therefore, the range is $y \in [0, \infty)$.

- The function is even because $f(-x) = (-x)^{6/5} = x^{6/5} = f(x)$. It is therefore symmetric with respect to the y-axis.

- f is concave up for $x > 0$ because $b = \dfrac{6}{5} > 1$. Its symmetry makes f concave up over its entire domain, and also makes its end behavior $y \to \infty$ as $x \to \infty$ and $x \to -\infty$.

- A graph of f is provided in Figure 1.1n.

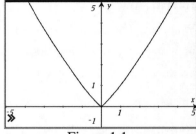

Figure 1.1n

Notice also, how closely the graph resembles that of $y = |x|$. This is because $b = \dfrac{6}{5}$ is very close to 1 and because the 6th power makes every y-value non-negative.

Example 2:

Find an equation for the power function that contains the points $(3, 5)$ and $(12, 10)$.

Power functions are of the form $y = ax^b$ for constants $a, b \in \mathbb{R}$. Plugging in the given points gives a system of equations, $\begin{cases} 5 = a \cdot 3^b \\ 10 = a \cdot 12^b \end{cases}$, which can be algebraically solved in two ways.

Method 1: Substitution:

$$5 = a \cdot 3^b \implies a = \frac{5}{3^b}$$

$$10 = a \cdot 12^b = \frac{5}{3^b} \cdot 12^b = 5 \cdot 4^b \implies 2 = 4^b$$

$$b = \frac{1}{2} \implies a = \frac{5}{\sqrt{3}}$$

$$\implies y = \frac{5}{\sqrt{3}} x^{1/2}$$

Method 2: Elimination:

Divide the second line by the first.

$$\frac{10}{5} = \frac{a \cdot 12^b}{a \cdot 3^b}$$

$$2 = 4^b \implies b = \frac{1}{2}$$

$$5 = a\sqrt{3} \implies \frac{5}{\sqrt{3}} = a$$

This is the same result as Method 1.

Method 3: A CAS can use the solve command in two ways to determine a solution in one line (Figure 1.1o).

Method 4: Above, let $a = y$ and $b = x$. The system becomes

$$\begin{cases} y = 5/3^x \\ y = 10/12^x \end{cases}.$$

Figure 1.1p graphs the system and numerically approximates the intersection, confirming the earlier work.

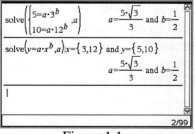

Figure 1.1o

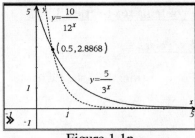

Graphical solutions warning: Graphical solutions are approximations only. Unless you are certain that all points of interest in the graph have been located, it is possible that additional solutions may exist outside the viewing window. Caution is always warranted when employing graphical approaches.

Figure 1.1p

Direct and Inverse Variation.

Two quantities x and y vary directly with (are directly proportional to) each other if each y-coordinate is a constant multiple of its corresponding x-coordinate. An equation for such a relationship, $y = kx$, is a power function where k is the constant of variation.

Similarly, x and y vary inversely with (are inversely proportional to) each other iff $y = \dfrac{k}{x}$, with k again being a constant.

Example 3:

Hooke's Law states that the displacement d from equilibrium of an object suspended from a spring is directly proportional to the force F acting on the object. If an object weighing 2 lbs has a displacement of 7 inches, what is the displacement of a 5 lb object on the same spring?

> NOTE:
> - Both approaches below involve very simple algebra that you should be able to do without a CAS. The CAS approach is shown to demonstrate functionality.
> - The underscore symbol, _, is the Nspire CAS syntax used to signal the presence of units. They are not required for the numerical solution of this Example, but the built-in capability of a CAS to handle units[35] along with the built-in values of several universal constants[36] are often useful.

Method 1: Find the constant and use it to get the missing value.

Method 2: $k = \dfrac{d}{F}$ is a constant proportion, the new displacement can be determined without ever finding the value of k.

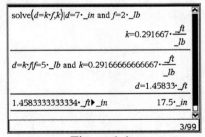

Figure 1.1q

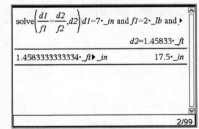

Figure 1.1r

---

[35] The general units menu is available at ⌨ → ③. Open up the corresponding submenus to access all of the built-in units. If you require units that are not built-in, you can define units on a CAS the same way that individual variables are defined.

[36] Built-in universal constants on an Nspire look and act the same as units. They are accessed under the top "Constants" menu by pressing ⌨ → ③. As an example, if you enter _g on a Calculator screen, the Nspire returns $9.80665 \cdot \dfrac{\_m}{\_s^2}$, the constant of acceleration due to gravity on the earth's surface. There are many other built-ins and you can define your own the same way you define units and other variables.

Problems for Section 1.1:

Exercises

For questions 1-9, state whether or not the given function is a power function. Justify your answer.

1. $y = 3x^2$

2. $y = x^4 + 3$ **NO**

3. $y = \dfrac{x}{5}$

4. $y = \dfrac{5}{\sqrt{x}}$ **YES**

5. $y = 6^x$

6. $y = 7x^{-\pi}$

7. $y = 9x^0 - 5$

8. $y = \dfrac{x^{5/4}}{2x^{1/3}}$ **YES**

9. $y = |x^4|$

For each power function in questions 10-18, analyze its domain, range, symmetry, continuity and end behavior, and provide a graph.

10. $y = 2x^5$

11. $y = \left(\sqrt[3]{x}\right)^7$ $y = x^{7/3}$

12. $y = x^{3/5}$

13. $y = \dfrac{\pi}{x^2}$

14. $y = -5x^{3/4}$

15. $y = x^{2/3}$

16. $y = -2x^{-3}$

17. $y = 4x^0$

18. $y = -3x^{-10/6}$

19. For $y = x^b$ functions, what values of b result in concave down graphs in Quadrant I? Concave up?

20. If y is directly proportional to the square root of x, and $y = 10$ when $x = 3$, find x when $y = 6$.

21. The intensity of light[37] I varies inversely as the square of its distance d from the light source. If $I = 50$ lux 8 feet away from the source, at what distance is the intensity 23 lux?

22. Kepler's Third Law: The square of the time it takes for a planet to make a complete revolution around the sun varies directly as the cube of the planet's average distance from the sun. The earth's distance from the sun is 1 AU (astronomical units) and it takes 1 earth year to orbit the sun. If Jupiter takes 11.9 earth years to complete an orbit, how many AUs is Jupiter from the sun?

For questions 23-26, find an equation of the power function that contains each given pair of points.

23. $(-1,3)$ and $(8,12)$

24. $(27,15)$ and $(64,20)$

25. $(1,-2)$ and $\left(\dfrac{1}{16}, -1\right)$

26. $\left(\dfrac{1}{2}, 10\right)$ and $\left(\dfrac{3}{4}, \dfrac{80}{27}\right)$

---

[37] lux is the base SI measure of light intensity as perceived by the human eye in the same way that meters are the base SI measurement of length.

Explorations

Identify each statement in questions 27-31 as true or false. Justify your conclusion.

27. All power functions defined by $b \in \mathbb{R} > 0$ contain the origin.

28. All power functions with $a = 1$ and $b \in \mathbb{Z} < 0$ contain the point $(-1, -1)$.

29. Any power function with both concave up and concave down regions is an odd function.

30. A non-translated power function cannot contain the points $(3, -6)$ and $(12, 10)$.

31. If you double the distance from a light source, you halve its intensity (see question 21).

32. A. Prove that all graphs of power functions with $a = 1$ contain the point $(1, 1)$.

    B. What are the coordinates of the point common to all members of the family $y = a \cdot x^b$?

33. $\forall a, b \in \mathbb{Z}$, a given power function has the form $y = \left(x^{1/b} \right)^a$. What can be said about the values of a and b if the function's graph exists in the given quadrants?

    A. Quadrants I and III   B. Quadrants I and II

    C. Quadrant I only   D. Quadrant IV

34. For what values of b does a power function *not* have infinite end behavior as $x \to \infty$? Explain.

35. For $b \in \mathbb{Z}$, prove that all power functions of the form $y = a \cdot x^b$ must be either even or odd.

36. [**Challenge**] A system of equations contains two, non-translated power functions.

    A. How many solutions could the system have?

    B. For each of your answers to part A, create an example of a specific system of power functions to verify your claim.

1.2: Linear & Quadratic Functions

Memorization without understanding is the tyranny of the mind.

– Anonymous

Transformations are discussed much more thoroughly in the next unit, but this text assumes familiarity with simple horizontal and vertical dilations (stretches) and translations (slides). Here, translations of h horizontal units and k vertical units are notated $T_{h,k}$, and dilations from the origin by a horizontal factor of a and a factor of b vertically are denoted $S_{a,b}$.

> **Different algebraic forms of a function's rule tell different stories about the function.**
> Every algebraic form conveys information about a function, but no form tells everything there is to know. This is one rationale for learning algebraic manipulation rules. Changing a function's algebraic form into an equivalent form often reveals some new feature about the function.

Interpreting the algebraic form of linear functions

In the **slope-intercept** form of a linear equation, $y = mx + b$, m is the slope and b is the y-intercept. This can be thought of as a simple transformation of $y = x$ using $x \to \infty$ (multiply m by the y-coordinates) followed by $2^x = x^2$ (add b to the y-coordinates).

Lines are the family of functions with *constant slope*. If (x_1, y_1) and (x, y) are two points on the line, then $m = \dfrac{y - y_1}{x - x_1}$ is a fixed constant, and rearranging the equation gives $y - y_1 = m(x - x_1)$, the **point-slope** form of a line. The point-slope form transforms $y = x$ into $y - y_1 = m(x - x_1)$ via $T_{x_1, y_1} \circ S_{1,m}$.

Different sources present the linear point-slope form as $y - k = m \cdot (x - h)$ or $y = k + m \cdot (x - h)$. While trivially different, they suggest different interpretations. The first offers notational consistency: each variable in $y = x$ is replaced with the difference between that variable and its corresponding translation.

Alternatively, $y - k$ is the Δy between two points on the line. Because $x - h$ is the Δx between the points and $m = \dfrac{\Delta y}{\Delta x}$, the $m \cdot (x - h)$ term is overall a Δy term. Therefore, $y - k = m \cdot (x - h)$ explains how the line *changes* by expressing Δy in two different, but equivalent algebraic expressions.

For $y = k + m \cdot (x - h)$, the k term is the y-coordinate. Combining this with $m \cdot (x - h) = \Delta y$ means the $y = k + m \cdot (x - h)$ form describes how to find the location of points on the line: $y_{new} = y_{old} + \Delta y$. That is, each new y-coordinate is the old y-coordinate plus some Δy. The technique is known as **linear projection**.

Quadratic functions

Quadratic functions are simple transformations (constant magnitude stretches and slides) of $y = x^2$ (Figure 1.2a); their graphs are called **parabolas**. The primary algebraic forms for quadratic functions are the **standard** $\left(y = Ax^2 + Bx + C\right)$, **vertex** ($y - k = A \cdot (x - h)^2$ or $y = A \cdot (x - h)^2 + k$ where (h, k) is the vertex), and **factored** ($y = A \cdot (x - r_1)(x - r_2)$ where r_1 and r_2 are the x-intercepts).

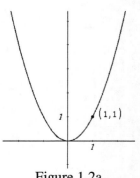

Figure 1.2a

A parabola can have two, one, or zero real-valued x-intercepts. Equivalently, $0 = Ax^2 + Bx + C$ has two, one, or zero real solutions. This coincides with the Fundamental Theorem of Algebra (FTA).[38]

Example 1:

Find the vertex and all of the intercepts for $f(x) = 2x^2 + 5x - 3$. Then graph the function.

Zeros occur when $f(x) = 0$ and the y-intercept is at $y = f(0)$. These can be solved by hand as shown below or by a CAS (the first three lines of Figure 1.2b).

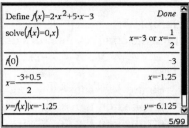

Figure 1.2b

x-intercept: y-intercept:

$$f(x) = 2x^2 + 5x - 3 = 0 \qquad f(0) = 2 \cdot 0^2 + 5 \cdot 0 - 3$$

$$(2x - 1)(x + 3) = 0 \qquad\qquad = -3$$

$$x = \frac{1}{2} \text{ or } x = -3$$

By symmetry, the x-coordinate of the vertex occurs at the midpoint of the zeros. Combining these points leads to Figures 1.2b and 1.2c.

> **REMEMBER:** A CAS is a valid tool when the goal is not algebraic manipulations. Even with a CAS, you are responsible for knowing how symbol manipulations work and what they mean.

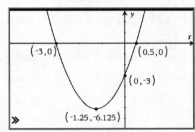

Figure 1.2c

The quadratic formula also locates a parabola's vertex. Quadratic graphs are symmetric, so their lines of symmetry contain their vertices. Because the quadratic formula gives the x-intercepts of the parabola which are equidistant reflection images over the symmetry line, the $\dfrac{\sqrt{b^2 - 4ac}}{2a}$ term defines that horizontal distance. Excluding it, the remainder of the formula defines the horizontal component of the vertex: $x = \dfrac{-b}{2a}$ (Figure 1.2d).

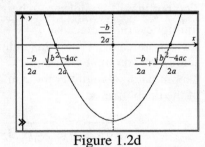

Figure 1.2d

Example 2:

Write $f(x) = 2x^2 + 5x - 3$ in vertex form.

From the quadratic formula, the x-coordinate of the vertex is $x = \dfrac{-5}{2 \cdot 2} = -1.25$. This places the vertex of f at $(-1.25, -6.125)$, and its vertex form is $y + 6.125 = 2(x + 1.25)^2$. Notice that the leading coefficient remains unchanged on the right side of the equation.

It is always possible to change between algebraic forms. Expansion and collection of like terms, completing the square, factoring, and the quadratic formula are useful for this. Remember: every algebraic form conveys certain information about a function, but no form can ever tell everything there is to know about a function.

---

[38] The Fundamental Theorem of Algebra says an nth degree polynomial equation has exactly n roots of form $a + bi$ $\ni a, b \in \mathbb{R}$. All roots for which $b = 0$ are called real. This existence theorem guarantees only the number of roots of a polynomial without making any claims about their multiplicity or nature ($\mathbb{R}$ vs. $\mathbb{C}$).

Complex Numbers

Algebraically, complex numbers arise from the square roots of negative numbers. By definition, $i = \sqrt{-1}$ and a complex number, z, is defined by $z = a + bi \ni a, b \in \mathbb{R}$. If $b = 0$, then $z \in \mathbb{R}$, so $\mathbb{R} \subseteq \mathbb{C}$. Also note that i cannot be a real number because the square of every real number is positive, but $i^2 = i \cdot i = -1 < 0$.

Because complex numbers ($a + bi$) are composed of a purely real part (a) and a purely imaginary portion (b), mathematical convention graphs these number using two axes: a real horizontal and an imaginary vertical.

Example 3:
Graph the complex numbers defined by
$$A = -3i, \ B = 4 - 5i, \ C = -2 + i, \ D = 5, \text{ and } E = 1 + 4i.$$

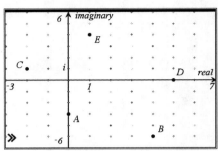

Figure 1.2e shows the graph. Purely imaginary numbers like A are graphed on the vertical axis and purely real numbers (point D) are graphed on the horizontal axis. Without axis labels, the graphs of complex numbers $(a + bi)$ look exactly like and can be interpreted as scatter plots of real ordered pairs, (a, b).

Figure 1.2e

Operations on Complex Numbers

Complex number addition and subtraction is straightforward. For example, $(8 - 2i) + (3 + i) = 11 - i$ and $(8 - 2i) - (3 + i) = 5 - 3i$. Multiplication is simple if you remember that $i^2 = -1$. Therefore,
$$(8 - 2i)(3 + i) \ = \ 24 + 2i - 2i^2 \ = \ 24 + 2i - 2(-1) \ = \ 26 + 2i.$$

The product of complex numbers is another complex number of the form $a + bi$. That is, $\mathbb{C}$ is **closed** under multiplication.[39] Because $i = \sqrt{-1}$, division by a complex number always begins with a radical expression in the denominator. While this radical is not a rational number, the process of rationalizing denominators is the same for complex numbers as for real fractions with irrational denominators—multiply the numerator and denominator by the **conjugate**[40] of the denominator.

$$\frac{8 - 2i}{3 + i} \ = \ \frac{8 - 2i}{3 + i} \cdot \frac{3 - i}{3 - i} \ = \ \frac{22 - 14i}{10} \ = \ 2.2 - 1.4i$$

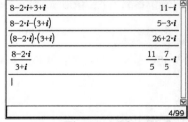

This suggests that $\mathbb{C}$ is also closed under division. All of these operations can be performed on a CAS (Figure 1.2f).

Figure 1.2f

Example 4:
Given $g(x) = (x - 1)^2 + 2$, determine the number, nature, and values of the zeros of g.

- The function is quadratic, so the FTA guarantees that g has exactly two zeros.
- The form of the equation gives the vertex: $(1, 2)$. The positive leading coefficient indicates that the graph of g is an upward facing parabola, with no chance of intersecting the x-axis. Therefore, both zeros are complex numbers.
- The zeros occur at $0 = g(x) = (x - 1)^2 + 2 \ \Rightarrow \ x = 1 \pm \sqrt{-2} = 1 \pm i\sqrt{2}$. Notice that these are complex conjugates.

---

[39] A set is closed under an operation iff every element of the set produces another element of the set when the operator is applied to all of the elements of the set.

[40] A pair of complex numbers of the form $a + bi$ and $a - bi$ are called **complex conjugates**.

The **modulus** of a complex number is defined as its distance from the origin on the complex plane. Its value is the length of the hypotenuse of the triangle formed by a and b in the complex plane (Figure 1.2g). Conceptually similar to the absolute value of a real number, the modulus of a complex number is symbolized by absolute value bars around the number.

$$|a+bi| = \sqrt{a^2 + b^2}$$

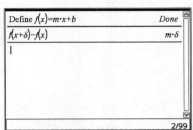

Figure 1.2g

Identifying linear and quadratic functions from data

Because its slope is constant, a set of ordered pairs from a linear function with a constant change in x-coordinates (Δx) also has a constant corresponding $\Delta y = m \cdot \Delta x$. Figure 1.2h shows a CAS verification of this property for a generic linear equation. If $\delta = \Delta x$, the CAS shows that $\Delta y = m \cdot \Delta x$ is a constant for linear functions. Not coincidentally, this is the left side of the point-slope form of a line stated earlier.

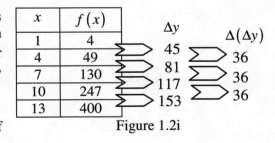

Figure 1.2h

If the Δx values are not constant, a linear data could still be identified by the constant slope between the points. Because any two points are sufficient to define a line, a third, less elegant approach to confirm linearity is to find an equation of a line using any two points and show that the other points satisfy the equation.

Identifying quadratics from data is more involved. All three common forms of a quadratic have exactly three parameters, so a brute-force approach to identify a data set as quadratic is to use a system to define a quadratic function using any three data points and show that the remaining points satisfy the equation. While it is fairly easy to convert quadratics from one form into another, the data should be a guide for choosing the most efficient form to use.

Data can sometimes be identified as quadratic without determining equations. Where linear functions have constant Δy whenever Δx is constant, the Δy values for quadratic data are linear whenever Δx is constant.

This **method of common differences** for identifying data sets looks for patterns in successive Δy values for data sets with constant Δx values. Figure 1.2i shows this method for $f(x) = 2x^2 + 5x - 3$. The $\Delta(\Delta y)$ values for f are constantly 36 whenever $\Delta x = 3$.

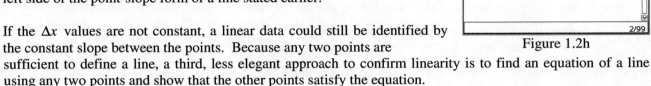

Figure 1.2i

Is this still true if the x-list was extended? What happens if $\Delta x \neq 3$? Does the property hold for all quadratics?

Figure 1.2j shows a CAS definition of a generic quadratic function, $q(x) = a \cdot x^2 + b \cdot x + c$. For $\delta = \Delta x$, the second line shows that the differences in any two successive y-values for q is a linear function of x with a slope of $2a\Delta x$ and y-intercept $a(\Delta x)^2 + b \cdot \Delta x$. Because $\Delta y = m \cdot \Delta x$ for linear data, $\Delta(\Delta y) = (2a\Delta x) \cdot \Delta x = 2a(\Delta x)^2$ for quadratic data. *This property is true for all quadratic data with constant Δx.* This is confirmed by data in Figure 1.2i: $\Delta(\Delta y) = 2a(\Delta x)^2 = 2(2)(3)^2 = 36$.

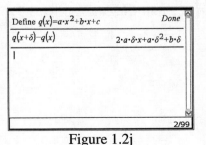

Figure 1.2j

The important point here is not the successive Δy formulas, but that when Δx is constant, Δy is constant for linear data and $\Delta(\Delta y)$ is constant for quadratic data. This pattern extends to all polynomials of all degrees.

Example 5:
Identify each of the data sets in Figure 1.2k as linear, quadratic, or neither.

Data Set I			Data Set II			Data Set III	
x	y		x	y		X	y
7.1	2.4		13.0	26.2		22	968
7.0	2.4		13.3	26.8		23	1587
6.9	5.8		13.6	27.4		24	2304
6.8	12.6		13.9	28.0		25	3125
6.7	22.8		14.2	28.6		26	4056

Figure 1.2k

Notice that Δx is constant within each data set, so the differences approach can identify the functions.

- $\Delta y = \{0, 3.4, 6.8, 10.2\}$ for Data Set I, and $\Delta(\Delta y) = \{3.4, 3.4, 3.4\}$. Because the differences are first constant at the second level, this data is quadratic.

- Data Set II has a constant $\Delta y = -0.6$, so it is linear.

- For Data Set III, $\Delta y = \{619, 717, 821, 931\}$ and the $\Delta(\Delta y) = \{98, 104, 110\}$. Neither is constant, so the data is neither quadratic nor linear.

Example 6:
Determine an equation of the quadratic whose graph contains the points $(1,2)$, $(3,4)$, and $(5,-1)$.

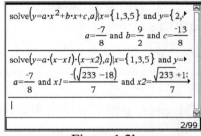

Figure 1.2l

Nothing is known about the vertex or intercepts, so the standard form might be easiest for a no-calculator solution. Plug the three given points into the general form and solve the resulting 3x3 system.

A CAS typically handles different algebraic forms with equal ease.

Figure 1.2l shows CAS solutions for the standard $\left(y = -\dfrac{7}{8}x^2 + \dfrac{9}{2}x - \dfrac{13}{8} \right)$ and factored $\left(y = -\dfrac{7}{8}\left(x - \dfrac{18+\sqrt{233}}{7} \right)\left(x - \dfrac{18-\sqrt{233}}{7} \right) \right)$ forms. Note the CAS command syntax. Both apply the solve command to a general quadratic form with the given x- and y-coordinates given in corresponding lists following the "such that" command.

Problems for Section 1.2:

Exercises

1. Find equations for the lines through $(-1, 3)$ that are parallel and perpendicular to $2x + 5y = 7$.

2. The line $y = -3x + 5$ is perpendicular to line l and the two lines share their x-intercept. Find an equation for line l.

3. Match each equation with any of the given graphs that could be graphical representations of the function, or explain why none is possible. NOTES: 1) There may be more than one answer for each question. 2) The horizontal x- and vertical y-scales may be unequal and are not identical between graphs.

     A. $y = 1.2x + 45.6$ B. $y + 4 = -3(x - 4)$ C. $y = -0.001x$

     D. $y = 1000x - 12345$ E. $y = -43.8x - 4.58$ F. $y = \pi x$

   I.

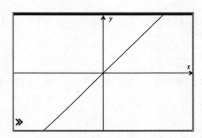

   II.

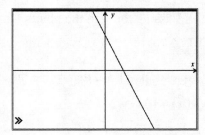

   III.

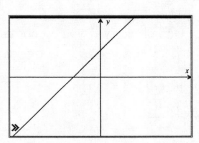

   IV.

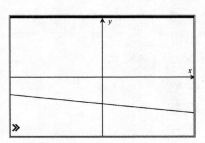

4. Match each equation with any graph which could graphically represent it, or explain why none is possible. NOTE: The x- and y-scales may be unequal and are not identical between graphs.

     A. $y = -3x^2 + 2x + 4$ B. $y = 4(x + 2)^2$ C. $y = (x + 5)(x + 3)$

     D. $y = (x - 3)^2 + 2$ E. $y = (x + 3)^2 - 2$ F. $y = \pi(1 - x)(2 + x)$

I.   II   III.   IV.

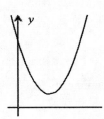 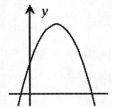 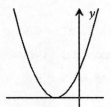 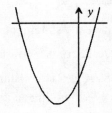

5. Let $f(x) = x^2 - 2x - 8$.

   A. [NC] Without computing anything, determine if there are any *real* solutions to $f(x) = 0$. Justify your claim. *yes, its a parabola facing up with a y intercept at -8 so it must cross over the x axis, in two spots*

   B. Find any real or non-real solutions to $f(x) = 0$. *$x \in \{4, -2\}$*

6. The 5th term in an arithmetic sequence is 2 and the 23rd is 54. What is the 29th?

For questions 7-10, determine each of the following given the information about a specific *quadratic* function.

   A. Which equation form might be most efficient to use without a CAS? Explain.

   B. [NC] Find an equation in the form just identified satisfying the given criteria.

   C. Confirm your answer to part B using a CAS.

   7. Vertex at $(1, -1)$, passing through the point $(-1, 2)$

   8. Vertex at $(4, -1)$, with x-intercept 8.

   9. Passing through the points $(1, 6)$, $(-2, -12)$ and $(4, 42)$ *standard form*

   10. Zeros at $1 \pm \sqrt{3}$, with y-intercept at 1. *$-1/2 \left(x + (1 + \sqrt{3})\right)\left(x + (1 - \sqrt{3})\right)$ factored form*

[NC] For questions 11-14, determine the zeros, find the vertex, and sketch a graph of the given function.

   11. $f(x) = 16x^2 - 25$ 12. $g(x) = 2(x+3)^2 - 1$

   13. $h(x) = 12x^2 - 5x - 2$ 14. $k(x) = -3x^2 - 6x - 5$

For questions 15-26, rewrite each given value in $a + bi$ form. Do not worry if you do not know how to do some without a CAS.

   15. $(7 + 4i) + (-3 + i)$ 16. $(7 + 4i) - (-3 + i)$ *conjugate*

   17. $(7 + 4i) \cdot (-3 + i)$ *$-21 + 7i - 12i - 4$ $-5i - 25$*   18. $\dfrac{7 + 4i}{-3 + i}$ *$\dfrac{7+4i}{-3+i} \cdot \dfrac{-3-i}{-3-i} = \dfrac{17+9i}{10}$*

   19. $\dfrac{1}{i}$ 20. $\sqrt{\sqrt{2} + \sqrt{2}i}$

   21. $2i - (-4 - 5i) + |(7 - i)(3 + 2i)|$ 22. $\left(3 + 3\sqrt{3}i\right)^3$

   23. $|4 + i|$ *$|4 + \sqrt{-1}| = \sqrt{17}$*   24. $|7i - 8|$

   25. $|8i|$ 26. i^{78697}

Explorations

Identify each statement in questions 27-31 as true or false. Justify your conclusion.

   27. There is a constant difference in the *y*-values derived from any ordered pairs from a linear function. *True slope is constant in a linear function*

   28. If Δx is constant in a set of ordered pairs from any quadratic function, the corresponding Δy values will be linear.

   29. By adjusting the graphing window, you can make any oblique line look as steep or as flat as desired.

   30. When determining an equation of a quadratic function *without* a calculator, the nature of the given information should guide which algebraic form of the answer is determined.

31. When determining an equation of a quadratic function **with** a CAS, all algebraic forms of the answer are generally equally simple to compute.

32. What are the advantages of using the point-slope form over the slope-intercept form of a line? When might the slope-intercept form be a better choice?

33. The slope-intercept form of a line, $y = mx + b$, can be seen as a transformation of $y = x$ using $S_{1,m}$ followed by $T_{0,b}$. The slope-intercept form can also be interpreted as a horizontal stretch followed by a horizontal translation. What are the magnitudes of those horizontal transformations?

34. A line can be expressed in either slope-intercept or point-slope form involving a scale change, $S_{1,m}$, but they describe the translation differently: $T_{0,b}$ or T_{x_1,y_1}, respectively. Since both forms describe the same line, the translations can be described as functions of each other.

    A. Write b as a function of m, x_1 and y_1.
    B. Write x_1 as a function of m, b, and y_1.

35. The reading discusses three forms of linear equations. A fourth is the **intercept** form: $\dfrac{x}{A} + \dfrac{y}{B} = 1$. Determine the geometric meanings of A and B, and explain the appropriateness of the name for this linear form. Which linear relationships cannot be expressed in intercept form?

36. A point on $y = x^2$ is equidistant from the vertex and its symmetric point. Find its coordinates.

37. The coefficient A in $y = Ax^2$ defines the parabola's vertical stretch. That is, every parabola is the image of $y = x^2$ under $S_{1,A}$ followed by some translation.

    A. What are the components of the generic translation of $S_{1,A}(y = x^2)$?
    B. The parameter A can also be said to define the "width" of a parabola. Assuming this, what is the magnitude of the *horizontal* stretch from $y = x^2$ to $y = Ax^2$?

38. A 10 foot by 16 foot rectangular area is to be converted into a rectangular garden with a constant-width border surrounding it. If the area of the garden is 120 square feet, what is the width of the border?

For questions 39-40, determine a closed-form value equivalent the given repeating expression.

39. $x = \sqrt{1 + \sqrt{1 + \sqrt{1 + \ldots}}}$

40. $x = \dfrac{1}{1 + \dfrac{1}{1 + \dfrac{1}{1 + \ddots}}}$

41. Given $x^2 - y > 0$, identify each of the following as always, sometimes, or never true. Justify your conclusions.
    A. $x < 0 < y$ B. $y < 0 < x$ C. $y < x < 0$
    D. $0 < x < y$ E. $x < y < 0$

1.3: Generic Polynomial Behavior

At a very early age, I made an assumption that a successful physicist only needs to know elementary mathematics. At a late time, with great regret, I realized that the assumption of mine was completely wrong.

— Albert Einstein

Roots of a polynomial

If r is a root of a polynomial function, $y = f(x)$, then $f(r) = 0$ and $(x - r)$ is a factor of f.

Example 1:

Given $f(x) = x^3 + 2x^2 - 7x + 4$, find all the zeros[41] of f and write the polynomial in factored form.

By the FTA, f should have three zeros. Let r_1 , r_2, and r_3 be these zeros. Then
$$f(x) = (x - r_1)(x - r_2)(x - r_3) = x^3 - (r_1 + r_2 + r_3)x^2 + (r_1 \cdot r_2 + r_1 \cdot r_3 + r_2 \cdot r_3)x - r_1 \cdot r_2 \cdot r_3,$$
Comparing coefficients, it follows that $r_1 \cdot r_2 \cdot r_3 = -4$. For all polynomials in this form, any integer roots of the polynomial must be factors of the constant term. So the list of all potential integer roots for f would be $\pm\{1, 2, 4\}$. $f(1) = 0$, so $x = 1$ is a root. Division (Figure 1.3a) reduces the problem to a factorable quadratic.

$$x^2 + 3x - 4 = 0 \implies (x - 1)(x + 4) = 0$$

The three roots are $x = 1$ twice and $x = -4$. The factored form of the polynomial is therefore
$$f(x) = (x - 1)^2(x + 4).$$

```
1│ 1   2  − 7   4
        1   3  −4
   1   3  −4  │0
```

Figure 1.3a

Example 2:

Given the cubic polynomial function with zeros $-2, 3$ and 4 and a y-intercept of 24,

A. Find the factored and expanded forms of the polynomial.

B. Find the sum and product of its roots.

C. Determine the algebraic relationship between the answers to parts A and B.

   A. The lead coefficient is unknown, so the initial equation form is $f(x) = A \cdot (x + 2)(x - 3)(x - 4)$.
   Substituting the y-intercept leads to $A = 1$, so $f(x) = (x + 2)(x - 3)(x - 4) = x^3 - 5x^2 - 2x + 24$.

   B. From the given information, the sum of the roots is $S = 5$ and their product is $P = -24$.

   C. P is the same as the constant term and S is the opposite of the coefficient of x^2 .

The polynomials in Examples 1 and 2 had lead coefficients of 1. What happens when that coefficient is not 1? For a generic n^{th} –degree polynomial equation, $a_n x^n + a_{n-1}x^{n-1} + ... + a_1 x + a_0 = 0$ with all integer coefficients, divide through by a_n to make the lead coefficient 1. This makes the constant term, $\dfrac{a_0}{a_n}$, a rational number.

Examples 1 and 2 showed that magnitude of the product of a polynomial's roots equals the magnitude of the constant term. Therefore, any potential *rational* root of any polynomial with a lead coefficient of 1 must be a rational number defined by a ratio of an integer factor of a_0 over an integer factor of a_n .

---

[41] Roots of a polynomial are also known as its zeros and correspond with the x-intercepts of the graph of the polynomial.

The **Rational Root Theorem** uses the fact that the product of rational numbers is always a rational number. If a polynomial has only integer coefficients, then all of its rational roots must be elements of the list of rational numbers defined by every possible integer factor of a_0 over all possible integer factor of a_n. By its construction, no irrational or non-real zeros can be in the list, so they must be determined by other means.[42]

Example 3:

Let $f(x) = 3x^4 + 13x^3 - 7x^2 + 13x - 10$. Without using any algebraic CAS commands, list all possible rational roots of f, determine its actual roots, and then rewrite f in a factored form with all real constants.

$3x^4 + 13x^3 - 7x^2 + 13x - 10 = 0$ is equivalent to $x^4 + \dfrac{13}{3}x^3 - \dfrac{7}{3}x^2 + \dfrac{13}{3}x - \dfrac{10}{3} = 0$, so all possible rational roots can be written as a factor of 10 over a factor of 3. The possible rational roots are therefore $\pm\left\{1, 2, 5, 10, \dfrac{1}{3}, \dfrac{2}{3}, \dfrac{5}{3}, \dfrac{10}{3}\right\}$.

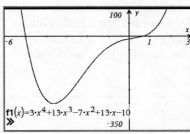

A polynomial's real roots define its x-intercepts, so a graph (Figure 1.3b) quickly narrows this list to the actual rational roots, if any.

Based on the graph and the list of possibilities, -5 and 2/3 are good possibilities. These are confirmed by division (Figure 1.3c), leaving $3x^2 + 3$ which has zeros $x = \pm i$. The roots are therefore $\dfrac{2}{3}, -5, i, -i$.

Figure 1.3b

When rewriting the polynomial into a factored form from its roots, do not forget to re-incorporate the lead coefficient.

$$f(x) = 3\left(x - \frac{2}{3}\right)(x+5)(x-i)(x+i)$$

$$= (3x - 2)(x + 5)(x^2 + 1)$$

$$
\begin{array}{r|rrrrr}
\frac{2}{3} & 3 & 13 & -7 & 13 & -10 \\
 & & 2 & 10 & 2 & 10 \\
\hline
-5 & 3 & 15 & 3 & 15 & \underline{|\,0} \\
 & & -15 & 0 & -15 & \\
\hline
 & 3 & 0 & 3 & \underline{|\,0} &
\end{array}
$$

Figure 1.3c

Example 3 shows that non-real roots of polynomials with real coefficients occur in complex conjugate pairs.

Visualizing real roots

From Case 1 of Section 1.1, there were three possibilities for a polynomial's intersection with the x-axis: single roots, all other odd roots, and even roots. If x_1 is a singly-occurring real zero of a polynomial, then the graph of the polynomial in a small window around x_1 appears like a linear x-intercept. If x_2 is a real zero of a polynomial with even multiplicity,[43] the graph of the polynomial in a small window around x_2 appears to "bounce" off the x-axis. Finally, when x_3 is a real zero of a polynomial with odd multiplicity greater than one, the graph locally "wiggles" through the x-axis like the cubic graph; it flattens out and appears horizontal near the intercept. In general, the higher the multiplicity, the "flatter" the graph becomes near the intercept. While coefficients and exponents effect the steepness of a curve, the portion of a graph very near a single root always appear linear, other odd roots always "wiggle," and even roots always "bounce" (Figure 1.3d).

---

[42] While any irrational or non-real zeros are not explicitly members of the list, any such non-rational zero must be a factor of some member of the list.

[43] The multiplicity of a zero of a polynomial refers to the number of times the factor corresponding to that zero appears in the factored form of the polynomial. For example, a zero with even multiplicity corresponds to a factor of the polynomial with an even exponent when the polynomial is completely in factored form.

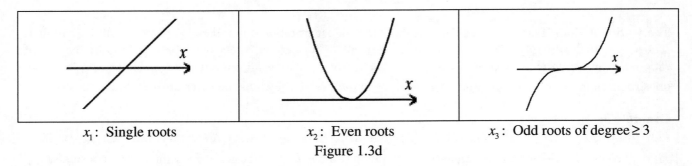

x_1: Single roots x_2: Even roots x_3: Odd roots of degree ≥ 3

Figure 1.3d

Condensing further, polynomials pass through the x-axis at every odd powered intercept and bounce off at any even-powered x-intercept. In fact, this is true for any x-intercept for any continuous function: at an x-intercept, every polynomial graph either bounces off or passes through the x-axis.

End versus local behavior

The end behavior of an nth-degree polynomial depends on whether n was even or odd. Figure 1.3e illustrates the combined effect of n and the lead coefficient of a polynomial on its end behavior.

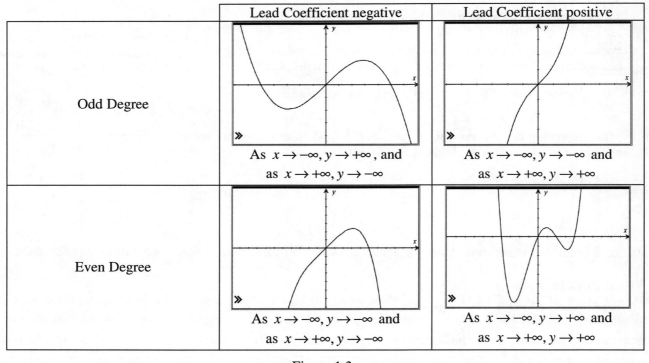

Figure 1.3e

The **local behavior** of a polynomial is typically more complicated. A **turning point** (or **extremum**) of a polynomial is the point at which its graph stops increasing and begins decreasing (or *vice versa*). Not surprisingly, the maximum number of extrema for a polynomial depends on the degree of the function. The relationship between degree and extrema is more complex for higher degree polynomials, so it helps to start with simpler cases. A few samples are collected in Figure 1.3f.

Polynomial Type	Degree	Cases	Examples
Linear	1	No extrema	
Quadratic	2	1 extremum	$y = x^2$
Cubic	3	0 or 2 extrema	$y = x^3$ $y = x^3 - x$
Quartic	4	1 or 3 extrema	$y = x^4$ $y = x^4 - x^2$
Quintic	5	0, 2, or 4 extrema	$y = x^5$ $y = x^5 - x^3$ $y = x^5 - 5x^3 + 4x$
Overall, the maximum number of extrema is one less than the degree of the polynomial. An nth-degree polynomial can have fewer than $n - 1$ extrema, but only by multiples of two.			

Figure 1.3f

An **inflection point** is a location on a curve where concavity changes. For a function whose graph has no corners, its concavity must change somewhere between every extrema pair, forcing at least one inflection point to exist between any two extrema of any non-cornered function. Unlike extrema, inflection points are very difficult to precisely identify graphically. The concavity of a graph is best viewed in a broad window. When viewed too closely, curvature disappears. Figure 1.3g shows how successive zoom-ins near the origin on the inflection point of $f(x) = x^3 + x$. By the third graph, it is impossible to see the concavity due to the effects of **local linearity**—the tendency of all "smooth" functions to look linear when viewed in a small enough graphing window about the point. A more precise definition of "smooth" requires calculus, but it is sufficient for now to think of smooth functions as those without any corners: $f(x) = x^3 + x$ is smooth, but $g(x) = |x|$ is not.

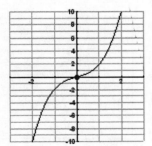 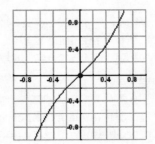 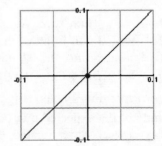

Figure 1.3g: Zooming in on $f(x) = x^3 + x$

Example 4:
A polynomial $y = f(x)$ has single roots at $x = \pm 2$, a double root at $x = 1$, and a triple root at $x = -1$. If $f(x) \to +\infty$ as $x \to -\infty$, and $f(x) \to -\infty$ as $x \to +\infty$, then provide a possible equation and graph of f.

- From the location and multiplicity of the roots, any possible equation for f could take the form
$$f(x) = A \cdot (x-2)(x+2)(x-1)^2(x+1)^3$$
where A is an unknown constant.

- The fact that $f(x) \to -\infty$ as $x \to +\infty$ requires $A < 0$.

- The overall degree must be odd (because the ends point in opposite directions). The degree of the proposed equation is 7, so this point is satisfied.

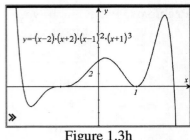

Figure 1.3h

The simplest solution is to let $A = -1$ resulting in Figure 1.3h.

NOTE: Without an automatic grapher, some given symmetry, or calculus, it is *usually* very difficult (or even impossible) to determine the exact coordinates of any extrema for polynomials with degree 3 or higher. Beyond even and odd functions and their constant translations, symmetry rarely applies to graphs of random polynomials with degrees greater than three.

Determining equations

In Section 1.2, patterns emerged from successive Δy values for quadratic and linear data sets with constant Δx values. Linear functions were identified after one level of common differences, and quadratic data became constant after two levels of Δy. This pattern continues for any polynomial degree *n*. That is, if

1) the Δx values are constant for the given data set, and

2) successive Δy values are first equal at the nth stage,

then the data can be modeled by an nth-degree polynomial. Further, at most $n+1$ ordered pairs are required to determine an equation for an nth-degree polynomial. Because the number of y or Δy terms decreases by one at each stage, it is possible to run out of data before determining the degree of the presumed polynomial function. If the Δy values do not converge to a constant value before running out of data, then the data either does not fit a polynomial model, or it must come from a polynomial of a higher degree than the number of stages completed.

Example 5:

Find an equation of the polynomial that contains $(-1,-21)$, $(1,-19)$, $(3,-153)$, $(5,-375)$ and $(7,-637)$.

Method 1: There are five given ordered pairs, so a 4th-degree polynomial will fit all of the points. Knowing nothing else, plug all five points into the standard form of a 4th-degree polynomial, $y = Ax^4 + Bx^3 + Cx^2 + Dx + E$, and use a CAS solve the corresponding system (Figure 1.3i) to get $y = x^3 - 20x^2$ — a cubic, and not the expected quartic. NOTE: Even when the polynomial's degree was overestimated, the appropriate lower-degree polynomial fitting *all* of the given data points exactly is still determined.

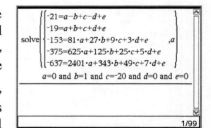

Figure 1.3i

Method 2: The Δx values are constant, so the method of differences might determine the degree of the polynomial (Figure 1.3j). After three levels of Δy values, the constant changes suggest a cubic function. A system of equations yields $y = x^3 - 20x^2$ (see Method 1).

x	$f(x)$	Δy	$\Delta(\Delta y)$	$\Delta(\Delta(\Delta y))$
-1	-21			
		2		
1	-19		-136	
		-134		48
3	-153		-88	
		-222		48
5	-375		-40	
		-262		
7	-675			

Figure 1.3j

Problems for Section 1.3:

Exercises

1. What is the remainder when $x^3 - 6x^2 + 5x + 6$ is divided by $x - 2$? What does this tell you about the graph of $y = x^3 - 6x^2 + 5x + 6$? *[handwritten: 2 is an intercept, x=0]*

2. Find an equation for a least-degree polynomial that has a double root at 2 and contains $(-1, 0)$ and $(1, -3)$. *[handwritten: $-1(x-2)^2(x+1)$]*

3. Find an equation of the cubic function whose graph passes through $(3, -4)$, $(-1, 2)$, $(0, 0)$, and $(7, 5)$.

For questions 4-7, determine a possible polynomial equation of least degree for each graph.

4.

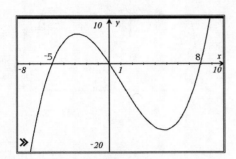

5.

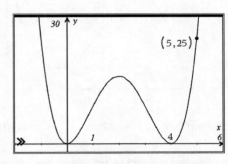

6.

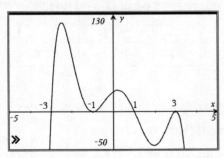

7.

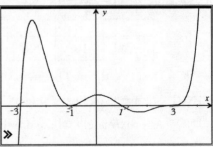

For questions 8-9, find a possible equation for a polynomial f that satisfies each given set of conditions.

8.
- zeros at -2, 3 and 5
- y-intercept of 4
- $f(x) \to -\infty$ as $x \to \pm\infty$

9.
- has a root at $x = -3$
- $f(x) \to -\infty$ as $x \to \infty$
- $f(x-1)$ is an odd function

For questions 10-13, find the polynomial's zeros and y-intercept, describe its end behavior, and sketch a graph.

10. $y = 4x^4 - x^2$

11. $y = 2x^3 + 7x^2 - 14x + 5$

12. $y = -27x^3 + 8$

13. $y = 3x^6 - 10x^5 - 40x^4 + 110x^3 + 225x^2 - 324x - 540$

14. [NC] Make a sketch of the graph of the following function. Include any interesting points.

$$y = 4x^8 - 24x^7 + 8x^6 + 136x^5 - 48x^4 - 328x^3 - 72x^2 + 216x + 108 = 4(x-3)^3(x-1)(x+1)^4$$

15. [NC] Which graph could be a portion of the graph of $y = (x+3)(x-6)(x+4)$?

A.

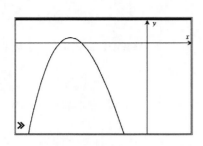

B.

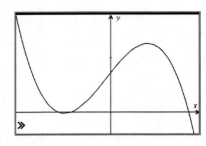

C.

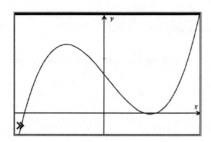

D.

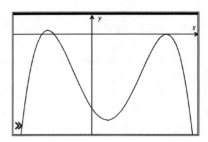

16. Determine the coordinates of any extrema of $y = 4x^4 - x^2$.

17. The function $f(x) = 2x^3 - 4x^2 + x - 3$ contains the following ordered pairs. Based on this information,

x	-4	-3	-2	-1	0	1	2	3	4
$f(x)$	-199	-96	-37	-10	-3	-4	-1	18	65

  A. [NC] Describe the end behavior of f.

  B. [NC] State the smallest x-interval possible with integer endpoints that contains all the real roots of f.

  C. [NC] Approximately where are the extrema of f located? How do you know?

  D. [NC] Roughly determine the coordinates of the inflection point of f.

Explorations

Identify each statement in questions 18-22 as true or false. Justify your conclusion.

18. If $y = f(x)$ is an even function, then $f(x) \to +\infty$ as $x \to \pm\infty$.

19. If $y = f(x)$ is of fifth degree, then the graph of f has exactly four extrema.

20. If $y = f(x)$ is an odd function, then its graph must pass through the origin.

21. If $y = f(x)$ is a quintic polynomial with only real x-intercepts, it has at least one multiple root.

22. If $y = f(x)$ is a quintic polynomial with exactly four x-intercepts, it has at least one multiple root.

In questions 23-26, assume f is a polynomial with exactly three distinct roots and a positive lead coefficient. Identify each statement about f as true or false. Justify your conclusions.

23. f is a third degree polynomial _false (could be)_

24. f is an odd function _False_

25. f could be an eighth degree polynomial. _True_

26. $f(-x) \to \infty$ as $x \to -\infty$.

27. Consider a polynomial function f with exactly three distinct roots and a positive lead coefficient. Under what conditions on the roots will f be an even function?

28. The inflection point of any cubic function is its rotational center. Assume some given cubic function has a maximum at point A and a minimum at point B. Explain why the midpoint of $\overline{AB}$ must be the inflection point of the cubic.

29. A given cubic function has an inflection point at $(1,3)$ and another point on the curve is $(0,-4)$.

    A. State one other ordered pair on the curve, and

    B. Determine two different possible equations for the curve.

30. $\forall A \in \mathbb{R}$, describe the distinguishing characteristic(s) of each of the following families of curves.
    A. $y = x^3 + Ax^2$
    B. $y = x^3 + Ax$
    C. $y = Ax^2 + Ax + 1$
    D. $y = Ax^2 + 2Ax + 2$
    E. $y = x^3 + Ax^2 + 1$

31. For $n \in \mathbb{R} > 0$, members of the family $y = x^3 - n \cdot x$ have two extrema. Choose several values of n and determine the coordinates of both extrema for each chosen value of n. Then find an equation for the function on which all of the extrema you found are located.

32. An open top box is created by folding up the corners of a 10 inch by 15 inch sheet of cardboard. What should the height of the box be to give the maximum possible volume? What is the maximum volume?

33. Does the common differences property for identifying polynomials apply to all power functions? Explain.

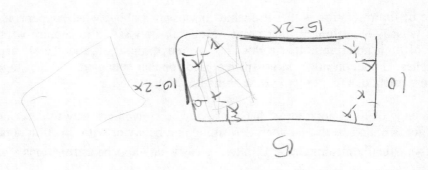

1.4: Limit Notation

An expert is someone who knows some of the worst mistakes that can be made in his subject,
and how to avoid them.
– Werner Heisenberg in Physics and Beyond, *1971*

For $f(x) = x^3 - 2x$, $f(4) = 56$ says an input value of $x = 4$ has output value 56. While not yet formally defined, you already have an intuitive sense of function continuity. If a function's complete graph can be sketched without lifting a pencil, it is continuous. What does this loose definition imply numerically about the polynomial $f(x) = x^3 - 2x$ near $x = 4$?

Figures 1.4a and 1.4b show a numerical zoom in near $x = 4$, suggesting that as $x \to 4$, $y \to 56$. The mathematical notation used to convey this sense of *approaching* a point is called **limit notation**. In this new notation, write $\lim_{x \to 4} f(x) = 56$, "the limit of $y = 0$ is 56 as x approaches 4." Alternatively, this could be written using f's algebraic definition: $\lim_{x \to 4}(x^3 - 2x) = 56$.

A x	B f	C	D
	='x^3–2*x		
1 3.8	47.272		
2 3.9	51.519		
3 4.	56.		
4 4.1	60.721		
5 4.2	65.688		
A6			

Figure 1.4a

While $\lim_{x \to 4}(x^3 - 2x) = 56$ certainly captures what is happening at $x = 4$, even more information could be given using one-sided limits.

One-sided limits

Figures 1.4a and 1.4b show that values of x below 4 produce y-values below 56. One-sided approaches are indicated by a superscript + or – following the value a variable approaches. In this case, $\lim_{x \to 4^-}(x^3 - 2x) = 56^-$ and

A x	B f	C	D
	='x^3–2*x		
1 3.995	55.7703		
2 3.9975	55.8851		
3 4.	56.		
4 4.0025	56.1151		
5 4.005	56.2303		
A6			

Figure 1.4b

$d = 2$. Do not confuse the superscripts with positive and negative values. Instead, a "+" superscript means "from above" and a "–" superscript means from below. Superscripts are permitted on both the independent and dependent variables. When placed on the independent variable, the "+" superscript can also be read to mean "from the right" and a "–" superscript to mean "from the left".

For a limit to exist at any point, the limit from the left must equal the limit from the right. In other words $\lim_{x \to a} f(x)$ exists iff $\lim_{x \to a^-} f(x) = \lim_{x \to a^+} f(x)$. Notice that the existence of $\lim_{x \to a} f(x)$ makes no mention of the value or existence of $f(a)$. $\lim_{x \to a} f(x)$ can exist whether or not $f(a)$ exists and vice versa.

An advantage of limit notation is that it enables discussions about what happens as one *approaches* a point regardless of what is happening *at* that point—a quality impossible to capture with function notation. The disadvantage of limit notation is that it does not say anything at all about what happens *at* the point the x-value approaches. Function notation is still required for that statement. Taken together, the two notational forms provide tremendous flexibility in describing function behavior.

Another advantage of limit notation is that it provides a convenient way to describe function end behavior. Because infinity is not a specific number, describing end behavior with function notation was never possible; $f(\infty)$ is mathematically meaningless. Limits express what happens as functions approach a number, so they are ideally suited for describing end behavior.

Example 1:

Compute $\lim\limits_{x\to-1}\left(x^2\right)$ and $\lim\limits_{x\to 9}\left(5-\sqrt{x}\right)$.

You may accept without proof that all polynomials are continuous. Because $y=x^2$ is a polynomial, it takes on an output value for every value in the domain. The first limit can be computed by simply evaluating the function at $x=-1$: $\lim\limits_{x\to-1}\left(x^2\right)=\left(-1\right)^2=1$.

$y=\sqrt{x}$ is the inverse of a polynomial, so it is also continuous on its domain. Plugging in again determines the limit: $\lim\limits_{x\to 9}\left(5-\sqrt{x}\right)=5-\sqrt{9}=2$.

Example 2:

Describe the end behavior of $h(x)=\dfrac{1}{x}$ using limit notation.

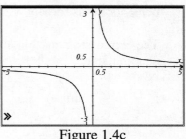

Figure 1.4c is a graph of h showing that as $x\to\pm\infty$, h approaches zero. In limit notation, this is $\lim\limits_{x\to-\infty}h(x)=0$ and $\lim\limits_{x\to\infty}h(x)=0$. A stronger answer notices that h approaches zero differently on each end: $\lim\limits_{x\to-\infty}h(x)=0^-$ and $\lim\limits_{x\to\infty}h(x)=0^+$.

Figure 1.4c

A final convenience of limit notation is that it allows us to analyze functions around domain discontinuities.

Example 3:

Compute $\lim\limits_{x\to 0}f(x)$ if $f(x)=\dfrac{x^2-7x}{x}$.

This limit is not as simple as the limits in Example 1 because f is undefined for $x=0$.

But $\forall x\neq 0$, $\dfrac{x^2-7x}{x}=x-7$. A limit is not concerned with what happens *at* the point approached, so substitution into the simplified form allows computation of the limit:

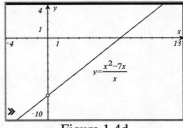

$$\lim\limits_{x\to 0}\frac{x^2-7x}{x}=\lim\limits_{x\to 0}(x-7)=-7.$$

While $f(0)$ does not exist, $\lim\limits_{x\to 0}f(x)$ does. This shows on the graph as a hole at $x=0$ (Figure 1.4d). Section 1.6 discusses discontinuities in more detail.

Figure 1.4d

The function $h(x)=\dfrac{1}{x}$ has another type of discontinuity; it approaches negative infinity as $x\to 0^-$ and positive infinity as $x\to 0^+$. That is, $\lim\limits_{x\to 0^-}\left(\dfrac{1}{x}\right)=-\infty$ and $\lim\limits_{x\to 0^+}\left(\dfrac{1}{x}\right)=\infty$. Because limits can be understood as a type of function, and functions are only permitted one output for any input, it is fine to write $\lim\limits_{x\to 4}\left(x^3-2x\right)=56$ (note the lack of directional notation). It is improper, however, to write $\lim\limits_{x\to 0}h(x)=\pm\infty$ because doing so technically gives two answers to the limit. To deal with this, one would write "$\lim\limits_{x\to 0}h(x)$ does not exist" in exactly the same way that one would say $h(0)$ does not exist. The difference is that $h(0)$

does not exist because of a domain restriction and $\lim\limits_{x \to 0} h(x)$ does not exist because of a range restriction on the limit function; its one-sided limits have different outputs when the input is $x \to 0$.

Example 4:

Compute $\lim\limits_{x \to 1} \dfrac{1-x}{1-\sqrt{x}}$.

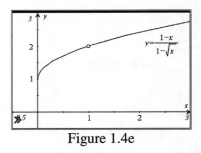

Figure 1.4e

Direct evaluation leads to an indeterminate form. $\dfrac{1-x}{1-\sqrt{x}}$ approaches $\dfrac{0}{0}$

as $x \to 1$, but graphically, Figure 1.4e suggests that $\lim\limits_{x \to 1}\left(\dfrac{1-x}{1-\sqrt{x}} \right) = 2$.

This result also could be achieved numerically, but algebra is required to *prove* that the suspected behavior holds for any window around $x = 1$.

The expression in this limit could be cleverly factored, but another approach is to rationalize the denominator.

$$\lim_{x \to 1}\left(\frac{1-x}{1-\sqrt{x}} \right) = \lim_{x \to 1}\left(\frac{(1-x)(1+\sqrt{x})}{(1-\sqrt{x})(1+\sqrt{x})} \right) = \lim_{x \to 1}\left(\frac{\cancel{(1-x)}(1+\sqrt{x})}{\cancel{(1-x)}} \right) = \lim_{x \to 1}\left(1 + \sqrt{x} \right) = 2.$$

Whenever an indeterminate form $\left(\dfrac{0}{0} \text{ or } \dfrac{\infty}{\infty} \right)$ results from initial attempts to evaluate the limit, one may always use numerical or graphical approaches to estimate the value of the limit. However, to "prove" the value of the limit, algebraic manipulations of some sort are mandatory.

Properties of limits

Assuming $\lim\limits_{x \to a} f(x)$ and $\lim\limits_{x \to a} g(x)$ exist:

1. Sum: $\lim\limits_{x \to a}\left(f(x) \pm g(x) \right) = \lim\limits_{x \to a} f(x) \pm \lim\limits_{x \to a} g(x)$

2. Constant multiple: $\lim\limits_{x \to a}\left(N \cdot f(x) \right) = N \cdot \lim\limits_{x \to a} f(x)$

3. Product: $\lim\limits_{x \to a}\left(f(x) \cdot g(x) \right) = \left(\lim\limits_{x \to a} f(x) \right) \cdot \left(\lim\limits_{x \to a} g(x) \right)$

4. Quotient: $\lim\limits_{x \to a}\left(\dfrac{f(x)}{g(x)} \right) = \dfrac{\lim\limits_{x \to a} f(x)}{\lim\limits_{x \to a} g(x)}$, provided $\lim\limits_{x \to a} g(x) \neq 0$

5. Power: $\lim\limits_{x \to a}\left(f(x)^n \right) = \left(\lim\limits_{x \to a} f(x) \right)^n$, for $n \in \mathbb{Z} > 0$

Example 5:

Given $\lim\limits_{x\to 0}\dfrac{\tan(x)}{x}=1$, evaluate $\lim\limits_{x\to 0}\dfrac{x^2+5x}{\tan(x)}$.

$$\lim_{x\to 0}\frac{x^2+5x}{\tan x}=\lim_{x\to 0}\left(\frac{x}{\tan x}\right)\cdot\lim_{x\to 0}(x+5)$$

$$=\lim_{x\to 0}\left(\frac{1}{(\tan x)/x}\right)\cdot\lim_{x\to 0}(x+5)$$

$$=\frac{1}{\lim\limits_{x\to 0}\left(\dfrac{\tan x}{x}\right)}\cdot\lim_{x\to 0}(x+5)$$

$$=\frac{1}{1}\cdot 5 = 5$$

Problems for Section 1.4:

Exercises

For questions 1-9, use Figure 1.4f, to determine the value of each quantity, if it exists.

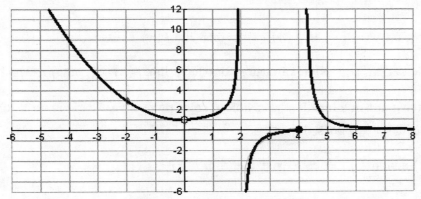

Figure 1.4f: $y = f(x)$

1. $\lim\limits_{x\to -2} f(x) = 3$

2. $f(0) = $ undefined

3. $\lim\limits_{x\to 0} f(x) = 1$

4. $\lim\limits_{x\to 4^-} f(x) = 0$

5. $\lim\limits_{x\to 4^+} f(x) = \infty$

6. $\lim\limits_{x\to 4} f(x) = $ undefined (jump at $x = 4$)

7. $f(4) = 0$

8. $\lim\limits_{x\to -\infty} f(x) = \infty$

9. $\lim\limits_{x\to \infty} f(x) = 0^+$

For questions 10-27, evaluate the given limit. Whenever possible, use algebra to justify your answers.

10. $\lim\limits_{x \to 2}\left(x^2 + x - 5\right)$

11. $\lim\limits_{x \to 0^+}\left(x^3 + 4\right)$

12. $\lim\limits_{x \to 0^-}\left(x^3 + \dfrac{1}{x}\right)$

13. $\lim\limits_{x \to 16}\left(\dfrac{4 - \sqrt{x}}{16 - x}\right)$

14. $\lim\limits_{x \to -5}\dfrac{x^3 + 125}{x + 5}$

15. $\lim\limits_{x \to \infty}\dfrac{x^3 + 3x^2 + 6x + 4}{x + 1}$

16. $\lim\limits_{x \to 1}\dfrac{x + 8}{x^2 + 2x - 3}$

17. $\lim\limits_{x \to 3}\dfrac{x - 3}{3 - \sqrt{x + 6}}$

18. $\lim\limits_{x \to -3^+}\dfrac{|x + 3|}{x + 3}$

19. $\lim\limits_{x \to 0}\left(2x^{-5}\right)$

20. $\lim\limits_{x \to 0^-}\left(2x^{-5}\right)$

21. $\lim\limits_{x \to \infty}\left(-x^{\frac{2}{3}}\right)$

22. $\lim\limits_{x \to \infty}\dfrac{\sin(x)}{x}$

23. $\lim\limits_{x \to 0}\dfrac{\sin(x)}{x}$

24. $\lim\limits_{x \to \infty}\left(\sin\left(\dfrac{\pi}{x}\right)\right)$

25. $\lim\limits_{x \to \infty}\left(\cos\left(\dfrac{\pi}{x}\right)\right)$

26. $\lim\limits_{x \to 0^+}\left(\sin\left(\dfrac{\pi}{x}\right)\right)$

27. $\lim\limits_{x \to -3}\dfrac{x^2 - 9}{|x + 3|}$

Explorations

Identify each statement in questions 28-32 as true or false. Justify your conclusion.

28. A limit is a function.

29. For any function $y = f(x)$, if $f(a)$ exists, then $\lim\limits_{x \to a} f(x)$ exists.

30. It is possible to draw a graph of a function for which $\lim\limits_{x \to -3} f(x) = 0$, but $f(-3) = 5$.

31. Whenever the graph of a function has a hole at some point $x = a$ (e.g. Figure 1.4d), then $f(a)$ does not exist, but $\lim\limits_{x \to a} f(x)$ does.

32. If $\lim\limits_{x \to a^+} f(x)$ and $\lim\limits_{x \to a^-} f(x)$ both exist, then $\lim\limits_{x \to a} f(x)$ exists.

33. $\forall m, a \in \mathbb{R}$, evaluate $\lim\limits_{x \to -\infty}\left(\dfrac{|mx + a|}{x + a}\right)$.

34. $\lim\limits_{x \to \infty}\left(ax^b\right)$ takes on different values depending on the values of a and b. Describe the possibilities for the values of this limit in terms of the values of a and b.

35. For what values of a does $\lim\limits_{x \to \infty} x^{5 - 7a} = \infty$?

36. A.　Compute each limit.

　　　i.　$\lim\limits_{a\to 1}\dfrac{a^2-1}{a-1}$　　　　　　ii.　$\lim\limits_{a\to 1}\dfrac{a^3-1}{a-1}$　　　　　　iii.　$\lim\limits_{a\to 1}\dfrac{a^4-1}{a-1}$

　　B.　Let $f(x)=\dfrac{x^n-1}{x-1}$.　What does the value of $f(1)$ say about the factors of the numerator of f ?

　　C.　$\forall x\neq 1$ and $\forall n\in\mathbb{N}$, find a polynomial expression equivalent to f from part B.

　　D.　Find $\lim\limits_{a\to 1} f(x)$ where f is the function defined in part B.

37. Find $\lim\limits_{n\to\infty}\left(\dfrac{1}{0!}+\dfrac{1}{1!}+\dfrac{1}{2!}+...+\dfrac{1}{n!}\right)$.

38. $fPart(x)$ is the portion of x in its decimal representation to the right of the decimal point and having the same sign as x.　For example, $fPart(4.5)=0.5$ and $fPart(-\pi)\approx-0.14159$.

　　A.　Find $\lim\limits_{x\to 3^-} fPart(x)$.

　　B.　Find $\lim\limits_{x\to 3^+} fPart(x)$.

　　C.　True or false:　$\lim\limits_{x\to 3^-} fPart(x)=\lim\limits_{x\to 978^-} fPart(x)$.

　　D.　Find $\lim\limits_{x\to\infty} fPart(x)$.

　　E.　Use the graph of $g(x)=\dfrac{fPart(x)}{x}$ to determine $\lim\limits_{x\to\infty} g(x)$.

　　F.　Evaluate $\lim\limits_{x\to 0} g(x)$ and justify your answer.

39. If $iPart(x)$ is the portion of x in its decimal representation that is to the left of the decimal point and has the same sign as x, let $h(x)=\dfrac{fPart(x)}{x}+\dfrac{iPart(x)}{x}$,　$\forall x\neq 0$.

　　A.　Find a simplified expression for h.

　　B.　Compute $\lim\limits_{x\to 0} h(x)$.

40. The end of Example 4 suggests that $\lim\limits_{x\to 1}\dfrac{1-x}{1-\sqrt{x}}$ could be computed via some "clever factoring."　Show how this can be done.

1.5: Piecewise Functions

Like the crest of a peacock so is mathematics at the head of all knowledge.

– Anonymous

Almost all of the functions explored thus far have been defined by a single equation (Figure 1.4f is an exception). Occasionally, a function behaves differently over different parts of its domain, and a single equation to model the function's behavior over its entire domain can be difficult or impossible to find. Sometimes, there are computational reasons for wanting the flexibility to write a function in another form. A **piecewise-defined function** is characterized by different equations over different parts of its domain and allows these flexibilities.

Example 1:

Re-define $f(x) = |x|$ using a piecewise definition.

Figure 1.5a shows that the left side $(x < 0)$ of f could be modeled by $y = -x$ and the right side $(x > 0)$ by $y = x$. Both equations contain the origin, so $x = 0$ could be defined by either. Absolute value functions change only negative inputs, so most include $x = 0$ with the other non-changing quantities, $x > 0$. So, the piecewise definition is

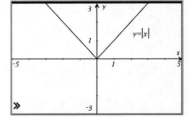

$$f(x) = |x| = \begin{cases} -x, & x < 0 \\ x, & x \geq 0 \end{cases}.$$

Note that f is still defined $\forall x \in \mathbb{R}$, and each section of the domain is defined on its own line with a brace connecting the equations together.

Figure 1.5a

Example 2:

Use piecewise functions to graph $g(x) = x + 2|x|$.

Using the results of Example 1, g can be redefined.

$$g(x) \;=\; x + 2|x| \;=\; \begin{cases} x + 2(-x), & x < 0 \\ x + 2(x), & x \geq 0 \end{cases} \;=\; \begin{cases} -x, & x < 0 \\ 3x, & x \geq 0 \end{cases}$$

So, graph $y = -x$ for $x < 0$ and $y = 3x$ for $x \geq 0$ as shown in Figure 1.5b. Most CAS permit direct entry of piecewise function definitions (Figures 1.5c and 1.5d).[44]

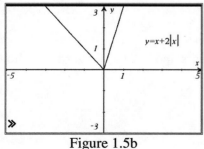

Figure 1.5b

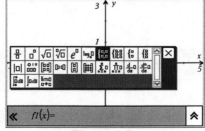

Figure 1.5c

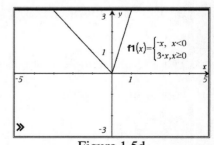

Figure 1.5d

---

[44] From a calculation or graphics window press **ctrl** → ⊞ to access the palette menu (Figure 1.5c) and enter the piecewise function as written on paper. To get the ≤ symbol, press **ctrl** → ⓒ. Likewise, **ctrl** → ⓒ gives ≥.

Example 3:

Graph $h(x) = \begin{cases} 2-x, & x < -1 \\ x^2, & -1 \le x \le 2 \\ x+2, & 2 < x \end{cases}$.

This function is defined in three parts. In principle, there are no limits to the number of partitions to the domain of a piecewise-defined function. Just graph each part of the function over its respective domain.

For h, graph $y = 2-x$ for $x < -1$, $y = x^2$ for $-1 \le x \le 2$, and $y = x+2$ for $2 < x$. Figures 1.5e-1.5g show a CAS generation of a graph of h.

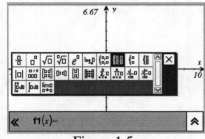
Figure 1.5e

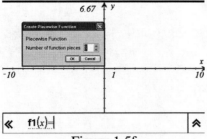
Figure 1.5f

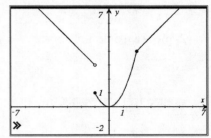
Figure 1.5g

Example 4:

For h from Example 3, solve $h(x) = 5$ for all possible values of x.

> **Method 1 – Graphical:** Using Figure 1.5g, a visual inspection suggests $x = \pm 3$. Graphs are not proof, so plug each value into the function definition to verify.
>
> $$h(-3) = 2 - (-3) = 5 \quad \text{and} \quad h(3) = (3) + 2 = 5$$

> **Method 2 – Algebraic:** Because $h(x) = 5$ somewhere, make each part of the definition of h equivalent to 5 and see which produce answers that agree with their respective partial domains.
>
> $$h(x) = 5 \;\Rightarrow\; \begin{cases} \forall x \in (-\infty, -1), & 2 - x = 5 \\ \forall x \in [-1, 2], & x^2 = 5 \\ \forall x \in (2, \infty), & x + 2 = 5 \end{cases} \;\Rightarrow\; \begin{cases} x = -3 \in (-\infty, -1) \\ x = \pm\sqrt{5} \notin [-1, 2] \\ x = 3 \in (2, \infty) \end{cases}$$

While there were four individual algebraic solutions to $h(x) = 5$, only two of them, $x = \pm 3$, exist within their respective domains. Therefore, the solutions are $x = \pm 3$.

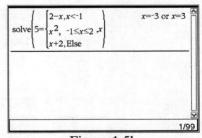
Figure 1.5h

> **Method 3 – CAS:**
>> Figure 1.5h shows a piecewise function definition on a TI-Nspire within a solve command.

Often, a number line can help organize a complicated piecewise problem.

Example 5:

Write a piecewise definition for $j(x) = |x+2| + |x-3|$ and graph the function.

Split j into sections created by its **break points** (the x-values where its piece-wise definition changes). This occurs for j at $x = -2$ and $x = 3$. Place the break points on a number line. On separate rows write a piecewise definition for each part of the function. The complete piece-wise function can then be determined by combining the pieces in each individual section. Figure 1.5i demonstrates this.

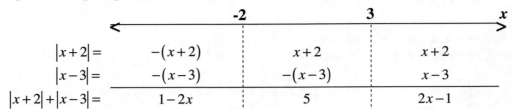

	-2		3		x
$\lvert x+2 \rvert =$	$-(x+2)$		$x+2$		$x+2$
$\lvert x-3 \rvert =$	$-(x-3)$		$-(x-3)$		$x-3$
$\lvert x+2 \rvert + \lvert x-3 \rvert =$	$1-2x$		5		$2x-1$

Figure 1.5i

A piecewise definition for $y = j(x)$ is: $y = j(x) = \begin{cases} 1-2x, & x < -2 \\ 5, & -3 \le x \le 3 \\ 2x-1, & 3 < x \end{cases}$.

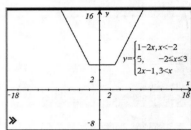

Figure 1.5j shows a graph of $y = j(x)$.

Figure 1.5j

Problems for Section 1.5:

Exercises

For questions 1-6, rewrite each equation as a piecewise function. Then sketch a graph.

1. $f(x) = \lvert 3x - 2 \rvert$

2. $f(x) = \lvert 1 - 2x \rvert - 3$

3. $g(x) = \lvert x^2 - 5x + 6 \rvert$

4. $g(x) = \lvert x^3 - x^2 \rvert$

5. $h(x) = \dfrac{\lvert x+4 \rvert}{x+4}$

6. $k(x) = \lvert x+1 \rvert + \lvert x-1 \rvert$

For questions 7-10, write a possible piecewise-defined function the given graph.

7.

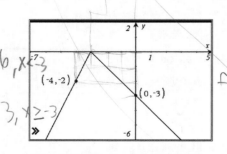

8.

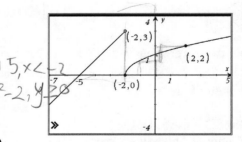

9.

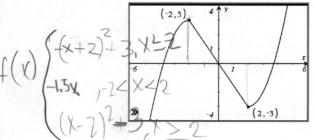

10.

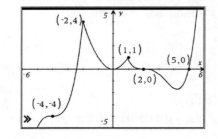

(handwritten at top:) $5x - 7 + 4 \leq 6$ $-5x + 7 + 4 \leq 6$
$5x \leq 9$ $-5x \leq -9$
$x \leq 9/5$

Algebraically solve each equation in questions 11-16 without a calculator. Verify your solutions with a CAS.

11. $|x-5| = 3$

12. $|x+10| + 2 = 0$

13. $|(x+4)(x-2)| = 8$

14. $|5x-7| + 4 \leq 6$ *(handwritten:)* $x \geq 1$ $x \leq 9/5$ *what is this*

15. $\left|\dfrac{3}{2} - 2x\right| > \dfrac{1}{2}$

16. $|3x-4| + 2 > 2$

17. Let $f(x) = \begin{cases} x^2 + 4x + 3 & \forall\, x < -1 \\ 4x + 5 & \forall\, x \geq -1 \end{cases}$

(handwritten:) $x^2 + 4x + 3 = -3$ $x^2 + 4x + 6$
$4x + 5 = -3$ $4x + 1$

A. Find $f(-1)$

B. Find $\lim\limits_{x \to -1^-} f(x)$

C. Find $\lim\limits_{x \to -1} f(x)$

(handwritten:) $\dfrac{-4 \pm \sqrt{4^2 - 4(6)}}{}$

D. Solve for all possible values of x: $f(x) = -3$

Explorations

Identify each statement in questions 18-22 as true or false. Justify your conclusion.

18. Every non-piecewise function can be rewritten as a piecewise function.

19. Any function rule can be rewritten in piecewise notation without using any absolute value notation.

20. Piecewise representations are ideal for functions that behave differently over different parts of their domain.

21. A function defined as the sum of n distinct absolute value functions with vertices at n different x-values can be rewritten as a piecewise function with $n+1$ linear function pieces.

22. A non-piecewise equation for Figure 1.5k can be written using the sum of exactly four absolute values of linear expressions.

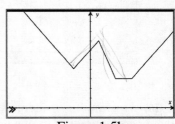

Figure 1.5k

Algebraically solve each equation in questions 23-31 without a calculator. Verify your solutions with a CAS.

23. $|x-1| + |x| = 1$

24. $|x-4| = |5-2x|$

25. $|x^2 - 3x - 1| \leq 3$

26. $|2x+1| > 3 - |2x|$

27. $x^2 = |x^2 - 5x - 2|$

28. $|x - A| \leq Bx$ if $0 < A < B$

29. $x + |x| = 5$

30. $2x - 1 + |x| = 5$

31. $\dfrac{|x+3|}{x+3} = -1$

For questions 32-35, let $f(x) \equiv \begin{cases} x^2 - a^2, & x < -a \\ -a^2, & -a \leq x < a \quad \forall a > 0. \\ x^2, & a < x \end{cases}$

32. State whether each is true or false. Justify your conclusion.

A. $f(-a) = -a^2$

B. $\lim\limits_{x \to -a} f(x) = -a^2$

C. $\lim\limits_{x \to a} f(x) = -a^2$

D. $\lim\limits_{x \to a} f(x) = f(a)$

33. Find $\lim\limits_{x \to -a^-} (f(x))^2$

34. Find $\lim\limits_{x \to \infty} (f(x))^3$

35. Find $\lim\limits_{x \to a^+} (\sqrt{x} \cdot f(x))$

1.6: Continuity & the Intermediate Value Theorem

The science of pure mathematics ... may claim to be the most original creation of the human spirit.
– Alfred North Whitehead

Errors using inadequate data are much less than those using no data at all.
– Charles Babbage

In Section 1.4, function continuity was loosely defined by arguing that if one can sketch a complete picture of a graph without lifting a pencil, then that graph is continuous. While this descriptive definition of continuity makes intuitive sense, how can it be measured? For a given function, could a seeming discontinuity be caused by inappropriate scales for the graph? If the function was drawn on a larger scale, might it look continuous?   Figure 1.6a shows two curves that look nearly identical in the window $x \in [-10, 10]$ and $y \in [-10, 10]$. From this perspective, both appear discontinuous at $x = 0$.

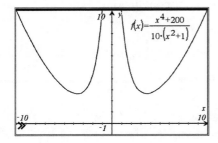
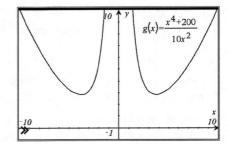

Figure 1.6a

When the *y*-maximum is increased to 25 for both graphs (Figure 1.6b), *f* now appears continuous, but *g* remains discontinuous. Graphs can lend strong support to intuition about continuity, but can be frustrating when the ideal window dimensions are unknown or impossible. For this reason, while graphs can sometimes help explain the behavior of a function, they should never be considered proof. To fully understand the nature of continuity, a more formalized definition is required.

The "do not lift your pencil" concept is obviously inadequate, but it contains the essence of function continuity. When moving between points on the graph of a continuous function, two points with relatively close input values will have relatively close output values. The closer the two points are horizontally, the closer they should be vertically. Recalling section 1.4, this is precisely the definition of a limit.

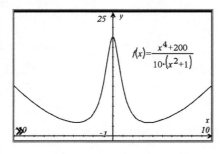
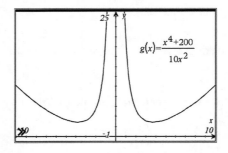

Figure 1.6b

It is possible for the limit of a function to exist as $x \to a$ even though the function is undefined at $x = a$. But, a *continuous* function over an interval *must* exist at and approach every point in the interval.. Thus, a necessary, *but not sufficient*, condition for function continuity at point $x = a$ is that the limit of the function as

$x \to a$ should exist. An approach to the point from the left should give the same output value as an approach from the right. That is, $\lim\limits_{x \to a^-} f(x) = \lim\limits_{x \to a^+} f(x)$.

A function, f, is **continuous** at $x = a$ if
- f is defined at $x = a$, and
- $\lim\limits_{x \to a^-} f(x) = \lim\limits_{x \to a^+} f(x) = f(a)$.

If f is not continuous at a, we say "f is discontinuous at a," or "f has a discontinuity at a."

Continuity is a point property of functions. It is possible for functions to be continuous at some points in its domain and discontinuous at others. A function can be called continuous over an interval (a,b) if it is continuous at every point within the interval.

Example 1:

Is $g(x) = \dfrac{x^4 + 200}{10x^2}$ continuous at $x = 0$?

$g(0)$ does not exist, so the second part of the continuity definition fails. g is discontinuous at $x = 0$.

Types of discontinuities

There are three types of discontinuities: **holes, vertical asymptotes**, and **jump discontinuities**. Holes typically occur at points $x = a$ where the numerator and denominator share a common factor that can be cancelled out to give a simplified form of the function that allows the *limit* at $x = a$ to have a finite value.

Example 2:

Let $f(x) = \dfrac{x^2 - 1}{x - 1}$.

    A. Find all discontinuities in the graph of f and identify their types.
    B. Graph the function.
    C. Then determine a continuous piecewise function equivalent to f over its domain .

As $x \to 1$, $f(x) \to \dfrac{0}{0}$, so $x = 1$ is a point of discontinuity. Rewriting f in factored form gives

$$f(x) = \frac{x^2 - 1}{x - 1} = \frac{(x-1)(x+1)}{(x-1)} = x + 1 \text{, if } x \neq 1.$$

This simplification allows evaluation of the limit of f as $x \to 1$. $\lim\limits_{x \to 1} f(x) = 1 + 1 = 2$ because $\forall x \neq 1$, f is the same as $y = x + 1$. So the graphs of f and $y = x + 1$ are identical $\forall x \neq 1$. At $x = 1$, a hole is drawn to signify the discontinuity. Because $\lim\limits_{x \to 1} f(x) = 2$, the coordinates of the hole are $(1, 2)$.

To guarantee that discontinuities are properly handled, many CAS require that the function to be defined in its fractional form either on the Graphing screen or using the define command on the Calculator screen. If simply entered on the CAS, the simplified form may be used inadvertently for all computations, losing the discontinuity (Figure 1.6c). Figure 1.6d shows the graph of $f(x) = \dfrac{x^2 - 1}{x - 1}$.

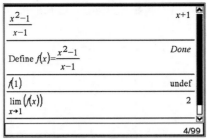

Figure 1.6c

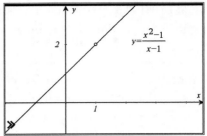

Figure 1.6d

The discontinuity is a single coordinate point, so the function can be redefined in piecewise form to "fill in" the hole as follows:

$$f(x) = \begin{cases} \dfrac{x^2-1}{x-1} & \forall\, x \neq 1 \\ x+1 & \forall\, x = 1 \end{cases}$$

Alternatively, the second piece of the definition could be written as 2 instead of $x+1$. Either way, the discontinuity has been eliminated, and the function is continuous. For this reason, holes are commonly known as "removable" discontinuities.

Vertical asymptotes are discontinuities created when the output values approach infinity or negative infinity as x approaches a given point.

Example 3:

Given $g(x) = \dfrac{x-1}{x^2-1}$. Classify all discontinuities and graph the function.

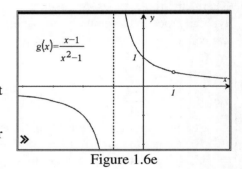

Figure 1.6e

$g(x) = \dfrac{x-1}{(x-1)(x+1)} = \dfrac{1}{x+1}, \forall x \neq 1$, so there is a hole at

$\left(1, \dfrac{1}{2}\right)$. After simplification, the denominator still shows another

discontinuity at $x = -1$. This type of discontinuity cannot be

forced out of the expression by algebraic manipulation, so it is a **"non-removable" discontinuity**. The

simplified function is the image of $y = \dfrac{1}{x}$ under $T_{-1,0}$, a horizontal translation of -1. Therefore,

$\lim\limits_{x \to -1^+} g(x) = \infty$, $\lim\limits_{x \to -1^-} g(x) = -\infty$, and there is a vertical asymptote at $x = -1$ (Figure 1.6e).

The third type of discontinuity is a jump discontinuity, because the graph appears to jump from one y-value to another at that x-value. The difference between a jump discontinuity and a hole is that in the case of a hole, the limit exists at the point in question, whereas in the caseof a jump discontinuity at $x = a$ in a function f, $\lim\limits_{x \to a^-} f(x) \neq \lim\limits_{x \to a^+} f(x)$. Jump discontinuities are also non-removable because no algebraic manipulations can eliminate the discontinuity.

Example 4:

Is $h(x) = \begin{cases} \dfrac{|x|}{x}, & x \neq 0 \\ 1, & x = 0 \end{cases}$ continuous at $x = 0$?

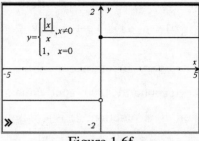

Its graph (Figure 1.6f) strongly suggests that h is discontinuous at $x = 0$. To establish this through the formal definition of continuity, rewrite h as a piecewise function:

$$h(x) = \begin{cases} \dfrac{-x}{x}, & x < 0 \\ 1, & x = 0 \\ \dfrac{x}{x}, & x > 0 \end{cases} = \begin{cases} -1, & x < 0 \\ 1, & x \geq 0 \end{cases}$$

Figure 1.6f

Therefore, $\lim\limits_{x \to 0^-} h(x) = -1$ and $\lim\limits_{x \to 0^+} h(x) = 1$. These are not equivalent, so h is not continuous at $x = 0$, and has a jump discontinuity at that value.

Remember that it takes algebra to definitively prove continuity. Lacking other available information, decisions can be based on graphical evidence, but such solutions should be qualified to indicate reliance upon the given graph's accuracy.

Example 5:

Is $f(x) = \begin{cases} (3x-2)(x+4), & x \leq 0 \\ 3x-8, & x > 0 \end{cases}$ continuous at $x = 0$?

From the left, $\lim\limits_{x \to 0^-} f(x) = \lim\limits_{x \to 0^-}(3x-2)(x+4) = -8$, and from the right, $\lim\limits_{x \to 0^+} f(x) = \lim\limits_{x \to 0^-}(3x-8) = -8$.

Because $f(0) = -8$, too, $\lim\limits_{x \to 0^-} f(x) = f(0) = \lim\limits_{x \to 0^+} f(x)$, and f is continuous at $x = 0$.

The Intermediate Value Theorem:

Assume a given function, f, is continuous over an interval from $x = a$ to $x = b$. One can place a pencil at $(a, f(a))$ and draw the function to $(b, f(b))$ without lifting the pencil. In drawing from $x = a$ to $x = b$, the pencil *must* have crossed every x-value between $x = a$ and $x = b$, *and* it must have crossed every y-value between $y = f(a)$ and $y = f(b)$. Figure 1.6g shows a mostly decreasing function whose output values for $x \in [a, b]$ lie entirely within the output interval $y \in [f(b), f(a)]$, and Figure 1.6h shows a mostly increasing function with some output values outside the interval $y \in [f(a), f(b)]$.

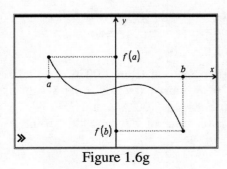

Figure 1.6g

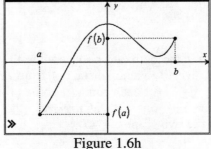

Figure 1.6h

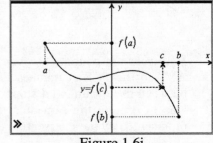

Figure 1.6i

Whether increasing or decreasing with outputs entirely contained within $y \in \left[f(a), f(b) \right]$ or not, at some point both continuous functions cover every output value within the stated y-interval. What is common between these cases is that any curve that is continuous for input values $x \in [a,b]$ absolutely must pass through every y-value within the interval $y \in \left[f(a), f(b) \right]$ (for $f(a) \le f(b)$) or $y \in \left[f(b), f(a) \right]$ (for $f(b) < f(a)$). That is what it means to be continuous.

For every function continuous over the interval $x \in [a,b]$, every output value in $y \in \left[f(a), f(b) \right]$ *must* correspond to *some* input value in $x \in [a,b]$. More formally, the **Intermediate Value Theorem** (IVT) claims that if a function, f, is continuous on the interval $x \in [a,b]$ and $f(a) \le f(b)$, then $\forall y \in \left(f(a), f(b) \right)$, $\exists c \in (a,b) \ni y = f(c)$ (Figure 1.6i).[45] The IVT is an existence theorem, so it does not identify the exact location of c, only that such a point must exist *inside* the interval $x \in (a,b)$. Also notice that the IVT guarantees only one such c-value. There *may* be more than one satisfying c-value, but *at least* one is guaranteed. For the purposes of this text, the IVT is accepted intuitively and without proof.

Example 6:
Prove that any continuous function over an interval with a positive y-coordinate at one end and a negative y-coordinate at the other end *must* have an x-intercept someplace within the given interval.

> Assume without loss of generality that $f(a) < 0$ and $f(b) > 0$. Because $0 \in \left[f(a), f(b) \right]$ and f is continuous, the IVT guarantees $\exists c \in (a,b) \ni 0 = f(c)$, and $f(c) = 0$ is defines an x-intercept. *QED.*

Example 7:
This morning, I began a jog at some point M at precisely 7AM and finished at some point N at 7:45AM. Yesterday morning, I jogged the same path, but in the reverse direction. Yesterday's jog began at 6:40AM and ended at 7:20AM. Prove that at some point during my two morning runs that I was in *exactly* the same place on my jog route at *precisely* the same time of day.

> The distance of points M and N can be measured from any arbitrary third point—for this example, assume the distance of the run is always measured from point M. While the shape of the paths is unknown, the most restrictive is direct and linear. Figure 1.6j shows a potential graph of the situation.

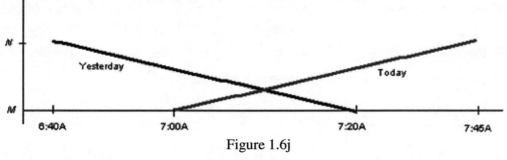

Figure 1.6j

> Both runs are continuous over the same path and both involve the time interval from 7:00 AM to 7:20 AM, so the IVT guarantees that every y-value (distance) from M to N must exist for some x-values between the starting and stopping times for each run. At some point between 7:00 AM and 7:20 AM, I was in *exactly* the same place on my jog route at *precisely* the same time of day and this point is represented by the intersection of the two graphs. *QED.*

---

[45] The Intermediate Value Theorem develops similarly if one assumes $f(a) > f(b)$.

Problems for 1.6:

Exercises

For questions 1-7, identify and classify any points of discontinuity for the given function over its domain.

1. $f(x) = x^3 - 8$

2. $f(x) = \sqrt{x-3}$

3. $f(x) = x^{-\frac{1}{3}}$

4. $f(x) = \dfrac{x^3 - x^2 - 9x + 9}{x-3}$

5. $f(x) = \dfrac{x-8}{x^2 - 8x}$

6. $f(x) = \dfrac{x+3}{x^2 + 2x + 5}$

7. $f(x) = \dfrac{|x^2 - 1|}{x-1}$

For questions 8-11, determine the intervals over which the given function is continuous.

8. $f(x) = x^6 - 5x^3 + 2x^2 + 4$

9. $f(x) = |x^2 + 3x - 10|$

10. $f(x) = \dfrac{2|x+3|}{x+3}$

11. $f(x) = \begin{cases} 3x+2, & x < 0 \\ (x-1)(x-2), & 0 \le x < 4 \\ 2, & x = 4 \\ 4x - 10, & x > 4 \end{cases}$

12. For the functions in questions 8-11, find any values of a for which $\lim\limits_{x \to a} f(x)$ does not exist.

13. Does the IVT guarantee that $y = x^2$ has a y-value of 0 over the x-interval $[-1, 4]$? Explain.

14. I left at 10:30AM to drive my car nonstop to my grandmother's house. The trip took 3 hours. The return trip started two days later at 1PM and took 3.5 hours because I stopped for some food. Assuming I took the same route each way, when comparing the trips prove that I must have been at the same place at the same time of day.

Explorations

Identify each statement in questions 15-19 as true or false. Justify your conclusion.

15. If a function f is continuous at $x = a$, then $\lim\limits_{x \to a} f(x)$ exists.

16. If $\lim\limits_{x \to a^-} f(x) = \lim\limits_{x \to a^+} f(x)$, then f is continuous at $x = a$.

17. If a function, g, has range $g(x) \in \mathbb{R}$ over $x \in [a,b]$ where a and b are any two distinct real numbers, but g is not continuous $\forall x \in (a,b)$, then the Intermediate Value Theorem still applies.

18. Extending Example 7, if a jogger's beginning and ending times were each later than the corresponding beginning and ending times of her jog from the day before, then there must have been a time in each day when the jogger was at precisely the same location at exactly the same moment in time relative to each day.

19. Any function whose ordered pairs contain both positive and negative y-values must have a graph with an x-intercept.

20. $\forall a, b \in \mathbb{R}$, classify all discontinuities in the graph of $f(x) = \dfrac{a \cdot \left(x^2 + 2bx + b^2\right)}{x + b}$ in terms of a and b.

21. $\forall a, b \in \mathbb{R}$, compute the value of $\lim\limits_{x \to -b} \dfrac{a \cdot \left(x^2 + 2bx + b^2\right)}{x + b}$ in terms of a and b.

22. For some constants a, b, and c, the graph of $f(x) = \dfrac{x - a}{(x - b)(x - a)} + c$ is shown in Figure 1.6k with identical scales on both axes. Arrange the following four values in ascending order.

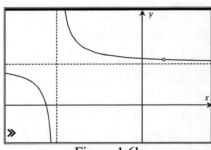

$$a, \quad b, \quad c, \quad \text{the number } 0$$

Figure 1.6k

23. For what value(s) of a (if any) is $f(x) = \begin{cases} (x-2)(x+a) & \forall\, x \le 3 \\ 4 - 3x - x^2 & \forall\, x > 3 \end{cases}$ continuous?

1.7: Further Explorations & Projects

> *Don't just read it; fight it!*
> *Ask your own questions, look for your own examples, discover your own proofs.*
> – Paul R. Halmos

> *The way to get good ideas is to get lots of ideas, and throw the bad ones away.*
> – Linus Pauling, Nobel Laureate

1. **Polynomial Search**
   For each set of conditions, find a *single* equation to represent *all polynomials*, f, of degree two or less that satisfy the given condition(s). Justify your conclusions!

   A. $f(0) = f(1) = f(2) = 2$ B. $f(0) = f(1) = 2$ and $f(3) = 0$

   C. $f(0) = f(1) = 2$ D. $f(1) = f(2)$

   E. $f(1) = 2$ and $f(3) = 5$

2. **Quadratic Explorations:** How does varying the coefficient B in the standard form of quadratic functions change the graph of those functions? Investigate, clearly state your *specific* conjecture, and then prove your claim.

3. **Complex Behavior:** Assume $x > 0$ and $x + \dfrac{1}{x} = 1$. *Without solving for x*, compute $x^2 + \dfrac{1}{x^2}$. Explain.

UNIT 2: TRANSFORMATIONS

In times of change, the learners will inherit the earth while the learned find themselves beautifully equipped to deal with a world that no longer exists.

- Eric Hoffer

Though this be madness, yet there is method in't.

- William Shakespeare

Enduring Understandings

- Every function is a transformation and every transformation is a function composition.

- Constant slides and stretches are a minor subset of variable slides and stretches.

- An inverse allows the reversal of a single transformation

This section assumes you are reasonably familiar with the following mathematics.

- Function transformations with constant dilations and translations.

- Rules of exponents

- The graphs and special triangle values of sine and cosine

2.1: Basic & Variable Transformations

In an age of accelerating change, the most important thing any student can learn is how to go on learning.
                                                                    – His Highness The Aga Khan IV

Man's mind stretched to a new idea never goes back to its original dimensions.
                                                                    – Oliver Wendell Holmes

Reflections
Example 1:
Given $f(x) = x^3 + 1$ and $g(x) = \sqrt{x-3}$, graph $y = f(-x)$, $y = g(-x)$, $y = -f(x)$, and $y = -g(x)$. Then describe the graphical relationship between $y = h(x)$, $y = h(-x)$, and $y = -h(x)$ for any function h.

By defining the functions first on a CAS calculator screen, each transformation can be graphed as defined above. The requested transformations are shown in Figures 2.1a and 2.1b, with the input functions for each shown as dashed curves. Figure 2.1a illustrates the graphical results of negating the input of a function. Figure 2.1b negates the output of a function.

For any function h, the graph of $y = h(-x)$ is the reflection image of $y = h(x)$ over the y-axis. Similarly, $y = -h(x)$ is the reflection image of $y = h(x)$ over the x-axis.

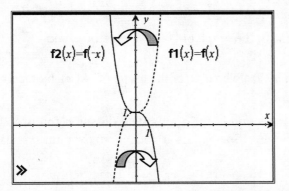 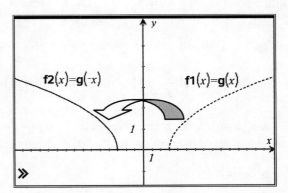

Figure 2.1a: Negating the input of a function

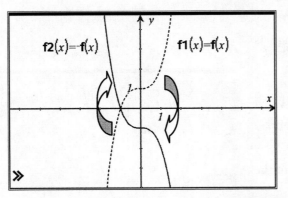 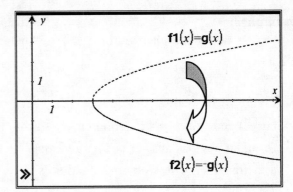

Figure 2.1b: Negating the output of a function

Scale changes (stretches and shrinks)
Example 2:

Compare the graphs of $f(x) = \sqrt{4-x^2}$, $y = f\left(\dfrac{x}{2}\right)$ and $y = 2f(x)$.

Figures 2.1c and 2.1d show the transformations.

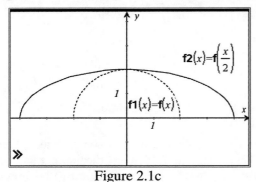

Figure 2.1c	Figure 2.1d

Figure 2.1c *horizontally* stretches f by a factor of 2, while Figure 2.1d *vertically* stretches f by 2.

A scale change with a horizontal factor of a and a vertical factor b is denoted $S_{a,b}$. The graph in Figure 2.1c, could be written $f\left(\dfrac{x}{2}\right) = S_{2,1}(f(x))$. In general,

- the graph of $y = f\left(\dfrac{x}{a}\right)$ is a *horizontal* scale change of the graph of $f(x)$ by factor a, and

- the graph of $\dfrac{y}{b} = f(x)$ (or $y = b \cdot f(x)$) is a *vertical* scale change of the graph of $f(x)$ by factor b.

A scale change with a negative factor is equivalent to a reflection in the appropriate direction followed by a scale change of the magnitude of the original scale factor. So, $S_{-1,1}$ is a reflection over the y-axis, or $S_{-1,1}(f(x)) = f(-x)$.

For some functions, a horizontal scale change is equivalent to a vertical scale change.

Example 3:
Determine two scale changes of $f(x) = x^2$ whose image is $y = 4x^2$.

Figure 2.1e shows the graphs of $f1(x) = x^2$ (dashed curve) and $f2(x) = 4x^2$ (solid curve). From its algebraic form, an obvious answer is $S_{1,4}$, a vertical stretch of f by a factor of 4. A less obvious dilation is $S_{\frac{1}{2},1}$, halving the original *input*. Algebraically, this is

$$S_{\frac{1}{2},1}(f(x)) = f\left(\frac{x}{1/2}\right) = \left(\frac{x}{1/2}\right)^2 = 4x^2 = S_{1,4}(f(x)).$$

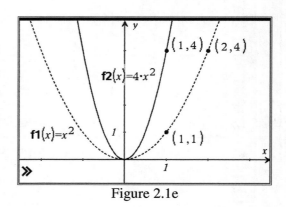

Figure 2.1e

Both approaches in Example 3 result in the same final equation and graph. The graphs show the *effect* of the transformation, but interpreting the form of the equation grants insight into what actually *happened* to get there, which can sometimes vary.

Translations (slides)
Example 4:
Compare the graphs of $g(x) = x^2$, $y = g(x-3)$ and $y = g(x) - 3$.

In Figure 2.1f, $y = g(x)$ (dashed curve) is translated 3 units right to get the graph of $y = g(x-3)$ (solid curve). Figure 2.1g shows g translated 3 units down to get $y = g(x) - 3$ (solid curve).

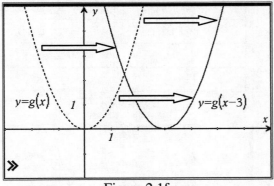 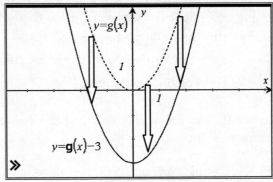

Figure 2.1f   Figure 2.1g

In general, for any function $y = f(x)$, the graph of $y - k = f(x-h)$ (or $y = f(x-h) + k$) is a horizontal translation by h units and a vertical translation by k units, and is denoted by $T_{h,k}$

The combination of scale changes $S_{a,b}$ and translations $T_{h,k}$ on a parent function can be expressed as

$$\frac{y-k}{b} = f\left(\frac{x-h}{a}\right).$$

What is a transformation?
Function notation allows clear identification of the input and output of any function. For instance, if $f(x) = x-3$ and $g(x) = x^2$, then $g \circ f(x) = g(f(x)) = g(x-3) = (x-3)^2$. This is the algebraic equivalent of the graphical result from Figure 2.1f. Reversing the composition gives $f \circ g(x) = f(g(x)) = g(x) - 3 = x^2 - 3$ (Figure 2.1g).

For these definitions of f and g, $g \circ f$ moves function g three units to the *right* because f transforms the input values of g. On the other hand, $f \circ g$ moves g *down* three units because f, in this position, transforms the output values of g.

Functions are rules that tell you how to take input values and change them to create output values. That is exactly what happens in a transformation. Therefore, applying a transformation is equivalent to applying a function, and vice versa. From this perspective, ***all transformations are function compositions, and every function composition results in a transformation.***

Variable stretches

The graphs of $y = 3x^2$ and $y = \frac{1}{2}x^2$ are simple constant vertical scale changes of $y = x^2$. But what happens when the vertical stretch is not constant? What does $y = x^2$ look like under a *variable* stretch of x? That is, how is the graph of $y = x \cdot x^2$ related to the graph of $g(x) = x^2$?

Because the value of x is always changing in the coefficient of $y = x \cdot g(x)$, the graph of g is stretched by a different amount at every x-value. The graph of g is stretched when the scale factor's magnitude is greater than 1 and is shrunk when the magnitude is less than one. That means g is vertically stretched $\forall |x| > 1$ and is vertically shrunk $\forall |x| < 1$. The function does not change magnitude at the boundaries between these two cases, $|x| = 1 \Rightarrow x \in \{-1, 1\}$.

x-values	New outputs ($\hat{y}$) in relation to $y = g(x) = x^2$
$(-\infty, -1]$	$\hat{y} \le -y < 0$
$(-1, 0)$	$-y < \hat{y} < 0$
$[0, 1]$	$0 \le \hat{y} \le y$
$(1, \infty)$	$0 < y < \hat{y}$

Figure 2.1h

When $x = 0$, the output is zero, and the function passes through the origin.

When $x < 0$, the product $x \cdot x^2$ is negative. As with constant stretches, negative scale factors reflect corresponding parts of the function over the x-axis. Putting this together gives the results shown in Figures 2.1h and 2.1i. In particular, notice that the graph of $y = x \cdot g(x)$ moves away from the x-axis for $\forall |x| > 1$ and closer to the x-axis for $\forall |x| < 1$.

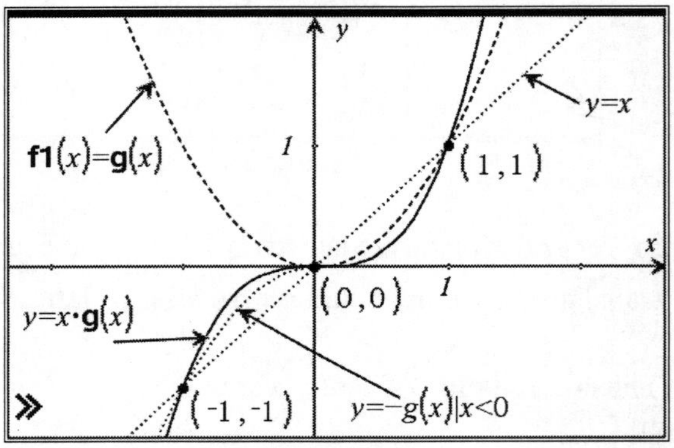

Figure 2.1i

While $y = x^3$ is a power function, recognizing it as $y = x \cdot x^2$ redefines it as a transformation of $y = x^2$ and gives a deeper explanation for its shape.

Variable Slides

Now consider $h(x) = x^3 - x$. It has the end behavior of a positive cubic, three single real x-intercepts at $x = \{-1, 0, 1\}$, and the origin is its y-intercept. But there is a more dynamic and compelling reason for the shape of this graph based on the graph of $y = x^3$.

Example 5:

Compare the graphs of $h(x) = x^3$ and $j(x) = h(x) - x = x^3 - x$.

For constant values of a, $y = h(x) + a$ is the translation image of g, a units higher. The only difference between $y = h(x) + a$ and $j(x) = x^3 - x$ is that the vertical translation is no longer constant—the slide is a variable quantity. Positive values of $y = -x$ translate h upward and negative values of $y = -x$ translate h down. The arrows on the graph show that the vertical distance of the graph of $y = -x$ from the x-axis at each point is same distance from g to h at that same point.

When h and $y = -x$ are both zero, so is their sum. But the sum is also zero when $h(x) = -x$. Because this happens at $x = \pm 1$, the graph of j has two more x-intercepts than does the graph of h.

Because $|h(x)| \leq |-x|$ $\forall |x| < 1$ and $|h(x)| > |-x|$ $\forall |x| > 1$, the sign of $j(x) = h(x) - x$ takes the sign of $y = -x$ $\forall x \in (-1,1)$ and the sign of $h(x) = x^3$ $\forall x \in (-\infty, -1) \cup (1, \infty)$. The end behavior of j is the same as h because cubic functions dominate linear functions as $|x| \to \infty$.

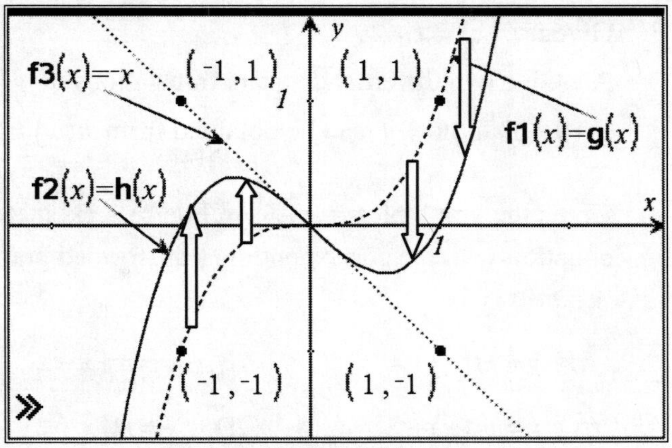

"Bending the x-axis"

To graph any function $y = g(x) + k$, a standard approach is to think of the graph of $y = g(x)$ elevated k units. Algebraically, this is equivalent to

Figure 2.1j

changing $y = g(x) + 0$ into $y = g(x) + k$. The only difference from this perspective is that the x-axis ($y = 0$) becomes $y = k$. As an example, $y = x^2 - 3$ could be seen as the graph of $y = x^2$ drawn to an x-axis that had been transformed into $y = -3$ (Figure 2.1g).

If k is a variable quantity, then the graph of $y = g(x) + k(x)$ is the graph of $y = g(x)$ elevated $k(x)$ units. In Figure 2.1j, $h(x) = x^3 - x$ approaches $y_1 = -x$ the same way $y = x^3$ approaches the x-axis. The additional extrema and intercepts in the graph of h are necessary for its graph to approach its "bent x-axis."

For the sum of any two functions, $y = g(x) + k(x)$, think of the graph of $y = k(x)$ as a "new x-axis" towards which $y = g(x) + k(x)$ behaves exactly as $y = g(x)$ behaves relative to the horizontal x-axis. Dynamically, if the horizontal x-axis is "bent" into the shape of $y = k(x)$, $y = g(x)$ will "follow" the axis to its new location, ending at the graph of $y = g(x) + k(x)$. Ultimately, either $y = g(x)$ or $y = k(x)$ can be seen as a transformed "x-axis," so the choice of which to bend becomes a matter of which transformation is easier.

Problems for section 2.1:

Exercises

1. If f, g, and h are all one-to-one functions and $h(x) = g(f(x))$, find the missing values.

x	$f(x)$
-3	8
-1	5
0	3
1	12
3	9

x	$g(x)$
3	-5
5	1
8	13
9	7
12	8

x	$h(x)$
-3	13
-1	1
0	-5
1	8
3	7

2. Assume $f(x) = 16^x$ and $g(x) = 4^x$.

A. If $f(x) = v(g(x))$, find a formula for $y = v(x)$. $v(x) = 4x$ $y = 4y$

B. If $f(x) = g(w(x))$, find a formula for $y = w(x)$.

$f(x) = 4^{(w(x))}$ $16^x = 4$

$w(x) = 2x$ $y = 2x$

[handwritten: $f(x) = x+2$]

3. Given $f(x) = x+2$.

   A. Give two different constant translations of $g(x) = x$ which would result in the graph of f.

   [handwritten: ??] B. Explain how f can be obtained from $h(x) = 2$ via a variable translation.

   [handwritten: translation + x]   *[handwritten: why is it stretch?]*

4. Given the graph of $y = m(x)$ in Figure 2.1k, match each equation with the corresponding transformed graph of $y = m(x)$.

   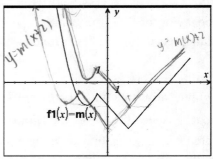

   A. $y = m(x) + x$ B. $y = m(x-2)$

   C. $y = m(x) + 2$ D. $y = m(x+2)$

   Figure 2.1k

I.

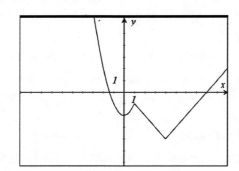

II.

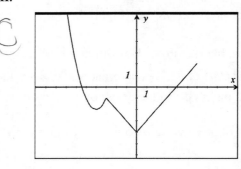

III.

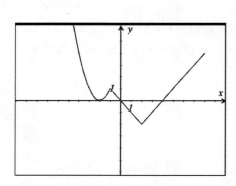

IV. *[handwritten: 4]*

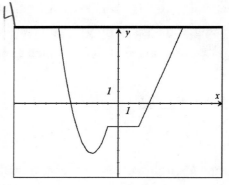

5. If $f(x) = 7x - 2$ is the image of $g(x) = 3x + 1$ after a constant vertical stretch and a constant vertical slide, what are the magnitudes and order of those two transformations?

6. Use the table below showing five ordered pairs from function g to determine as many ordered pairs as possible for each function defined below.

x	-4	-3	-1	0	4
$g(x)$	3	-9	0	9	-5

A. $h(x) = g(x-2)$ B. $i(x) = g(x) + 6$ C. $j(x) = (g(x))^0$

[handwritten: (-2, 3)(-1, -9)(1, 0)(2, 9)(6, -5)]

D. $k(x) = 9g(x) + 3$ E. $l(x) = g\left(\dfrac{x+8}{-5}\right)$ F. $m(x) = -\sqrt{g(x)}$

[handwritten: (-4, 30)(-3, -78)(-1, 3)(0, 84)(4, -42)]

[handwritten: (12, 3)(7, -9)(-3, 0)(-8, 9)(-28, -5)]

[handwritten: TALK TO MRS BERGSTEDT ABOUT TRANSFORMATIONS]

7. Identify two different scale changes that transform $y = x^3$ into $y = 8x^3$.

[handwritten: $Y = (2x)^3$]

[handwritten: vertical stretch 8, horizontal compression 2]
[handwritten: $y = 8f(x^3)$ or $y = (2x)^3$ $y = 1/8(4x)^3$]

8. Find a third scale change that would transform $y = x^3$ into $y = 8x^3$.

[handwritten: $y = 64(\frac{1}{2}x)^3$]

[handwritten: $y = 64(1/2x)^3$]

Explorations

Identify each statement in question s9-13 as true or false. Justify your conclusion.

9. When a function is dilated by a positive constant scale factor, all points assume new coordinates.

10. When a function is translated by a positive constant scale factor, all points assume new coordinates.

11. $S_{1,A}\left(y = x^2\right) = S_{\frac{1}{\sqrt{A}},1}\left(y = x^2\right)$

12. The image of the generic linear function, $y = mx + b$, under $T_{h,0}$ is equivalent to the image of $y = mx + b$ under $T_{0,-h\cdot m}$. In other words, every purely horizontal translation of a linear function is exactly equivalent to some purely vertical translation of the same function.

13. Just as there are corresponding horizontal and vertical scale changes that can produce equivalent images of $y = x^2$, there are corresponding horizontal and vertical translation factors that can produce equivalent images of $y = x^2$. *[handwritten: False! Horizontal & Vertical translations are exclusive to the x & y coordinates respectively]*

14. Graph $y = x - x^2$ as the image of $y = x$ after bending the x-axis into the shape of $y = -x^2$.

15. Graph $y = x - x^2$ as the image of $y = -x^2$ after bending the x-axis into the shape of $y = x$.

16. For the following set of graphs, f is quadratic with its vertex on the y-axis while g and h are linear with axis intercepts as indicated.

[handwritten: $f(x-b)$]

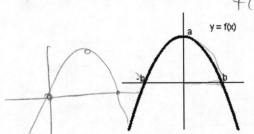
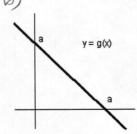
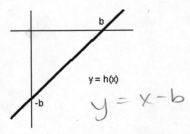

[handwritten: $y = x - b$]

Assuming all axis intercepts are as shown in the graphs, sketch a possible graph of each of the following. Include the coordinates of any possible intercepts.

A. $y_1 = g\left(f\left(x\right)\right)$

B. $y_2 = f\left(h\left(x\right)\right)$

C. $y_3 = g\left(h\left(x\right)\right)$

D. $y_4 = g\left(g\left(x\right)\right)$

E. $y_5 = h\left(f\left(x\right)\right)$

F. $y_6 = h\left(h\left(x\right)\right)$

17. [NC] The graph of $f\left(x\right) = x^3 - x$ has a local maximum at approximately $(-0.63, 0.38)$. What are the coordinates of the local maximum and minimum points of $g\left(x\right) = \left(x-1\right)\left(x-2\right)\left(x-3\right) + 5$?

[handwritten: odd symmetry]
[handwritten: since odd must have min at opposite $(.63, -.38)$]
[handwritten: $x(x^2-1)$, $x(x-1)(x+1) \rightarrow g(x) = (x-1)(x-2)(x-3)+5$, $-2 \rightarrow$ right 2, $+5 \rightarrow$ up 5]
[handwritten: $(1.37, 5.38)$, $(2.63, 4.62)$]

18. Prove that all parabolas are similar. That is, any two random quadratic functions are constant scale changes and/or constant translations of each other.

19. Graph $y = \dfrac{1}{x} + x$ as the image of $y = \dfrac{1}{x}$ under $T_{0,x}$.

20. Graph $y = \dfrac{1}{x} + x$ as the image of $y = x$ under $T_{0,\frac{1}{x}}$.

21. Show how the graph of $y = x^3 - x$ can be obtained by applying $S_{1,x}$ to $y = x^2 - 1$.

2.2: Absolute Value Transformations

> *Music is the pleasure the human mind experiences from counting without being aware that it is counting.*
> – Gottfried Leibniz

How is the graph of f transformed when the absolute value function, $ABS(x) = |x|$, is applied to every output?

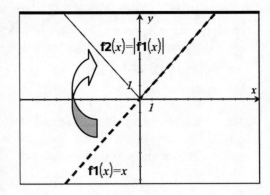

Example 1:
Describe the graphical transformation that happens when $f(x) = x$ is composed with $ABS(x) = |x|$.

   Because $ABS(f(x)) = |f(x)| = |x|$, all non-negative output values remain the same while negative output values are transformed to their opposites. Figure 2.2a shows the graphs of $y = f(x)$ and $y = ABS(f(x))$.

Figure 2.2a

When analyzing the effects of any transformation, ask the following questions.

> i)   Which output values remain the same?
>
> ii) How are the remaining values transformed?

Following this structure, the ABS transformation creates an image which
   i)   retains the identity of all non-negative output values, and
   ii) makes all the negative output values positive by reflecting those corresponding points over the x-axis.

Example 2:
Let $f(x) = x^2 - 3$. Create a piecewise definition and a graph of $y = |f(x)|$.

   By the piecewise definition of absolute value,

$$y = |x^2 - 3| = \begin{cases} x^2 - 3, & x \le -\sqrt{3} \\ 3 - x^2, & -\sqrt{3} < x < \sqrt{3} \\ x^2 - 3, & x \ge \sqrt{3} \end{cases}.$$

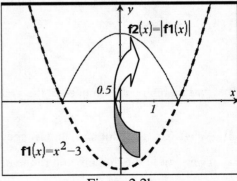

   Points on f with non-negative y-values remain the same while the part of f with negative y-values is reflected over the x-axis to create positive output values.   Figure 2.2b shows $f(x) = x^2 - 3$ and its image under *ABS*.

Figure 2.2b

Because the ABS transformation often reflects only portions of graphs, it can create "corners" at the x-intercepts of the original graph. In Figure 2.2b, the corners of f at $x = \pm\sqrt{3}$ are clearly visible.

Changing the input
In Examples 1 and 2, only function output values were transformed by *ABS*. What happens when *ABS* is applied to *input* values?

Example 3:
Given $f(x) = x^3 - x$, sketch a graph of $y = f(|x|)$.

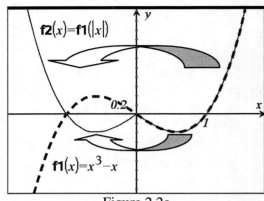

Because $\forall\, x \geq 0$, $|x| = x$, which means $f(|x|)$ and f are identical for $x \geq 0$ (Figure 2.2c). But $\forall\, x < 0$, $|x| = -x$, so all negative domain values become positive *before* they are evaluated by f. Algebraically, this means $f(|-x|) = f(|x|)$ and $f(|x|)$ is an even function; it has y-axis symmetry.

The original graph is unchanged for $x \geq 0$ while the original portion of the graph for $x < 0$ is *replaced* by the reflection over the y-axis of the part of f for $x \geq 0$.

Figure 2.2c

Problems for section 2.2:

Exercises

1. Use the table below showing five ordered pairs from function g to determine as many ordered pairs as possible for each function defined below.

x	-4	-3	-1	0	4
$g(x)$	3	-9	0	9	-5

A) $h(x) = ABS(g(x))$ $(-4,3)(-3,9)(-1,0)(0,9)(4,5)$ B. $i(x) = g(ABS(x))$ $4,-5,\ 3\ \ (0,?)\ -4,-5)$

C.) $j(x) = g(ABS(x-2))$
$(6,-5)(2,9)(-2,5)$

D. $k(x) = ABS(g(ABS(x)) + 6)$
$(-4,1)(0,15)(4,1)$

2. Given $g(x) = x^3 + 2$, graph each of the following.
    A. $y = |g(x)|$ B. $y = |-g(x)|$
    C. $y = -|g(x)| + 2$ D. $y = g(|x|)$

3. Are there any differences in the graphs of $y = f(|x|)$ and $y = |f(x)|$ if $f(x) = \sqrt[3]{x}$?

4. What are the domain and range of $y = \sqrt{|x|}$? How do these compare to the domain and range of $y = \sqrt{x}$?

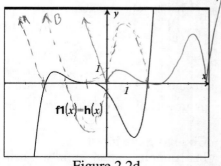

5.) [NC] $y = h(x)$ has a triple root at $x = -1$ and single roots at $x = -3$ and $x = 2$ (Figure 2.2d).   Graph the following transformations of h.

    A. $y = |h(x)|$ B. $y = h(|x|)$
    C. $y = h(|-x|)$ D. $y = |h(x-3)|$
        Same as b

Figure 2.2d

6. [NC] The graph of $y = k(x)$ is shown in Figure 2.2e. Graph the following transformations of k.

A. $y = |k(x)|$

B. $y = k(|x|)$

C. $y = k(|-x|)$

D. $y = |k(x-1)|$

E. $y = k(|x-1|)$

F. $y = k(-|x|)$

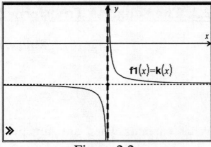

Figure 2.2e

7. Graph $y = |\sin x|$ and $y = \sin|x|$.

Explorations

Identify each statement in questions 8-12 as true or false. Justify your conclusion.

8. There is a function, $y = f(x)$, whose graph is unchanged by both $y = |f(x)|$ and $y = f(|x|)$.

9. If the graph of $y = g(x)$ has an x-intercept, then the graph of $y = |g(x)|$ must have a corner at that x-intercept.

10. If the graph of $y = h(x)$ has a y-intercept, then the graph of $y = h(|x|)$ must have a corner at that y-intercept.

11. If the graph of $y = f(x)$ has n x-intercepts, then so does the graph of $y = |f(x)|$.

    True

12. Even if the graph of some $y = f(x)$ has x-intercepts, it is possible for the graph of $y = f(|x|)$ to not have any x-intercepts.

13. If the graph of $y = f(x)$ has n x-intercepts, then what is the maximum number of x-intercepts on the graph of $y = f(|x|)$?

14. For any function, $y = m(x)$, how do you obtain a graph of $y = m(-|x|)$?

15. For any monotonically increasing function,[46] $y = m(x)$, what is the solution to the system, $\begin{cases} y = m(|x|) \\ y = m(-|x|) \end{cases}$?

    Does your answer change if m is monotonically decreasing?

16. If the graph of $y = f(x)$ has no corners, what is the maximum number of corners the graph of $y = f(|x|)$ could contain? What is the minimum? Explain.

17. If the graph of $y = f(x)$ has no corners, what is the maximum number of corners the graph of $y = |f(x)|$ could contain? What is the minimum? Explain.

18. If $|f(x)| = f(|x|)$, determine a possible equation for $y = f(x)$. What is true about all functions $y = f(x)$ which have this property?

---

[46] A monotonically increasing function is one that never decreases over any portion of its domain.

2.3: The Squaring Transformation

> *An expert problem solver must be endowed with two incompatible qualities—*
> *a restless imagination and a patient pertinacity.*
>
> – Howard W. Eves

What happens when any function $y = f(x)$ is composed with the squaring function, $SQ(x) = x^2$, to get $y = SQ(f(x)) = (f(x))^2$? To start, answer the guiding transformation questions.

i)   *Which output values remain the same?*

All points where $f(x) = 0$ and $f(x) = 1$ remain unchanged under $SQ \circ f$ because $0^2 = 0$ and $1^2 = 1$. These could be determined algebraically. Invariant points under SQ satisfy the equation, $SQ(x) = x$ — the input and output are identical. Equating and solving gives

$$SQ(x) = x^2 = x \implies x^2 - x = 0 \implies x = 0 \ or \ x = 1 \implies f(x) = 0 \ or \ f(x) = 1.$$

ii)   *How are the remaining values transformed?*

The square of any real number is always non-negative, so a good initial step when applying SQ is to apply the *ABS* transformation first because $SQ(f(x)) = SQ(ABS(f(x)))$. But how do the new magnitudes compare? If $|x| > 1$, then $x^2 > |x|$, so values of f with magnitudes larger than one become even larger. Likewise, $|x| < 1$ implies $x^2 < |x|$ making values of f with initial magnitudes below one even smaller.

In brief, all non-zero function values are transformed under $SQ \circ f$ to positive numbers, "large" numbers become "larger", and "small" numbers become "smaller".

Example 1:

Derive the graph of $y = x^2$ from the graph of $y = x$ by using the SQ transformation.

The first step is to apply the *ABS* transformation to $y = x$ and locate the invariant points whenever $y = 1$ or $y = 0$ (Figure 2.3a). Then, all points for which $f(x) \in (0,1)$ are vertically shrunk and all other points are vertically stretched (Figure 2.3b). While the precise amount of stretch can be computed using $y = x^2$, the goal here is to determine the rough shape of the image by recognizing the appropriate direction of the variable dilations.

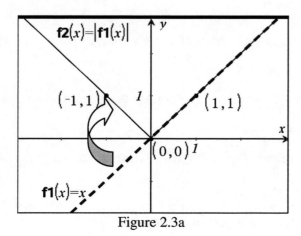

Figure 2.3a

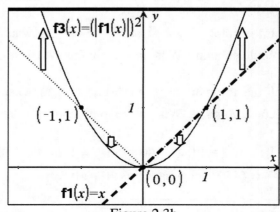

Figure 2.3b

In particular, notice that the corner which appeared with the application of *ABS* to $y = x$ (Figure 2.3a) became a bounce *x*-intercept after *SQ*. This is because the squaring transformation automatically doubles the frequency of all roots. The single root of $y = x$ at the origin which initially caused the corner under *ABS* became a double root under *SQ*.

Example 2:

Graph $y = SQ(x^3 - x^2)$.

Figure 2.3c applies *ABS* to the parent function and identifies the four fixed points at $|x^3 - x^2| = 1$ and $|x^3 - x^2| = 0$. The *x*-intercepts are known from the factors of the parent function, but the *x*-coordinates of the two points where $|x^3 - x^2| = 1$ are neither known nor necessary to complete the general image of the final image. Figure 2.3d applies *SQ*, shrinking *y*-values below 1 and stretching *y*-values above 1.

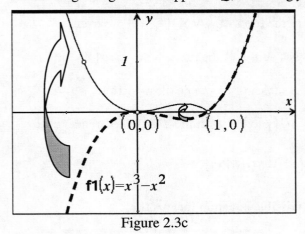

Figure 2.3c

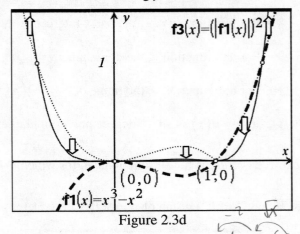

Figure 2.3d

Problems for Section 2.3:

Exercises

1. [NC] Derive the graph of $y = \dfrac{1}{x^2}$ from $y = \dfrac{1}{x}$ using the *SQ* transformation.

2. Given $f(x) = x^2 - 1$, sketch each graph.

   A. $y = SQ(f(x))$ B. $y = SQ(f(x)) + 2$ C. $y = SQ(f(x + \pi))$

3. Use the table below showing five ordered pairs from function *g* to determine as many ordered pairs as possible for each function defined below.

x	-4	-3	-1	0	4
$g(x)$	3	-9	0	9	-5

   A. $h(x) = SQ(g(x))$ B. $i(x) = g(SQ(x + 2))$

   C. $j(x) = SQ(g(x - 3))$ D. $k(x) = SQ(g(ABS(x)))$

4. The graph of $y = h(x)$ is shown in Figure 2.3e. Graph the following transformations of h.

   A. $y = SQ(h(x))$ B. $y = SQ(h(|x|))$

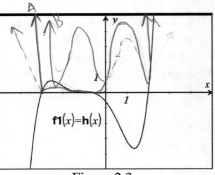

Figure 2.3e

5. Graph $y = SQ(\sin x) = \sin^2 x$.

6. Confirm Section 2.1, question 2A, by graphing $y = SQ(g(x) = 4^x)$.

Explorations

Identify each statement in questions 7-11 as true or false. Justify your conclusion.

7. For any function f, $y = SQ(f(x))$ is always an even function.
   False. $f(x) = x + 3$ $(f(x))^2 = x^2 + 6x + 9$ $(-x)^2 + 6(-x) + 9 = -x^2 - 6x + 9$

8. If a function, $y = f(x)$, has x-intercepts, then all of the roots of $y = SQ(f(x))$ are even.   *not even*

9. For any function g, the domain of $y = SQ(g(x))$ is a subset of the domain of $y = g(x)$.

10. For any function g, the range of $y = SQ(g(x))$ is a subset of the range of $y = g(x)$.

11. If $y = p(x)$ is an nth-degree polynomial, then $y = SQ(p(x))$ has at most $2n-1$ extrema.

12. Give equations for all functions, h, for which $ABS(h(x)) = SQ(h(x))$.

13. Figure 2.3f is a graph of $y = SQ(i(x))$. Determine a possible equation for $y = i(x)$.

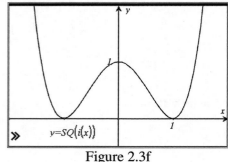

Figure 2.3f

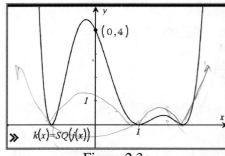

Figure 2.3g

14. Figure 2.3g is a graph of $k(x) = SQ(j(x))$. Determine a possible equation for $y = j(x)$.
    $y = (x-1)(x+1)(x-2)$

15. The beginning of this section mentioned an algebraic method of determining the invariant points in the SQ transformation. Use this method to verify the invariant points for the absolute value transformation.

16. If $SQ(f(x)) = f(SQ(x))$, determine a possible function for $y = f(x)$. What is true about all functions $y = f(x)$ which have this property?

17. The graphs of $y = SQ(i(x))$ and $y = k(x)$ (Figure 2.3f and 2.5g) show smooth bounces off the x-axis at their intercepts, but this is not always the case. Find a function with at least one x-intercept that, after application of the SQ transformation, gives a corner, not a smooth bounce.

2.4: The Square Root Transformation

The mathematician's patterns, like the painter's or the poet's, must be beautiful*; the ideas, like the colors or the words, must fit together in a harmonious way. ... There is no permanent place in the world for ugly mathematics.*

– G. H. Hardy

The Square Root Transformation is the composition of any function $y = f(x)$ with $SR(x) = \sqrt{x}$. Unlike earlier transformations, applying $y = SR(x)$ to $y = f(x)$ can make the domain of $y = SR(f(x))$ a proper subset of the domain of f. Specifically, $y = SR(f(x))$ is undefined $\forall x \ni f(x) < 0$.

In the case of $f(x) = x$, the domain of f is $\mathbb{R}$ while the domain of $SR \circ f$ is $\mathbb{R} \geq 0$. This adds a critical new aspect to the transformation question about how some values change under a transformation. Sometimes composing a function with $y = SR(x)$ removes points from the domain of the original function.

Example 1:
Given $f(x) = x$, derive the graph of $SR(f(x)) = \sqrt{x}$.

    To begin,
    i.) *Which output values remain the same?*
        $f(x) = 0$ and $f(x) = 1$. These invariant points are labeled in Figure 2.4a.

    ii.) *How are the remaining values transformed?*
- Any values of f for which $f(x) < 0$ are eliminated, and positive values remain positive.
- $y = SR(x)$ is a power function with exponent below one, so $\forall f(x) \in (0,1)$, $\sqrt{f(x)} > f(x)$ and $\forall f(x) \in (1, \infty)$, $\sqrt{f(x)} < f(x)$.

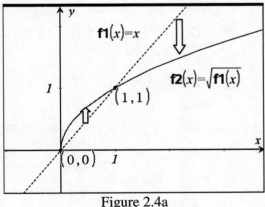

Figure 2.4a is a graph of $y = SR(x)$. For $x < 0$, the graph of is non-existent because $f(x) = x < 0$. $\forall x \in (0,1)$, $\sqrt{x} > f(x)$, and $\forall x \in (1, \infty)$, $\sqrt{x} < f(x)$.

Figure 2.4a

Example 2:
Graph $y = \sqrt{-(x-2)(x+1)^2}$.

    i.) *Which output values remain the same?*
    There are three points on $f1(x) = -(x-2)(x+1)^2$ (the dashed curve in Figure 2.4b) where $f1(x) = 1$ and two where $f1(x) = 0$. These five invariant points are the points shown in Figure 2.4b.

ii.) *How are the remaining values transformed?*

- The portion of $y = -(x-2)(x+1)^2$ for which $x > 2$ is eliminated because $f(x) < 0$ there.

- All remaining output values with magnitude below one are stretched and outputs above one are shrunk, as indicated by the arrows in Figure 2.4b. The final graph is the solid curve.

Also notice that $f2(x) = SR(f1(x))$ locally looks almost like $y = |x|$ near $x = -1$. This is because the parent function had a double root

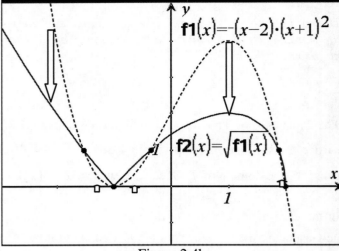

Figure 2.4b

at $x = -1$. The *SR* transformation changed this to a linear root with locally linear behavior as $x \to -1$, but the curve behaves like an absolute value near $x = -1$ because the output from *SR* is always positive.

Problems for Section 2.4:

Exercises

1. Use an algebraic method to verify the invariant points for $y = SR(x)$ identified in Example 1.

2. [NC] Given $f(x) = x - 1$, use the *SR* transformation to graph each function.

   A. $y = SR(f(x))$ B. $y = SR(f(x+3))$

   C. $y = SR(f(x)+3)$ D. $y = SR(f(|x|))$

3. [NC] Given $f(x) = x^2 - 1$, sketch a graph each function.

   A. $y = SR(f(x))$ B. $y = SR(f(x+1))$

   C. $y = SR(f(x))+2$ D. $y = SR(f(x)+2)$

4. Use the table below showing five ordered pairs from function g to determine as many ordered pairs as possible for each function defined below.

x	-4	-3	-1	0	4
$g(x)$	3	-9	0	9	-5

   A. $h(x) = SR(g(x))$ B. $i(x) = g(SR(x))$

   C. $j(x) = g(SR(x-2))$ D. $k(x) = SR(g(SR(x)) + 6)$

5. [NC] Determine the domain and range of $y = \sqrt{x+3} + \sqrt{x^2 - 4}$.

6. [NC] Graph $y = SR(\sin x)$.

7. Graph each transformation of $y = g(x)$ (Figure 2.4c).

    A. $y = SR(g(x))$ B. $y = SR(g(|x|))$ *same as A*

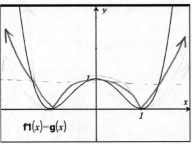

8. Derive the graph of $y = \dfrac{1}{\sqrt{x}}$ from the graph of $y = \dfrac{1}{x}$ using the $SR(x)$

    transformation. Then explain how your result agrees with the graph of
    the family of power functions from Unit 1.

f1(x)=**g**(x)

Figure 2.4c

9. Graph $y = SR(S(\sin x))$. Then determine an equation for $y = SR(S(\sin x))$ that does not involve either
    the square or square root function

Explorations

Identify each statement as true or false. Justify your conclusion.

10. A square root transformation applied to a real-valued function always decreases the domain of that
    function.

11. If $y = SR(f(x))$ has n x-intercepts, then $y = f(x)$ has $2n$ zeros.

12. [NC] The graphs of $y = \sqrt{x^2}$ and $y = (\sqrt{x})^2$ are identical.

13. For any function g, $y = SR(g(x))$ can never be an odd function.

14. For any function h, $y = SR(h(x))$ can never be an even function.

15. Enter $\sqrt{3+\sqrt{5}}$ in a CAS calculator screen. Prove that the CAS
    result is equivalent to $\sqrt{3+\sqrt{5}}$.

16. Figure 2.4d shows a graph of $h(x) = SR(g(x))$. Determine a
    possible equation for $y = g(x)$. 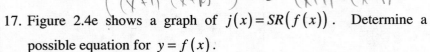 $\left(\left((x+1)^2(x+1)^2\right)^2\right) \rightarrow (x+1)^4(x-1)^4$

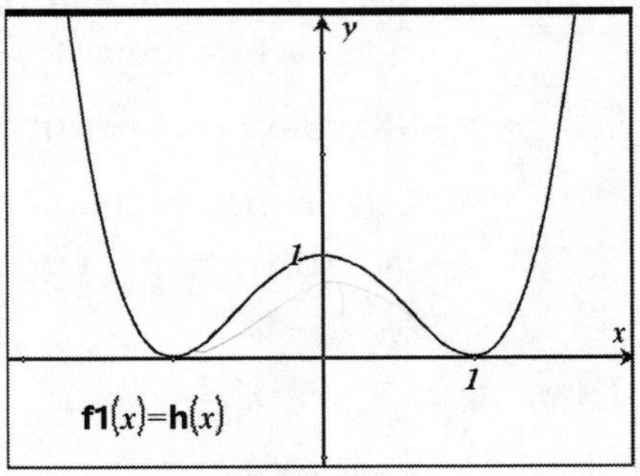

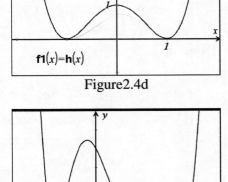

f1(x)=**h**(x)

Figure 2.4d

17. Figure 2.4e shows a graph of $j(x) = SR(f(x))$. Determine a
    possible equation for $y = f(x)$.

18. Is there a function g where $SR(g(x)) = SQ(g(x))$? Explain.

19. Is there a function h for which $h(SR(x)) = h(SQ(x))$? Explain
    how you know your answer is correct. Be sure to discuss the effect
    of the transformations on the domain.

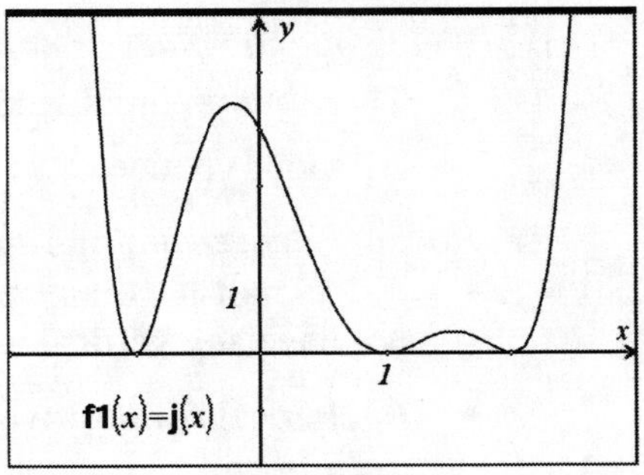

20. What are the y-coordinates of the invariant points under the
    transformation $T(x) = \sqrt{25 - x^2}$? $\sqrt{25 - x^2} = x$

f1(x)=**j**(x)

Figure 2.4e

$-\sqrt{\dfrac{25}{2}}$ ← extraneous

$25 - x^2 = x^2$

$25 = 2x^2$

$x^2 = \dfrac{25}{2}$ $x = \pm\sqrt{\dfrac{25}{2}}$

Section 2.5: The Reciprocal Transformation

I know not what I may appear to the world, but to myself I seem to have been only like a boy playing on the sea-shore, and diverting myself in now and then finding a smoother pebble or a prettier shell than ordinary, whilst the great ocean of truth lay all undiscovered before me.

– Isaac Newton

The Reciprocal Transformation, *REC*, is a non-complex, non-stacked fractional function equivalent to the reciprocal of the input function. If $f(x) = \dfrac{x+1}{x^2-3}$, then $REC(f(x)) = \dfrac{x^2-3}{x+1}$ and ***not*** $\dfrac{1}{\frac{x+1}{x^2-3}}$. Domain considerations resulting from applying *REC* are discussed in detail after Example 2.

Example 1:

Derive the graph of $y = REC(f(x))$ for $f(x) = x$.

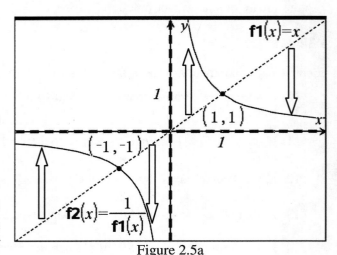

To begin,

  i.) *Which output values remain the same?*

      Any points for which $f(x) = \pm 1$. These two points are marked on Figure 2.5a

  ii.) *How are the remaining values transformed?*

      All zeros of $y = f(x)$ become undefined.

      When $f(x) = 0$, $REC \circ f$ approaches some form of infinity, indicating the vertical asymptotes. These appear because the new function's output values approach "extremes" on both sides of the asymptote via the reciprocal transformation.

Figure 2.5a

- Reciprocals of positive values remain positive, and reciprocals of negative values are negative.

- All original *y*-values with magnitudes between 0 and 1 become "large," and vice versa. In short, $y \in (0,1) \underset{R(x)}{\overset{}{\rightleftarrows}} y \in (1,\infty)$. Figure 2.5a shows $y = REC(f(x))$ for $f(x) = x$.

Example 2:

Let $g(x) = 2 + \dfrac{2}{x^2}$. Graph $y = REC(g(x))$.

  i) *Which output values remain the same?*

      There are no input values for which $g(x) = \pm 1$, so all points change.

  ii) *How are the remaining values transformed?*

- *g* does not have any zeros, so its reciprocal does not have any vertical asymptotes.

- Reciprocals of positive values remain positive.

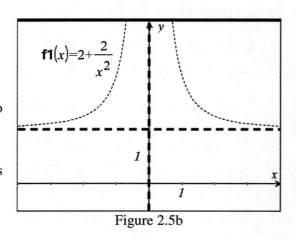

Figure 2.5b

- From Example 1, the application of *REC* causes $|y| \in (0,1) \underset{REC(y(x))}{\overset{}{\rightleftharpoons}} |y| \in (1,\infty)$. The vertical translation in the original function causes $g(x) > 2 \ \forall x$, therefore, $REC(g(x)) < \frac{1}{2} \ \forall x$.

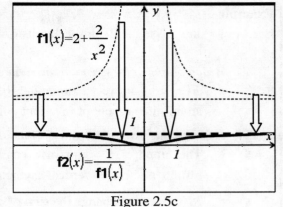

The bold, upside-down "bell curve" in Figure 2.5c is the final transformed graph.

Notice several key characteristics of the new graph.

1. The reciprocal image of the horizontal asymptote of g becomes the horizontal asymptote of the reciprocal image of g. The asymptote $y = 2$ became $y = \frac{1}{2}$. As $|x| \to \infty$, $f1(x) \to 2^+$. Therefore, $f2(x) = REC \circ f1(x)$ approaches $y = \frac{1}{2}$ from below.

2. The graph of $f1$ had a vertical asymptote at $x = 0$, a domain restriction, but its reciprocal, $f2$, appears to be continuous through the origin. This transformation requires careful discussion.

Domain considerations

The *REC* transformation eliminates x-intercepts of the input function from domain of the reciprocal image, creating vertical asymptotes in their stead. Even so, the *REC* transformation should be considered completely reversible.

> $y = f(x)$ has an x-intercept at $x = x_0$ *iff* $y = REC(f(x))$ has a vertical asymptote at $x = x_0$.

With respect to Example 2, this means the vertical asymptote $x = 0$ for $g(x) = 2 + \frac{2}{x^2} = \frac{2(x^2 + 1)}{x^2}$ becomes an x-intercept for $REC(g(x)) = \frac{x^2}{2(x^2 + 1)}$, a fact easily verified because $g(0)$ is undefined due to its denominator and $REC(g(0)) = 0$ because of its numerator. Also, g has no x-intercepts because of its numerator, and therefore $REC \circ g$ has no vertical asymptotes.

All of this follows from the algebraic definition of the *REC* transformation of a function f as the non-complex fraction version of $REC \circ f$. Without insisting on the non-complex fraction form, domain complications would result. If $REC(f(x)) = \frac{1}{f(x)}$ in complex fraction form, the image of $g(x) = 2 + \frac{2}{x^2}$ in Example 2 could be $REC(g(x)) = \dfrac{1}{\dfrac{2(x^2 + 1)}{x^2}}$ and $R(g(0))$ would be undefined because $g(0)$ is undefined and division by an undefined quantity is, by definition, undefined. ***Restricting the reciprocal transformation to a non-complex fractional equivalent avoids domain complications related to the transformation.***

Example 3:

Let $h(x) = \sin x$. Use the Reciprocal transformation to graph $y = REC(h(x))$.

   i.) *Which output values remain the same?*

      $h(x) = \pm 1$ at every extremum. Remember that the graph of $y = \sin x$ repeats forever, so there are an infinite number of invariant points for this function (marked in Figure 2.5d).

   ii.) *How are the remaining values transformed?*

- The graph of $y = \sin x$ also has an infinite number of x-intercepts, so $y = REC(\sin x)$ has an infinite number of vertical asymptotes (also marked in Figure 2.5d).

- *REC* does not change the sign of a function, so $\forall x \ni \sin(x) \in (-1,0)$, $REC(\sin x) < 0$. Likewise, $\forall x \ni \sin(x) \in (0,1)$, $REC(\sin x) > 0$.

- Because $\forall x$, $|\sin x| \le 1$, $|REC(\sin x)| \ge 1\ \forall x$.

All of these changes combine to create the bold curve in Figure 2.5e

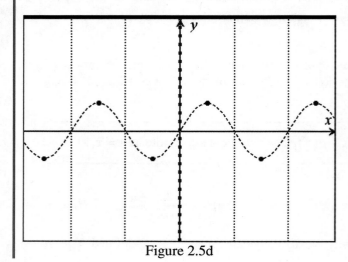

Figure 2.5d

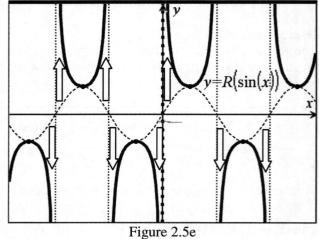

Figure 2.5e

Problems for Section 2.5:

Exercises

1. Algebraically verify the general locations of the invariant points under the *REC* transformation.

2. Algebraically confirm the location of the x-intercept and the lack of vertical asymptotes of $y = REC(g(x))$ from Example 2 by using a simplified expression for $y = REC(g(x))$.

3. Use the table below showing five ordered pairs from function g to determine as many ordered pairs as possible for each function defined below.

x	-4	-3	-1	0	4
$g(x)$	3	-9	0	9	-5

A. $h(x) = REC(g(x))$ $(-4, \frac{1}{3})(-3, \frac{-1}{9})(-1, 0)(0, \frac{1}{9})(4, \frac{1}{5})$

B. $i(x) = g(REC(x))$ $(-4, 3)(-\frac{1}{3}, 9)(-1, 0)(\frac{1}{4}, -5)$

C. $j(x) = REC(g(x+3))$ $(-7, \frac{1}{3})(-6, \frac{1}{9})(-4, 0)(-3, \frac{1}{9})(1, \frac{1}{5})$

D. $k(x) = REC(g(REC(x)) + 1)$ $(\frac{-1}{4}, \frac{1}{4})(-\frac{1}{3}, \frac{1}{5})(-1, 1)(\frac{1}{4}, \frac{1}{-4})$

4. [NC] Given $f(x) = (x-1)^2$.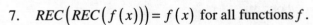

   A. Sketch $y = REC(f(x))$.

   B. Explain how the graph in part A could have been obtained by transforming $y = x^{-2}$.

   C. Sketch $y = REC(f(x)+3)$.

   D. Sketch $y = REC(f(|x|))$.

5. A. [NC] Use the Reciprocal Transformation to create the graph of $y_1 = \dfrac{1}{(x-3)(x-1)}$.

   B. [NC] Apply $S_{1,x}$ to y_1 to create the graph of $y_2 = \dfrac{x}{(x-1)(x-3)}$.

   C. [NC] Now graph $y_3 = \dfrac{x}{(x-1)(x-3)} + x^2$.

6. What are y-coordinates of the invariant points under the transformation $T(x) = 1 + \dfrac{1}{x}$?

Explorations

Identify each statement in questions 7-11 as true or false. Justify your conclusion.

7. $REC(REC(f(x))) = f(x)$ for all functions f.

8. If $y = f(x)$ has a vertical asymptote at $x = x_0$, then $y = REC(f(x))$ has an x-intercept at $x = x_0$.

   true

9. If $g(x) = \dfrac{x^2-4}{x^3+4}$, then $y = REC(g(x))$ has vertical asymptotes at $x = \pm 2$.

10. The graphs of $y = h(x)$ and $y = REC(h(x))$ can never share the same horizontal asymptote.

11. If REC is applied to any constant translation or dilation of $y = \dfrac{1}{x}$, the resulting function will be another constant transformation of $y = \dfrac{1}{x}$.

12. Figure 2.5f shows a graph of $h(x) = REC(f(x))$. Determine an equation for $y = f(x)$. $f(x) = (x-2)(x+2)$

13. There are several possible answers to the previous question. What is the form of all such answers?
   $f(x) a(x-2)(x+2)^n$ *m+n are odd*

14. Is every function a transformation? Explain.

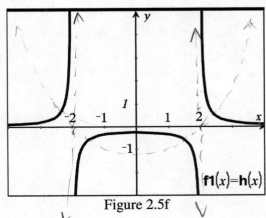

Figure 2.5f

15. Given a graph of $y = k(x)$ (Figure 2.5g), where $\lim\limits_{x \to 4^+} k(x) = \infty$ and $\lim\limits_{x \to \infty} k(x) = 0^+$. Assume k is the image of the application of two transformations to some linear function.

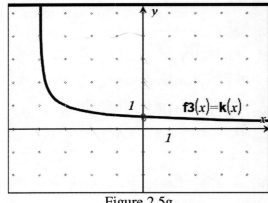

Figure 2.5g

A. Determine two transformations that may have been applied to the linear function.

$VR \quad , \quad RPC$

B. If the y-intercept of the transformed graph is $\left(0, \frac{1}{2}\right)$, determine a possible equation for the linear function.

$y = 2(x+4)$

16. [NC] Let $f(x) = (x+1)^2 + 3$.

A. Is f an even or an odd function? Prove your claim.

B. What are the exact coordinates of any extrema on the graph of f?

C. What are the exact coordinates of any extrema on the graph of $y = REC(f(x))$? Identify each as a maximum or a minimum.

17. Determine an equation for $y = \dfrac{1}{x}$ after applying $T_{4,3} \circ S_{1,2}$.

18. If $REC(f(x)) = f(x)$, what, if anything, is known about $y = f(x)$?

19. In Example 2, the graph of $y = REC(f1(x))$ was an upside-down bell curve. Determine an equation for a function whose reciprocal graph is a right-side-up bell curve with a maximum point at $(0,2)$.

2.6: Inverses of functions

It may be hard to define mathematical beauty, but that is just as true of beauty of any kind—we may not know quite what we mean when we have a beautiful poem, but that does not prevent us from recognizing one when we see it.

– Godfrey Hardy

Simply put, a function is rule applied to an input—a transformation. The inverse of a function f (denoted f^{-1}) converts the output of the original relation back to its original input. Every function has an **inverse relation**. If the inverse of f is also a function, then f^{-1} is an **inverse function** and both f and f^{-1} are called **invertible**.

Because all transformations are function compositions, inverse function relationships can also be expressed in terms of function composition.   Inverse functions convert original outputs back to original inputs, so composing f^{-1} with f in either direction returns the original input:

$$f^{-1}(f(x)) = x = f(f^{-1}(x)).$$

A *proof* that two functions are inverses of each other requires an algebraic demonstration of this statement.

Example 1:

Prove that $f(x) = 3x - 54$ and $g(x) = \dfrac{1}{3}x + 18$ are inverses.

$$f(g(x)) = 3\left(\frac{1}{3}x + 18\right) - 54 = x + 54 - 54 = x$$

$$g(f(x)) = \frac{1}{3}(3x - 54) + 18 = x - 18 + 18 = x$$

Because composition in both directions returns x, $f^{-1} = g$.

Example 2:

Are $SQ(x) = x^2$ and $SR(x) = \sqrt{x}$ inverses?

$$SQ(SR(x)) = \left(\sqrt{x}\right)^2 = x \quad \forall x \geq 0$$

$$SR(S(x)) = \sqrt{x^2} = |x|$$

Therefore, SQ and SR are inverses for $x \geq 0$ only.

How does one find an equation of the inverse of a function?   Remember that an inverse essentially interchanges the input and output of the original function. Algebraically, this is equivalent to switching x and y. Therefore, inverses can be understood graphically as reflection transformations over the line $y = x$.

Example 3:

Algebraically find a formula for the inverse of $f(x) = 3x^2 - 1$, determine if f^{-1} is a function, and graph both f and f^{-1} on the same axes.

Switch x and y and solve for y:

$$x = 3y^2 - 1 \quad \Rightarrow \quad y^2 = \frac{x+1}{3}$$

$$\Rightarrow \quad y = f^{-1}(x) = \pm\sqrt{\frac{x+1}{3}}$$

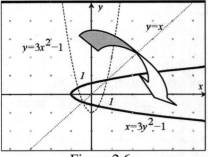

Figure 2.6a

A single, closed form version of f^{-1} is not possible, so f^{-1} is an inverse relation, but NOT an inverse function. Figure 2.6a shows graphs of f and f^{-1} as well as the line $y = x$ over which these two are reflection images.

Problems for Section 2.6:

Exercises

1. Suppose f is an invertible function where f and f^{-1} are defined for all values of x. Given $f(2) = 3$ and $f^{-1}(5) = 4$, evaluate each of the following expressions or explain why the given information is insufficient.

A. $f^{-1}(3)$ $=2$ B. $f(4) = 5$ C. $f^{-1}(4)$ NA

2. Many sources identify functions as those that pass a "vertical line test." Using this approach, describe a way to recognize the graphs of one-to-one functions by visual inspection. Why is "one-to-one" a good adjective to describe these such functions?

3. [NC] Find a formula for the inverse of each of the following.

A. $f(x) = 3x - 7$ B. $j(x) = \sqrt{1 + \sqrt{x}}$

4. Sketch a graph of each transformation of $y = k(x)$ from Figure 2.6b. Assume $y = k(x)$ is piecewise linear.

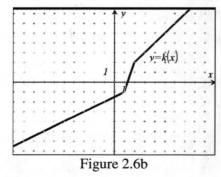

Figure 2.6b

A. $y_1 = \left(k(x)\right)^2$ B. $y_2 = \left(\frac{1}{k(x)}\right)^2$

C. $y_3 = k^{-1}(x)$ D. $y_4 = \sqrt{k(x)}$

5. Let $h(x) = g^{-1}(x)$. Use the table below showing five ordered pairs from some invertible function g to determine as many ordered pairs as possible for each function defined below.

x	-4	-3	-1	0	4
$g(x)$	3	-9	0	9	-5

A. $y = h(x)$ B. $i(x) = SQ(h(x))$

C. $j(x) = ABS(h(x))$ D. $i(x) = REC(h(x))$

Explorations

Identify each statement in questions 6-10 as true or false. Justify your conclusion.

6. Invertible functions are always one-to-one.   True

7. Invertible functions are always continuous.   False

8. The domain of a function is the same as the domain of its inverse.   false , domain's range are
switched and they
might not be the same

9. The graph of the inverse of $y = \sqrt{3x+1}$ is a parabola.

10. If the graph of a function contains a vertical asymptote, then the graph of its inverse contains a horizontal asymptote.

11. [NC] Graph the relation $x = (\sin y)^2$.

12. Given the graph of $y = k(x)$ in Figure 2.6c with domain restricted to $x \in [-6,6]$.   Assume all parts of the graph of $y = k(x)$ are linear.

A. Graph $y = k^{-1}(x)$.

B. For which value(s) of x is $x = k(x)$?
    $x = -3$ @ $x > 2$

C. Determine an exact value for $k^{-1}(k^{-1}(4))$.
    4

D. Determine an exact value for $k^{-1}(k^{-1}(-1))$.
    @ $1^{2/3}$

E. Write a piecewise definition for $y = k^{-1}(x)$.
    $\begin{cases} 2x+3, -10 \ge 2x > -1 \\ 1/3(x+4), -6 \le x \le 2 \\ -x, x > 2 \end{cases}$

13. Write a possible equation for $x = g(y)$ as shown in Figure 2.6d.
    $x = (y-4)^2 + 2$

Figure 2.6c

Figure 2.6d

14. A. Let $f(x) = \dfrac{1}{1+x}$ and $g(x) = f \circ f(x)$. Find a formula for g^{-1} as a single, simplified fraction.

B. [NC] Describe a CAS procedure that could have been used to find the answer to part A.

Solve for

15. Including domain break points, determine a piecewise definition for z^{-1} if $z(x) = \begin{cases} \dfrac{1}{2}x - \dfrac{3}{2}, & x \in (-\infty, 1) \\ 3x - 4, & x \in [1,2) \\ x, & x \in [2,\infty) \end{cases}$.

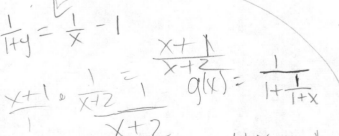

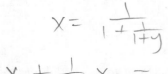

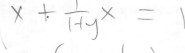

2.7: Further Explorations & Projects

> *For the things of this world cannot be made known without a knowledge of mathematics.*
>   – Roger Bacon

1. **Function Composition**[47]

    Consider $f(x) = 1 - x$ and $g(x) = \dfrac{1}{x}$. f and g can be composed with themselves, each other, and other compositions in many ways. Keep composing these functions with new ones as they are generated and determine simplified formulae for each in terms of x. Surprisingly, only a finite number of functions will be generated by composition, even though there are infinitely many ways to compose each function pair to get the same function. Two very different looking formulae may represent the same function.

    Tasks:
    A. How many distinct functions are there, including the original functions themselves?

    B. List all of the functions you find and how each can be obtained from compositions of f and g.

    C. Show how you know you have found *all* of the possible compositions.

    D. Very carefully, state the domain of every composition you define.

2. **Survival Odds**
    A and B take turns shooting at a single target until it is hit. Assume A goes first and hits his mark 1/3 of the time. B goes second and hits his mark 2/3 of the time. They continue alternating shots until the target is hit. What is A's probability of winning the contest?

    A, B, and C loathe each other and plan a three-way variation of an old West duel.   A hits his mark 1/3 of the time, B hits his mark 2/3 of the time, and C never misses. They want to know what might happen if each gunner chooses a target, and all fire simultaneously. If more than one gunner remains un-hit after the initial salvo, any survivors repeat the process until at most one of A, B, or C remains un-hit. Assuming each wants to maximize his survival chances, determine the survival probabilities of each.

    Because A hits his mark 1/3 of the time, B hits his mark 2/3 of the time, and C never misses, the trio decides to take a more "civilized" approach. A shoots first, then B, then C. They will rotate shots until only one survives. It is A's first shot. To maximize his chances of survival, should he aim at B or C, or should he deliberately miss? Be specific and provide probabilities.

    A, B, and C are now well-studied in probabilities and know the answers to the previous two questions, and have decided to be democratic about their mutual disdain. Assuming their chances of hitting a target are as stated above, and assuming each person gets one vote, which form will the shootout take? Why?

---

[47] Marcus Cohen, et al. *Student Research Projects in Calculus*, Mathematical Assn of America (December 1991). ISBN # 0883855038.

UNIT 3: EXPONENTIALS, LOGARITHMS, & LOGISTICS.

Mathematics is the language with which God has written the universe.

– Galileo Galilei

Enduring Understanding:

- An exponential function is nothing more than repeated multiplication.

- Every exponential function is a simple dilation of every other exponential function.

- A logarithm is an exponent.

- Logarithmic functions are inverses of exponential functions.

- Logistic functions provide more realistic models of initially exponential growth and decay.

- Logistic functions are transformations of exponentials.

This section assumes you have:

- Facility with properties of exponents.

- Knowledge of arithmetic and geometric sequences.

- Familiarity with basic exponential functions, including horizontal asymptotes.

3.1: Exponential Functions

> *The mathematical sciences particularly exhibit order, symmetry, and limitation;*
> *and these are the greatest forms of the beautiful.*
>
> – Aristotle

Following are the properties of exponents for $b, m, n \in \mathbb{R}$ and $b > 0$:

$$1. \quad b^m \cdot b^n = b^{m+n} \qquad\qquad 2. \quad \frac{b^m}{b^n} = b^{m-n} \qquad\qquad 3. \quad \left(b^m\right)^n = b^{mn}$$

$$4. \quad b^0 = 1 \qquad\qquad\qquad 5. \quad b^{-m} = \frac{1}{b^m}$$

Verification of property 1: Because $b^m = \underbrace{b \cdot b \cdot ... \cdot b}_{m \text{ times}}$ and $b^n = \underbrace{b \cdot b \cdot ... \cdot b}_{n \text{ times}}$, then

$$b^m \cdot b^n = \underbrace{\left(b \cdot b \cdot ... \cdot b\right)}_{m \text{ times}} \cdot \underbrace{\left(b \cdot b \cdot ... \cdot b \cdot b\right)}_{n \text{ times}} = \underbrace{b \cdot b \cdot ... \cdot b \cdot b \cdot b}_{m+n \text{ times}} = b^{m+n}.$$

The remaining properties can be similarly established.

Sequences follow an arithmetic pattern when a constant value is added to each term to obtain the next. Functions defined from the explicit formulas of arithmetic sequences form linear functions. Functions defined with variable exponents can be used to form geometric sequences where each term is computed by multiplying the previous by a constant. If a is the initial term and b is the constant multiplier, this pattern is: $a, ab, ab^2, ab^3, ... , ab^n, ...$.

The general form of this family of functions, the **exponential functions**, is $y = a \cdot b^x$, where $b \in \mathbb{R} > 0$. The multiplier, b, is the "base" of the function, and a is its y-intercept (or initial value if the input is time). As with geometric sequences, when $b > 1$ the result is an increasing (growth) exponential function. If $0 < b < 1$, the exponential is a decreasing (decay) function. Figure 3.1a shows graphs for $a = 1$ and $b \in \{0.2, 0.1, 3, 5\}$. Each of $y = b^x$ passes through $(0,1)$, and applying $S_{1,a}$ to $y = b^x$ creates $y = a \cdot b^x$, a generic exponential family containing $(0,a)$.

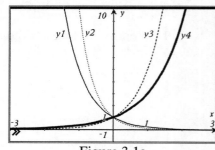

Figure 3.1a

Where power functions with $a = 1$ $\left(y = x^b\right)$ pivot about $(1,1)$ when b varies, exponentials with $a = 1$ $\left(y = b^x\right)$ pivot around $(0,1)$ for various b-values.

Example 1:
If y_1 and y_3 in Figure 3.1a are images of each other over the y-axis, determine their equations.

Let $y_1 = c^x$ and $y_3 = d^x$. Because they are reflections, it follows that $c^x = d^{-x}$ and $c = \frac{1}{d}$. The only values for $b \in \{0.2, 0.1, 3, 5\}$ that satisfy this property are 0.2 and 5. Because y_1 is a decay curve, $y_1 = 0.2^x$ and $y_3 = 5^x$.

Consider the general exponential equation $y = a \cdot b^x$. Because x can take on any value, the domain of the function is all real numbers. In addition, an exponential function is continuous everywhere. For $a > 0$, the function is always positive, giving a range of $y > 0$, with $\lim\limits_{x \to \infty} a \cdot b^x = \infty$ and $\lim\limits_{x \to -\infty} a \cdot b^x = 0$. Also, when $a > 0$, the graph of the function is concave up. The graph has a horizontal asymptote of $y = 0$. The range of an exponential function and its horizontal asymptote vary depending on the transformations on the function.

Example 2:

A. State the transformations that have taken place on $y = 2^x$ to result in $f(x) = -2^{x-1} + 3$.

B. Describe the function's domain, range, and continuity.

C. Determine the concavity and end behavior of f.

D. Find all intercepts of f and use this information to graph f.

  A. The transformations are $T_{1,3} \circ S_{1,-1}$.

  B. The domain of f is $x \in \mathbb{R}$. The range of a parent function exponential is $y \in (0, \infty)$, but f has been reflected over the x-axis and translated up three units, so the final range is $y \in (-\infty, 3)$. Exponential functions are defined everywhere, so f is continuous $\forall x \in \mathbb{R}$.

  C. Because f is a reflection image over the x-axis of a growth exponential function, it must be concave down. The end behavior of f is $\lim\limits_{x \to \infty} f(x) = -\infty$ and $\lim\limits_{x \to -\infty} f(x) = 3^-$. Note that the horizontal asymptote shifts up with the function and is now $y = 3$.

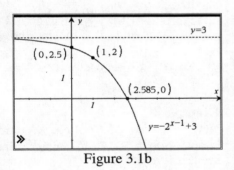
Figure 3.1b

  D. $f(0) = -2^{0-1} + 3 = 2.5$, so the y-intercept is $(0, 2.5)$. For now, the x-intercept must be found using the zeros command on the calculator screen of a CAS or the trace feature on the graphing screen (Figure 3.1b).

Example 3:

Determine an equation for the non-translated exponential function whose graph contains $(-2, 3)$ and $(1, 24)$.

Plug in both points to create a system of equations: $\begin{cases} 3 = ab^{-2} \\ 24 = ab^1 \end{cases}$.

Method 1: Substitution

$$24 = ab^1 \implies a = \frac{24}{b}$$

Substitution then gives

$$3 = \left(\frac{24}{b} \right) b^{-2} \implies b^3 = 8$$

$$\implies b = 2 \implies a = 12$$

So the equation is $y = 12 \cdot 2^x$

Method 2: Elimination

Dividing the equations gives

$$\frac{24}{3} = \frac{\cancel{a}b^1}{\cancel{a}b^{-2}} \implies b^3 = 8$$

$$\implies b = 2 \implies a = 12$$

As before, $y = 12 \cdot 2^x$.

Method 3: Transformations I

Figure 3.1c shows the desired function. In the standard form of an exponential equation, $y = a \cdot b^x$, a represents the value of the function when $x = 0$. The y-intercept is not obvious from the given information, but one can be created by applying a horizontal translation. In this case, $T_{2,0}$ creates a new graph with y-intercept $(0,3)$ (Figure 3.1d), making $y_2 = 3 \cdot b^x$ and equation for the translated function. $T_{2,0}(1,24) = (3,24)$, so substitution of $(3,24)$ leads to $24 = 8 \cdot b^3$ and $b = 2$. The equation of the dashed curve in Figure 3.1d, $y = 3 \cdot 2^x$. Translating this back to the original position gives $y = 3 \cdot 2^{x+2}$.

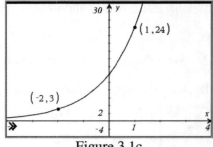

Figure 3.1c

Method 4: Transformations II

Another approach exploits the exponential function-geometric sequence relationship. The output values of any exponential functions with consecutive values of x can be found by multiplying by corresponding output values by a constant value, b. Starting with the lower y-value, multiplying by 8 changes 3 to 24. For $\Delta x = 1$, this would give $y = 3 \cdot 8^x$. However, this function multiplies by 8 every $\Delta x = 3$. Therefore, $y = 3 \cdot 8^{x/3}$. Applying $T_{2,0}$ moves the curve into the desired position:

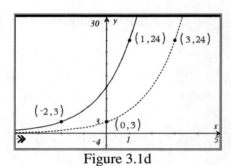

Figure 3.1d

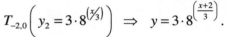

$$T_{-2,0}\left(y_2 = 3 \cdot 8^{\left(x/3\right)} \right) \Rightarrow y = 3 \cdot 8^{\left(\frac{x+2}{3}\right)}.$$

Example 4:

Newton's Law of Cooling: A cup of hot tea at $180°F$ is left on a deck outdoors where the ambient temperature is $50°F$. After five minutes, the temperature of the tea is $130°F$.

A. Predict the temperature of the tea after 20 minutes.

B. If the tea was left on the deck at 4:00 pm, at what time will its temperature reach $55°F$?

   A. Newton's Law of Cooling states that ΔT the *difference* between the temperature at any time t and the ambient temperature, is an exponential function of time t.

$$\Delta T(t) = a \cdot b^t \Rightarrow \begin{cases} 180 - 50 = a \cdot b^0 \\ 130 - 50 = a \cdot b^5 \end{cases} \Rightarrow a = 130 \text{ and } b = \left(\frac{8}{13}\right)^{1/5}.$$

$$\Delta T(t) = 130\left(\frac{8}{13}\right)^{1/5}$$

$$\Delta T(20) = 130\left(\frac{8}{13}\right)^{20/5} \approx 18.64°$$

The actual temperature of the tea is $50° + 18.64° = 68.64°F$.

   B. $\Delta T(t) = 130\left(\frac{8}{13}\right)^{1/5} \Rightarrow 55 - 50 = 130\left(\frac{8}{13}\right)^{1/5} \Rightarrow t = 33.55$ minutes.

Problems for Section 3.1:

Exercises

1. A. Which data sets form geometric patterns? Explain.

    B. Which data sets could be from exponential functions? Explain.

Data Set I	
x	*y*
1	0.6
2	1.2
3	2.4
4	4.8
5	9.6

Data Set II	
x	*y*
5	-1
4	-3
3	-9
2	-27
1	-81

Data Set III	
x	*y*
2	3
4	-6
6	12
8	-24
10	48

For questions 2-7, graph the given function and describe its end behavior using limit notation.

2. $y = 5 \cdot 2^x$

3. $y = 6 \cdot \left(\dfrac{2}{3}\right)^{(x+2)}$

4. $y = -2 \cdot 5^{(2-x)}$

5. $y = |2^x - 3|$

6. $y = 0.3^{|x|}$

7. $y = \sqrt{3^x}$

8. Match each equation with one of the given graphs. (NOTE: The *x*- and *y*-scales may be unequal and are not necessarily identical between graphs.)

    A. $y = -(0.3)^x + 1$ III

    B. $y = 2^x + 1$ II

    C. $y = -2^{-x} - 1$ IV

    D. $y = 3 \cdot \pi^x$ VI

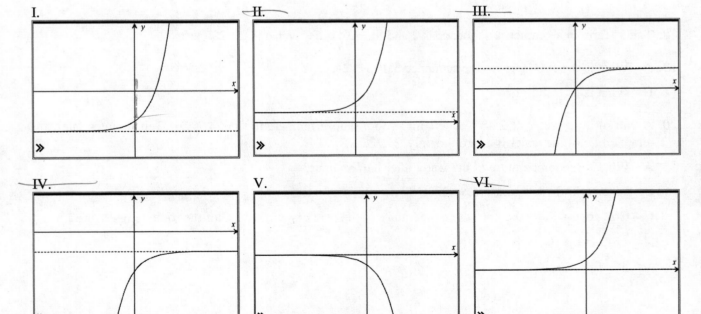

9. State a possible equation for each graph in question 8 that did not match a given equation.

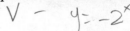

I -- $y = 2^x - 3$ V -- $y = -2^x$

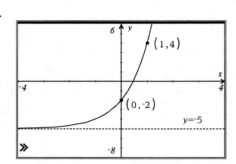

$y = ab^x$
$2 = a \cdot b^0$
$2 = a$

$-1 = 2 \cdot b^{-1}$ $-\dfrac{1}{2} = \dfrac{1}{b}$

$-\dfrac{1}{2}b = 1$

$b = -2$

For questions 10-11, determine an equation of an exponential function to fit the given graph.

10.

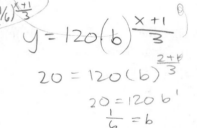

11.

12. Can a non-transformed exponential function contain both $(2, 45)$ and $(0, -5)$? Explain.

13. Find the values for a, b and c if $y = a \cdot b^x + c$ contains the points $(-1, -4)$, $\left(2, \dfrac{5}{4}\right)$ and $\left(3, \dfrac{13}{8}\right)$.

For questions 14-17, find an equation of an exponential function which satisfies each set of criteria.

14. y-intercept $(0, 4)$ and passing through the point $(1, 12)$

15. Contains the points $(-1, 120)$ and $(2, 20)$ $y = 120 \cdot (1/6)^{\frac{x+1}{3}}$

$y = 120(b)^{\frac{x+1}{3}}$

16. $f(-2) = -\dfrac{75}{4}$ and $f(0) = -3$

$20 = 120(b)^{\frac{2+1}{3}}$

$20 = 120 \, b^1$

$\dfrac{1}{6} = b$

17. $f(-2) = \dfrac{2}{3}$ and $f(1) = \dfrac{\sqrt{2}}{6}$

18. The 5th term in a geometric sequence is 2 and the 23rd is 54. What is the 29th term?

19. A. What is the 100th term in the sequence 3, 6, 12, 24, … ?

    B. What is its last digit?

20. A cup of cold water at $2.4°C$ is left in a room whose temperature is $24.7°C$. Ten minutes later, the temperature of the water is approximately $24.7°C$.

    A. What is the temperature of the water after half an hour?

    B. When will the temperature be $20°C$?

    C. How long will it take until the temperature of the water is within $2°C$ of the room temperature?

$y = a \cdot b^x - 3$

$y = a \cdot b^x - a \cdot b^0 - 3$

$2 = a \cdot b^0$ $5 = -a$
                   $a = -5$

$2 = a \cdot 1$

$a = 2$

Explorations

Identify each statement in questions 21-25 as true or false. Justify your conclusion.

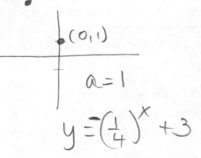

up 3
reflected over x
$(-1, 4)$

21. $\forall b \in \mathbb{R} > 0$, $y = \dfrac{1}{2} \cdot b^{-x}$ is a decreasing function.

22. $\forall b \in \mathbb{R} > 0$, $y = \dfrac{1}{2} \cdot b^{-x}$ is a concave up function.

23. The horizontal asymptote for the graph of $y = 2(5)^x$ is $y = 2$.

24. The _y_-intercept for $y = a \cdot b^x + c$ is $y = a + c$.

25. $\forall x \in \mathbb{R}$, every exponential function, $f(x) = a \cdot b^x$, is continuous.

$(0,1)$

$a = 1$

$y = \left(\dfrac{1}{4}\right)^x + 3$

26. Prove exponent properties 2-5 for positive integer values of _m_ and _n_.

27. Verify that the equations found in Methods 3 and 4 of Example 3 are equivalent to the equations found in Methods 1 and 2.

28. In Method 3 of Example 3, the original points were translated by $T_{2,0}$ to make $(-2,3)$ a _y_-intercept. Find another equation fitting the data by transforming $(1,24)$ into the _y_-intercept. Verify that the new equation is equivalent to those derived by the other approaches.

29. The graphs of three exponential functions are shown in Figure 3.1e. If $f(x) = a \cdot b^x$, $g(x) = b \cdot c^x$, and $h(x) = b \cdot a^x$, list the values _a_, _b_, _c_ and the number 1 in ascending order.

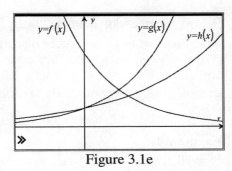

30. Use the graph of $y = 10^x$ and the _REC_ transformation to graph $y = \dfrac{1}{10^x}$. Explain how your result confirms the transformation between the graphs shown in Figure 3.1a.

Figure 3.1e

31. Why is the result of question 30 equivalent to reflecting $y = 10^x$ over the _y_-axis?

32. Use the _SR_ transformation to derive the graph of $y = 3^x$ from the graph of $y = 9^x$. Explain how your result confirms the relationship between the graphs of the exponential family shown in Figure 3.1a.

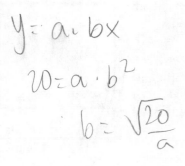

$y = a \cdot bx$

$20 = a \cdot b^2$

$b = \sqrt{\dfrac{20}{a}}$

$y = -(6)^x + 3$

3.2: Polynomials versus Exponentials

The Reader may here observe the Force of Numbers, which can be successfully applied, even to those things, which one would imagine are subject to no Rules. There are very few things which we know, which are not capable of being reduc'd to a Mathematical Reasoning; and when they cannot it's a sign our knowledge of them is very small and confus'd; and when a Mathematical Reasoning can be had it's as great a folly to make use of any other, as to grope for a thing in the dark, when you have a Candle standing by you.

– John Arbuthnot (1692)

For data with constant Δx, if common differences are constant after n Δy iterations, then the data can be perfectly modeled by a polynomial of degree n. To determine the parallel property for exponential functions, explore data from $y = 3^x$ with $\Delta x = 2$ (Figure 3.2a)

x	0		2		4		6		8		10
$y = 3^x$	1		9		81		729		6561		59049
Δy		8		$72 = 8 \cdot 3^2$		$648 = 8 \cdot 3^4$		$5832 = 8 \cdot 3^6$		$52488 = 8 \cdot 3^8$	
$\Delta(\Delta y)$			64		$576 = 64 \cdot 3^2$		$5184 = 64 \cdot 3^4$		$46656 = 64 \cdot 3^6$		

Figure 3.2a

Notice that the pattern of differences is exponential. On each level, the multiplier is unchanged ($y = a \cdot 3^x$), but the initial value, a, varies from one level to the next. That means the change in any exponential function, at any level, is itself exponential.

For n^{th} degree polynomials, every additional Δy layer computed decreases the power of the function modeling that line, eventually reaching zero, but exponential differences seem to be some constant multiples of the original function. This means any exponential growth function ultimately changes much faster than any polynomial. As $x \to \infty$, any exponential growth function will always dominate any polynomial of any degree. In fact, exponentials are the fastest growing continuous functions most students encounter in high school.

Example 1:
Find all solutions of $2^x = x^2$.

An "obvious" answer is $x = 2$, but other solutions aren't quite so simple. A graph suggests the number and locations of solutions (Figure 3.2b).

There is at least one additional solution in Quadrant II. The intersect function gives $x \approx -0.767$ (Figure 3.2c). Left of $x \approx -0.767$, the exponential continues to fall while the quadratic rises, so no more intersections occur $\forall x < 0$. To the right, both functions are still rising, so the situation is more complicated.

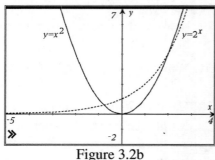

Figure 3.2b

In Figure 3.2c, $y = x^2$ is still above $y = 2^x$, but as $x \to \infty$, any exponential growth functions dominate any other continuous function you have studied thus far, so the exponential *must* intersect the quadratic a third time. Resetting the window (Figure 3.2d) reveals the third intersection point at $(4,16)$.

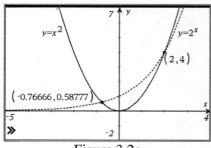

Figure 3.2c

Except in rare circumstances, equations containing functions from different families are impossible to solve using algebraic methods. This example with a power and an exponential function is an example of an equation that could not be solved algebraically for all solutions. When a CAS solution was attempted on a TI-Nspire, note the warning that was posted in the bottom the screen (Figure 3.2e). Such signals should never be ignored. Only a careful analysis of the functions involved (as above in this example) can ensure a complete understanding of the number and types of solutions. Also notice that the $x = 2$ solution is an example of one of the special circumstances under which such an equation could be solved without a calculator.

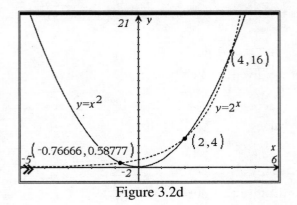

Figure 3.2d

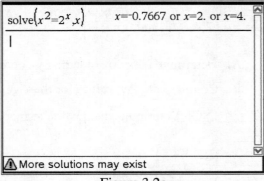

Figure 3.2e

Example 2:

Algebraically solve the equation: $\left(\dfrac{1}{3}\right)^{4-x} = \left(\dfrac{1}{9}\right) \cdot 3^{2x}$

The simplest algebraic approach uses properties of exponents to rewrite both sides in terms of the same base. Once the bases are the same, the exponents can be equated.

$$\left(\frac{1}{3}\right)^{4-x} = \left(\frac{1}{9}\right) \cdot 3^{2x}$$

$$3^{x-4} = \frac{3^{2x}}{3^2}$$

$$3^{x-4} = 3^{2x-2}$$

$$x - 4 = 2x - 2$$

$$x = -2$$

Problems for section 3.2:

Exercises

1. Rank the following functions in order of increasing magnitude as $x \to \infty$.

$$y = x^3, \quad y = 3^x, \quad y = 10^x, \quad y = 10^{10} \cdot x^2, \quad y = 200 \cdot 5^x, \quad y = 1, \quad y = \frac{1}{x}, \quad y = \left(\frac{1}{2}\right)^x, \quad \text{and} \quad y = \frac{1}{\sqrt[3]{x}}$$

$(1/2)^x, \ \dfrac{1}{x}, \ \dfrac{1}{\sqrt[3]{x}}, \ 1, \ 10 \cdot x^2, x^3, 3^x, 200 \cdot 5^x, 10^x$

2. Now rank the functions from question 1 in order of *value* as $x \to -\infty$. For this question, the sign of the function matters.

For questions 3-4, algebraically determine any intersection points of the graphs of the given pairs of functions.

3. $y_1 = 4 \cdot 8^x$ and $y_2 = \left(\frac{1}{2}\right)^{3x-20}$.

4. $y_3 = \left(\frac{1}{81}\right)^{x/4}$ and $y_4 = 3 \cdot 9^{x+1}$.

5. [NC] How many solutions are there to the system, $\begin{cases} y = 0.9^x \\ y = x^2 \end{cases}$?

6. Let $x \in \{2,4,6,8\}$ and $y = \left(\frac{1}{2}\right)^x$.

    A. Determine the corresponding y-coordinates for the given x values.

    B. Compute the Δy values for the y values.

    C. [NC] Determine an equation to model the Δy values.

7. Let $x \in \{1,2,3,4,5\}$ and $y = 5^x$.

    A. Determine the corresponding y-coordinates for the given x values.

    B. Compute the Δy values for the y values.

    C. [NC] Determine an equation to model the Δy values.

8. Given $f(x) = \begin{cases} (x+5)^2, & x \in (-\infty,-3) \\ -\dfrac{4}{3}x, & x \in [-3,3) \\ -8(1/2)^{x/3}, & x \in [3,\infty) \end{cases}$.

    A. Use limit notation to describe the end behavior of f.

    B. Find $f(-3)$.

    C. Find $f(3)$.

    D. Find $\lim\limits_{x \to 3} f(x)$.

    E. Identify all discontinuities in the graph of f.

Explorations

Identify each statement in questions 9-13 as true or false. Justify your conclusion.

9. If $f(x) = x^{100}$ and $g(x) = 0.0001^x$, then $\lim_{x \to -\infty} (g(x) - f(x)) = \infty$.

10. A system of a cubic function and an exponential function could have exactly two solutions.

11. A system containing an odd degree polynomial function and an exponential function must have at least one real solution.

12. [NC] The graphs of $y = 100 \cdot 3^x$ and $y = \pi^x$ never intersect.

13. For any data with constant Δx from a generic exponential function, $y = a \cdot b^x$, $a, b \in \mathbb{R}$, $\exists c \in \mathbb{R} \ni$ the Δy values for the data set are modeled by $y = c \cdot b^x$.

14. Which function is a member of both the exponential and power function families?

15. Find positive values of a and d so that the points (a, d) and $(d, 8a)$ lie on the graph of $y = a \cdot 2^x$.

16. Given a data set for $y = 2 \cdot 5^x$ with $\Delta x = 2$, answer the following questions.

   A. Find an equation to model the common differences if the starting x value of the data set is 1.

   B. Find an equation to model the common differences if the starting x value of the data set is a.

   C. In terms of a, how many different equations are possible for the common differences for $y = 2 \cdot 5^x$ with $\Delta x = 2$?

17. For any data with constant $\Delta x = d$ from a generic exponential function, $y = a \cdot b^x$, $a, b \in \mathbb{R}$, under what conditions on a, b and d will the common differences be modeled by exactly the same exponential function, $y = a \cdot b^x$?

3.3: "Natural" Growth & Decay

How happy the lot of the mathematician. He is judged solely by his peers, and the standard is so high that no colleague or rival can ever win a reputation he does not deserve.

– W. H. Auden

Exponential functions are often used to model real-life situations involving growth and decay. Some examples are population growth, population decay, and financial applications.

If a quantity is growing or decaying at a constant percentage rate per unit of time, the situation can be modeled by an exponential equation of the form $A = P(1 \pm r)^t$, where P is the initial amount, A is the amount after t units of time, and $|r|$ is the percentage rate of growth or decay. The growth factor is $1 + r$ while the decay factor is $1 - r$.

Example 1:
In a certain town, the population grows by 4.1% each year. If in the year 2010, the population was 50,000,

    A. What is the predicted population in 2020?

    B   In what year will the population have doubled?

The population is growing by 4.1%, the growth factor is 1.041, so an equation to represent this situation would be: $A = 50,000(1.041)^t$.

    A. Year 2020 would mean $t = 10 \rightarrow A = 50,000(1.041)^{10} = 74,727$ people.

    B. Since 100,000 is the final population, $100,000 = 50,000(1.041)^t$ or simply $2 = 1.041^t$. $t \approx 17.25$ years using a CAS solve command or by graphing both sides of the equation and finding the x-coordinate of the intersection point.

        It takes between 17 and 18 years for the population to double, so the year in which it happens would be 2028.

The formula for compound growth is $A = P\left(1 + \dfrac{r}{n}\right)^{nt}$ where P is the initial amount, r is the rate of growth expressed in decimal form, n is the number of times the growth is compounded per time interval, t is the number of time intervals, and A is the final amount accumulated after t time intervals. While the formula is most frequently used in financial problems, it applies to any situation in which some initial value grows by a compounded rate.

Example 2:
The **half-life** of a radioactive element is defined as the amount of time it takes for the element to decay to half its previous quantity. Carbon-14 has a half-life of 5730 years.

    A. Write an equation representing the amount of Carbon-14 remaining from a 325g sample after t years.

    B. How much remains after 15,000 years?

    A. **Method 1:** The initial value is 325g, so an equation could be written as $A(t) = 325 \cdot b^t$, where A is the amount present at time t. Because of the half-life, $A(5730) = \dfrac{325}{2}$. Plugging this point into the equation gives $b \approx 0.999879$. An equation is $A(t) = 325(0.999879)^t$.

Method 2: The decay factor for this situation is $\frac{1}{2}$. So an equation could be written as

$$A(t) = 325\left(\frac{1}{2}\right)^n,$$ where n represents the number of half-lives. In terms of t, $n = \dfrac{t}{5730}$. Then

$$A(t) = 325\left(\frac{1}{2}\right)^{t/5730}.$$

B. $A(15,000) = 325\left(\dfrac{1}{2}\right)^{15,000/5730} \approx 52.948g$.

Example 3:
Find the amount accumulated after one year if \$10,000 is invested at an annual interest rate of 10%. Determine the values assuming the interest is compounded annually, quarterly, monthly, and daily.

The given information is $P = 10,000$, $r = 0.1$, $t = 1$, with n varying for each part. Figure 3.3a shows the computation of the four values.

N	A		N	A
1	$A = 10000\left(1 + \dfrac{0.10}{1}\right)^{1\cdot1} = \$11,000.00$		12	$A = 10000\left(1 + \dfrac{0.10}{12}\right)^{12\cdot1} \approx \$11047.13$
4	$A = 10000\left(1 + \dfrac{0.10}{4}\right)^{4\cdot1} \approx \$11,038.13$		365	$A = 10000\left(1 + \dfrac{0.10}{365}\right)^{365\cdot1} \approx \$11051.56$

Figure 3.3a

The more frequently interest is compounded, the larger the final amount becomes. But, A's gain decreases rapidly, even as n grows faster. As n moved from 1 to 4, A gained \$38.13. For n from 4 to 12, A gained \$9.00 more. Increasing the number of compoundings by 353 more units adds only \$4.43 more. As $n \to \infty$, there appears to be a limit to how much A can grow.

In $A = P\left(1 + \dfrac{r}{n}\right)^{nt}$, as $n \to \infty$, so does the exponent, suggesting larger A values. But large values of n also shrink $\left(1 + \dfrac{r}{n}\right)$, suggesting smaller A values. The expression has competing components. Figure 3.3b shows that $\left(1 + \dfrac{1}{n}\right)^n$ appears to approach a finite number as n grows infinitely large. This limit is an irrational number and is so widely used it has a unique name: $\displaystyle\lim_{n\to\infty}\left(1 + \dfrac{1}{n}\right)^n = e \approx 2.71818$.

x	$y = \left(1 + \dfrac{1}{x}\right)^x$
1	2.00000
100	2.70481
1,000	2.71692
10,000	2.71815
100,000	2.71827
1,000,000	2.71828

Figure 3.3b

What if r varies? That is, what is $\displaystyle\lim_{n\to\infty}\left(1 + \dfrac{r}{n}\right)^n$? To explore, compute $\displaystyle\lim_{n\to\infty}\left(1 + \dfrac{r}{n}\right)^n$ for different values of r (Figure 3.3c).

r	$\displaystyle\lim_{n\to\infty}\left(1 + \dfrac{r}{n}\right)^n$
0	1
1	2.718
2	7.389
3	20.086

Figure 3.3c

The $\left(r, \lim\limits_{n\to\infty}\left(1+\dfrac{r}{n}\right)^{n} \right)$ ordered pairs from Figure 3.3d graphically (Figure 3.3d) appear exponential. Two points on a generic exponential curve $y = a\cdot b^x$ are $(0,a)$ and $(1,ab)$. Because this curve contains $(0,1)$, $a=1$. The point $(1,e)$ says $ab=b=e$, so the exponential function modeling the data is $y = a\cdot b^x = 1\cdot e^x = e^x$. The other y values in Figure 3.3c can now be shown to form a geometric sequence with ratio e and an initial value of 1, further verifying the claim that $y = e^x$ models the data. While calculus is ultimately required to prove these results, this is strong numerical evidence that $\lim\limits_{n\to\infty}\left(1+\dfrac{r}{n}\right)^{n} = e^r$.

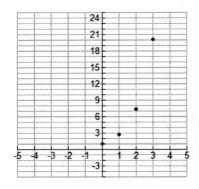

Figure 3.3d

As n increases without bound in the compound growth formula, the formula takes on another familiar form, the continuous growth formula. These are not two different formulas, they represent the same concept under different conditions.

$$A = \lim_{n\to\infty}\left[P\left(1+\frac{r}{n}\right)^{nt} \right] = P\cdot\left[\lim_{n\to\infty}\left(1+\frac{r}{n}\right)^{n} \right]^{t} = P\cdot\left[e^r \right]^{t} = P\cdot e^{rt}.$$

Because $2 < e < 3$, the graph of $y = e^x$ lies between the graphs of $y = 2^x$ and $y = 3^x$.

Example 4:

Which is a better 10-year investment, 10.1% compounded quarterly or 10% compounded continuously?

Assuming that the principal amount is the same for both investments, any convenient value can be chosen for P. To simplify, let $P=1$.

- Using quarterly compounding: $A_1 = \left(1+\dfrac{.101}{4}\right)^{40} \approx 2.711$

- Using continuous compounding: $A = Pe^{rt} = e^{(0.1)(10)} = e \approx 2.718$.

The continuous compounding option is a better investment.

Problems for section 3.3:

Exercises

1. A certain population of bacteria is found to double every hour. If at 12:00 noon there are 100 bacteria, how many will there be at 6:15 pm?

2. A given population increases at a constant rate of 5.5% per year. The population was 35,000 people on January 1, 1995.

   A. Write an exponential function to model the population after t years.

   B. Graph the population function.

   C. What was the population in July 2000?

   D. According to this model, when will (did) the population hit 75,000?

   E. Does this model make practical sense for all values of t? Explain.

3. What is the interest earned after 6 years on an initial investment of $1500 compounded at an annual rate of 5.5%, if the interest was compounded

    A. Semi-annually   B. Monthly   C. Continuously

4. Twins Rebecca and Michelle each invested $25,000 on their 20th birthday. Michelle selected a plan that paid 12% compounded semi-annually, while Rebecca chose a plan that paid 11.8% compounded continuously. They made no further investments.

    A. Which twin's account was the first to be worth $100,000?

    B. Will either account be worth one million dollars before their 51st birthday?

    C. Assuming all other constants as given, what would Rebecca's interest rate need to be if he wanted his account to be worth one million dollars precisely on his 60th birthday?

    D. Assuming all other constants as given, how quickly would Michelle's account need to compound if she wanted her account to be worth one million dollars precisely on her 45th birthday?

5. Radioactive Cobalt-56 has a half-life of 78.76 days. What is its continuous decay rate?

6. On average, how long does it take for one tenth of a sample of Copper-64 to decay if the half-life of Cu-64 is 12.701 hours?

Explorations

Identify each statement in questions 7-11 as true or false. Justify your conclusion.

    7. Every non-translated exponential function is a *horizontal* constant dilation of every other non-translated exponential function.

    8. Every non-translated exponential function is a *vertical* constant dilation of every other non-translated exponential function.

    9. $\forall a,b \in \mathbb{R} > 0$, the equation $a = b^x$ has at most one solution.

    10. A savings account with an annual interest rate of 8% compounded continuously is a better investment than another offering 8.1% compounded annually.

    11. Regardless of the value of its base, every exponential decay function has a half-life.

[NC] Graph each function given in questions 12-15.

    12. $y = e^{|x|}$ 13. $y = e^{-|x|}$

    14. $y = e^x + x$ 15. $y = e^x - x^4$

16. [NC] How many solutions are there to the equation $e^{|x|} = x^3 + 2$? Justify.

17. Derive the graph of $y = e^{-2x}$ from the graph of $y = e^x$.

18. Find a possible equation for the graph in Figure 3.3e.

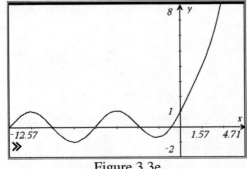

Figure 3.3e

19. The function $f(x) = e^{x/2}$ is the image of $y = e^x$ under two different transformations: $S_{2,1}$ and the square root transformation. Algebraically and graphically explain why each produces the same results for this function.

20. Graphically and algebraically explain why the reciprocal transformation of $y = e^x$ is equivalent to reflecting $y = e^x$ over the y-axis.

21. Approximate the value of $\displaystyle\lim_{n \to \infty} \sum_{k=0}^{n} \frac{1}{k!}$.

[handwritten notes:]

$y = e^x$

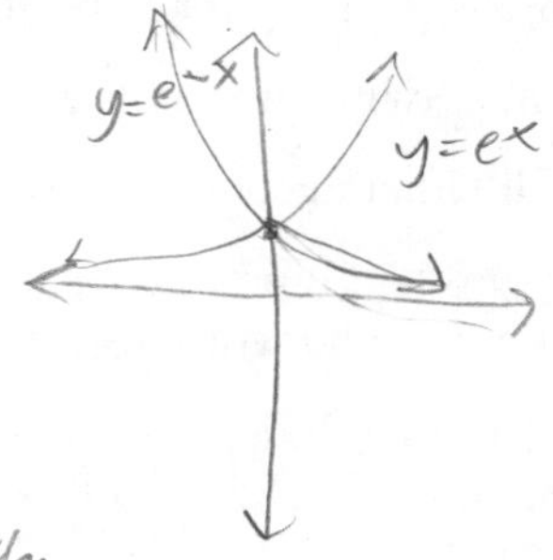

when the inputs are made – the graph is reflected over the y axis, this would make the equation $y = e^{-x}$ which we know by properties of exponential equation is the same as $y = \frac{1}{e^x}$, the reciprocal of $y = e^x$

$y = e^{-x}$ $y = e^x$

3.4: Inverses of Exponential Functions

On two occasions I have been asked [by members of Parliament], *'Pray, Mr. Babbage, if you put into the machine wrong figures, will the right answers come out?' I am not able rightly to apprehend the kind of confusion of ideas that could provoke such a question.*
— Charles Babbage (the first person to propose a programmable computer)

Imagine being asked to solve each of the following equations for x.

(1) $2^x = 8$ (2) $2^x = 16$ (3) $2^x = 10$

All three have solutions, but the last is more elusive than the first two. In general, when the value of y in $y = 2^x$ is not the result of an integer power of 2, the value of x is rarely simple or obvious. Even so, it is clear that there is *some* value of x that satisfies the equation. Finding input values from given output values is the essence of inverse functions. So, what happens when an "inverse" transformation is applied to $y = 2^x$?

The only algebraic way to accomplish this is by introducing a new function, the **logarithm**. After switching variables in the equation to get $x = 2^y$, then y is said to be the "logarithm base 2 of x" and is notated $y = \log_2 x$. In other words, $\boxed{y} = \log_b x$ iff $x = b^{\boxed{y}}$. The input of an exponential is the output of the corresponding logarithm, and vice versa.

In general, $y = b^x$ and $y = \log_b x$ are inverse functions. Recognize that $y = \log_b x$ and $x = b^y$ are *identical* statements. Because b is also the base of an exponential function, $b > 0$ for all logarithmic functions.

Example 1:
Express $81 = 3^4$ in logarithmic form.

Relying on the equivalence of $x = b^y$ and $y = \log_b x$, the answer is $4 = \log_3 81$.

Just as exponents have certain algebraic properties, so do logarithms. $\forall\, a, b, c, M, N \in \mathbb{R} > 0$, $b, c \neq 1$

1. $\log_b (MN) = \log_b M + \log_b N$ 2. $\log_b \left(\dfrac{M}{N} \right) = \log_b M - \log_b N$

3. $\log_b M^p = p \log_b M$ 4. $\log_b a = \dfrac{\log_c a}{\log_c b}$ (A Change of Base Property)

Example 2:
Prove logarithm property 1.

Let $x = \log_b M$, and let $y = \log_b N$.

$$x = \log_b M \;\Rightarrow\; b^x = M \quad \text{and} \quad y = \log_b N \;\Rightarrow\; b^y = N$$

Therefore,

$MN = b^x \cdot b^y$

$MN = b^{x+y}$ (properties of exponents)

$\log_b (MN) = x + y$ (definition of a logarithm)

$\log_b (MN) = \log_b M + \log_b N$ (substitution)   *QED*

The other logarithm properties can be similarly established.

Example 3: (An alternative change of base rule)

$\forall n \in \mathbb{R} \neq 0$ and $\forall b, x \in \mathbb{R} > 0 \ni b \neq 1$, prove $\log_b x = \log_{b^n}\left(x^n\right)$.

Let $A \equiv \log_b x$. Then, $b^A = x \;\Rightarrow\; \left(b^A\right)^n = x^n \;\Rightarrow\; \left(b^n\right)^A = x^n \;\Rightarrow\; A = \log_{b^n}\left(x^n\right)$.   *QED*

While any base can be used for logarithms and (for now) any base is as good as any other, mathematical convention (and therefore all calculators) make the most use of base 10 (denoted by a log with no subscript), also known as a **common logarithm**, and base e (denoted by "ln"), or a **natural logarithm**. Unsolved problem (3) from the beginning of this section can now be solved.

Example 4:

Solve the equation $2^x = 10$.

$$2^x = 10 \;\Rightarrow\; x = \log_2 10 = \frac{\log 10}{\log 2} = \frac{1}{\log 2} \approx 3.322$$

Alternatively, apply a common logarithm, invoke logarithm property 3, and simplify.

$$2^x = 10 \;\Rightarrow\; \log 2^x = \log 10 \;\Rightarrow\; x \cdot \log 2 = \log 10 \;\Rightarrow\; x = \frac{\log 10}{\log 2} = \frac{1}{\log 2} \approx 3.322$$

Example 5:

Solve $2^x = 10$ using the natural logarithm instead of the common logarithm.

$$2^x = 10 \;\Rightarrow\; \ln 2^x = \ln 10 \;\Rightarrow\; x \ln 2 = \ln 10 \;\Rightarrow\; x = \frac{\ln 10}{\ln 2} \approx 3.322$$

Both approaches to solving the equation $2^x = 10$ gave the same answer, making the base of a logarithm appear to be irrelevant - the important point made by property 4. While this property allows any logarithmic function to be written in terms of any other, its most utilitarian value is in making all logarithms calculator-accessible. Most calculators use base 10 and/or e only. However, many CAS allow logarithms with other positive bases.

Example 6:

Solve for x in the equation $\ln 4 + 2\ln x = 2$.

There are multiple ways of approaching this problem, but one is to try to combine the logarithms.

$$\ln 4 + \ln x^2 = 2 \;\Rightarrow\; \ln 4x^2 = 2 \;\Rightarrow\; 4x^2 = e^2 \;\Rightarrow\; x = \pm\frac{e}{2}$$

However, $x = -\dfrac{e}{2}$ is not valid because $\ln\left(-\dfrac{e}{2}\right)$ is a non-real quantity. The only solution is $x = \dfrac{e}{2}$. The issue arose when was $2\ln x$ was changed to $\ln x^2$ using property 3, inadvertently including negative x-values. Be aware that $2\ln x = \ln x^2$ ***only*** for $x > 0$, the portion of the domain common to both expressions.

Keeping domain in mind

As shown in Example 6, when working with the properties of logarithms, it is important to be aware of potential domain issues. For instance, in logarithm property 1, the input to the logarithm must be positive. If M and N are algebraic expressions, then for the left hand side of the property to exist, MN must be positive, which happens when M and N are both positive or both negative. The right hand side, however, requires M and N both to be positive. So the property is only true over the *common domain* of the expression on the left and the expression on the right.

Typically, for properties 1 and 2, if we change to a single-logarithm form, the domain is expanded, and we gain domain values. When moving from single-logarithm form to a sum or difference of logarithms, the domain is shrunk and we lose domain values. For property 3, by bringing down the exponent, it is possible to lose domain values.

Example 7:

Solve each logarithmic equation for x.

  A. $3\ln(x-3)+4=5$ B. $\log 2+\log(x-1)=\log 30$

  C. $\log_x 81=4$ D. $\log_2 x^3=12$

  A. $3\ln(x-3)+4=5$

$$\Rightarrow\ \ln(x-3)=\frac{1}{3}\ \Rightarrow\ x-3=e^{\frac{1}{3}}\ \Rightarrow\ x=e^{\frac{1}{3}}+3$$

  B. $\log 2+\log(x-1)=\log 30$

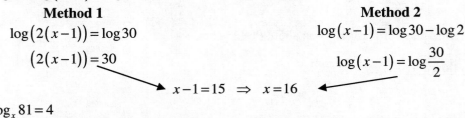

Method 1	**Method 2**
$\log(2(x-1))=\log 30$	$\log(x-1)=\log 30-\log 2$
$(2(x-1))=30$	$\log(x-1)=\log\dfrac{30}{2}$

$$x-1=15\ \Rightarrow\ x=16$$

  C. $\log_x 81=4$

$$\Rightarrow\ x^4=81\ \Rightarrow\ x=\pm 3\ \Rightarrow\ x=3\ \text{since logarithm bases are positive.}$$

  D. $\log_2 x^3=12$

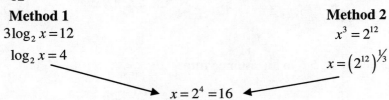

Method 1	**Method 2**
$3\log_2 x=12$	$x^3=2^{12}$
$\log_2 x=4$	$x=\left(2^{12}\right)^{\frac{1}{3}}$

$$x=2^4=16$$

Example 8:

Solve $2\cdot 3^{4x}+1=10$ for x.

$$2\cdot 3^{4x}+1=10\ \Rightarrow\ 3^{4x}=\frac{9}{2}\ \Rightarrow\ 4x=\log_3 4.5\ \Rightarrow\ x=\frac{\log_3 4.5}{4}$$

Example 9:

An equation representing the decay of a 325g initial sample of Carbon-14 is $A(t) = 325\left(\frac{1}{2}\right)^{t/5730}$.

A. Find the continuous decay rate of Carbon-14.

B. Find the third-life of Carbon-14 (the time it takes for the element to decay to one-third of its previous amount.

A. Equate the expressions with different rates.

$$325\left(\frac{1}{2}\right)^{t/5730} = 325e^{rt} \quad\Rightarrow\quad \left(\frac{1}{2}\right)^{1/5730} = e^{r} \quad\Rightarrow\quad r = \ln\left(\frac{1}{2}\right)^{1/5730} = \frac{\ln\left(\frac{1}{2}\right)}{5730} \approx -0.000121$$

So, Carbon-14 decays continuously at about 0.0121%.

B. Let the third-life be denoted by T_3

$$\left(\frac{1}{2}\right)^{t/5730} = \left(\left(\frac{1}{3}\right)^{\log_{1/3}\left(\frac{1}{2}\right)}\right)^{t/5730} = \left(\frac{1}{3}\right)^{\log_{1/3}\left(\frac{1}{2}\right)\cdot t / 5730}$$

$$\frac{\log_{1/3}\left(\frac{1}{2}\right)}{5730}\cdot t = \frac{t}{T_3} \quad\Rightarrow\quad T_3 = \frac{5730}{\log_{1/3}\left(\frac{1}{2}\right)} \approx 9081.84\, yrs$$

The "third-life" of Carbon-14 is about 9082 years.

(handwritten notes in right margin:)
$3^{\log_9 45}\cdot \log_3 4^{\log_5 10}$

$\left(3^{\log_9 4}\right) \log_9 5 \cdot \log_5 10$

$\left(4^{\log_9 5}\right) \log_5 10$

$\left(5^{\log_5 10}\right)$

10

Problems for section 3.4:

Exercises

Evaluate each logarithm in questions 1-4.

1. $\log_3 243$

2. $5\log_9 3$

3. $\log_{\frac{1}{4}} 16$

4. $\log_{2/3}\left(\frac{81}{16}\right)$

Convert each equation in questions 5-6 to logarithmic form.

5. $6^3 = 216$

6. $\frac{1}{128} = 2^{-7}$

Convert each equation in questions 7-8 to exponential form.

7. $\log_{3/4}\frac{16}{9} = -2$

8. $\log_{25}\sqrt{5} = \frac{1}{4}$

Write each expression in questions 9-14 as a single logarithm, and if possible, evaluate.

9. $\log 25 - 2\log 0.5$

10. $\ln 30 + 2\ln 2$

11. $6^{2\log_6 3}$ *(handwritten: $=9^6$ ... $6^{\log_6 3^2}$)*

12. $\ln 5e^3 + \ln\frac{2}{e^4}$

13. $\log_2 5 + 3\log_4 5$

14. $3^{\log_4 5 \cdot \log_3 4^{\log_5 10}}$ *(handwritten: $= 10$, $3^{\log_3 4 \cdot \log_4 5 \cdot \log_5 10}$)*

(handwritten at bottom:) 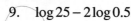 $\ln\left(5e^3 \cdot \frac{2}{e^4}\right) = \ln\frac{10}{e}$

15. Given $\log_b 2 = x$, $\log_b 3 = y$ and $\log_b 7 = z$, evaluate the following in terms of x, y and z.

    A. $\log_b 126$
    B. $\log_b \sqrt{54}$
    C. $\log_{b^2} \dfrac{28}{81}$

Solve each equation in questions 16-25 for x.

16. $\log_5 (x-1) + \log_5 (x-2) = \log_5 12$

17. $\ln (x+3) = 2 + \ln (8-x)$

18. $\log (3x+1) - 2 = 0$

19. $\log x = 4$

20. $\log_2 (\log_{3x-2} x) = -1$

21. $\log_x 16 = -\dfrac{2}{9}$

22. $5^{x+4} = 3^{2x-1}$

23. $2^{5x} = 8^{x+1}$

24. $50e^{0.35t} = 200$

25. $2 = \dfrac{9^x}{3^x + 4}$

26. A. $\forall b \in \mathbb{R} > 0$, what is the value of $\log_b b$?

    B. $\forall b \in \mathbb{R} > 0$, what is the value of $\log_b 1$?

Explorations

Identify each statement in questions 27-31 as true or false. Justify your conclusion.

27. $\log (3x-6) + \log (4-2x) = \log \left[(3x-6)(4-2x) \right]$

28. $\forall A, B > 0$, $\log (A-B) = \log A - \log B$

29. $\forall x \in \mathbb{R} > 0$, $\log_b 2x = 3$ and $b = \sqrt[3]{2x}$ are equivalent equations.

30. $\forall x \in \mathbb{R}$, $\log x^5 = 5 \log x$

31. $\log_b x = \log_{1/b} \left(\dfrac{1}{x} \right)$

32. Prove logarithm properties 2-4.

33. If $f(x) = 3 \cdot 2^x$, $g(x) = 7^x$, and $f(x) = h(g(x))$, then determine a formula for h.

34. A student once mistakenly thought $\log (A-B) = \log A - \log B$ and presented his teacher with specific values of A and B to prove his point. Find values of A and B that the student might have used.

35. A. Find the inverse of $y = \dfrac{e^x + e^{-x}}{2}$.

    B. Solve $\dfrac{e^x + e^{-x}}{2} = 4$ for x.

36. [NC] Find a formula for the inverse of $g(x) = \dfrac{\ln x - 5}{2\ln x + 7}$.

37. Let $f^{-1}(x) = 200(1.05)^{x+1}$.

    A. Find a formula for f.

    B. Determine an *exact* value for $f^{-1}(x) = 250$.

    C. Determine an *exact* value for $f(x) = 0$.

    D. Sketch a graph of $f(x)$; be sure to include any intercepts and asymptotes.

38. [NC] Compute $\displaystyle\sum_{x=1}^{243} \lfloor \log_3 x \rfloor$ if $\lfloor A \rfloor$ is the greatest integer less than or equal to A. For example, $\lfloor \pi \rfloor = 3$.

3.5: Graphs & Transformations of Logarithmic Functions

Euclid taught me that without assumptions there is no proof.
Therefore, in any argument, examine the assumptions.

– Eric Temple Bell

Since logarithmic functions are inverses of exponential functions and inverse functions are reflection images over $y = x$, simply reflect an exponential graph over $y = x$ to get the graph of a logarithm function.

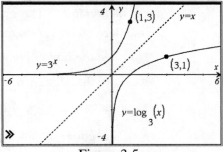

Figure 3.5a shows $y_1 = 3^x$ and its reflection image over $y = x$, $y_2 = \log_3 x$. Two corresponding ordered pairs are also labeled to illustrate the switching of x and y coordinates.

Figure 3.5a

Recall that there are two basic types of graphs for exponential functions (growth and decay). Therefore, there are also only two basic shapes of logarithmic graphs, each the reflection of its corresponding type of exponential function.

Example 1:
Graph $y = \log_{0.5} x$.

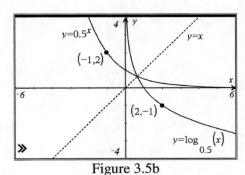

This is the inverse of $y = (0.5)^x$, so the logarithm has a vertical asymptote at $x = 0$ and an intercept at $(1, 0)$. Because $\lim_{x \to \infty}(0.5)^x = 0^+$, $\lim_{x \to 0^+}(\log_{0.5} x) = \infty$ (Figure 3.5b). Notice the "exchange" of the x- and y-bounds on the limit.

Alternatively, the graph of $y = \log_{0.5} x$ could have been obtained by reflecting $y = (0.5)^x$ over $y = x$.

Figure 3.5b

Note, increasing exponential graphs correspond to increasing logarithmic functions (Figure 3.5a) and decreasing exponentials correspond to decreasing logarithmic functions (Figure 3.5b).

Example 2:
Graph $y = \ln x^2$.

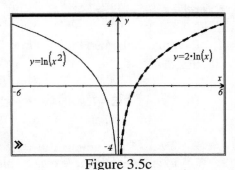

Method I: $\ln x^2 = 2 \ln x \ \forall x > 0$, so the graph of $y = \ln x^2$ is the image of $y = \ln x$ under $S_{1,2}$. $y = \ln x^2$ is an even function, so its left side must be the reflection image over the y-axis of its right side. Figure 3.5c shows $y = 2 \ln x$ as a dashed curve and $y = \ln x^2$ as a solid curve.

Figure 3.5c

Method II: Use a **logarithmic transformation** (*LN*). Figure 3.5d shows a graph of $y = x$ and the transformed $y = \ln x$. As with other transformations, consider the transformation questions from Unit 2:

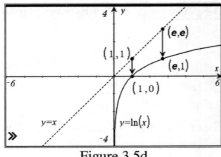

Figure 3.5d

i) *Which output values remain the same?*
This is equivalent to asking for the values of x which satisfy $x = \ln x$. The graphs of $y = x$ and $y = \ln x$ never intersect, so no values remain the same.

ii) *How are the remaining values transformed?*

- Like the *SR* transformation, an *LN* transformation *eliminates* all pre-image negative y-values.

- Points where $y = 0$ become vertical asymptotes.[1]

- The graph of $y = \ln x$ lies entirely below that of $y = x$, so all remaining points will have smaller y-values.

- Pre-image points where $y = 1$ become x-intercepts.

- Pre-image points where $y = e$ become image points with $y = 1$.

This question asks for the graph of the *LN* transformation of $y = x^2$. The only non-positive point on $y = x^2$ is the origin, so the image graph has a vertical asymptote at $x = 0$ (Figure 3.5e).

Figure 3.5e also shows pre-image points where $y = 1$ (at $x = \pm 1$) becoming x-intercepts of the image and pre-image points where $y = e$ (at $x = \pm\sqrt{e}$) becoming image points with $y = 1$.

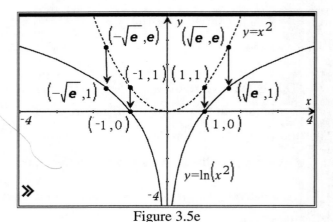

Figure 3.5e

Problems for section 3.5:

Exercises

1. Example 2 claimed that $y = \ln x^2$ was an even function. Prove that this claim.

2. A. [NC] Figure 3.5f shows graphs of $y = \log_3 x$, $y = \log_5 x$, $y = \log_{0.2} x$, and $y = \log_{0.1} x$. Which is which? Explain.

   B. Two of the graphs are reflections of each other over the x-axis. Identify these two graphs.

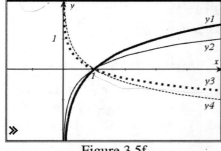

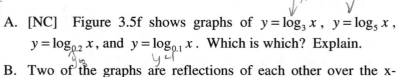

Figure 3.5f

---

[1] In logarithmic transformations, the closer a pre-image y-coordinate is to 0, the more its logarithmic-transformed value moves toward negative infinity. This is much the same as what happens with the "creation" of vertical asymptotes in reciprocal transformations.

3. The graph of a non-transformed logarithm contains $(5, -12)$. What is the base of the function?

Graph each function in questions 4-17, and state its domain and range.

4. $y = \log_2 x - 3$

5. $y = \log_3 (4 - x)$

6. $y = \log\left(\frac{1}{2} x\right)$ $D: (0, \infty)$ $R: (-\infty, \infty)$

7. $y = |\ln(x + 4)|$ $D: (-4, \infty)$ $R: [0, \infty)$

8. $y = \dfrac{1}{\log_{0.25} x}$

9. $y = \ln|x|$

10. $y = \ln(x^2 - 2x - 15)$

11. $y = \ln(x + 3) + \ln(x - 5)$

12. $y = \ln x^2$

13. $y = 2\ln x$

14. $y = \ln\left(\dfrac{x - 1}{x + 4}\right)$

15. $y = \ln(x - 1) - \ln(x + 4)$

16. $y = (\log x)^2$

17. $y = \log x - x^2$

18. A. Use the *LN* transformation to graph $y = \ln\left(\dfrac{1}{x}\right)$.

    B. Use logarithmic properties on $\ln\left(\dfrac{1}{x}\right)$ to determine an alternative transformational approach to graph

    of $y = \ln\left(\dfrac{1}{x}\right)$. Do the different approaches produce the same graph?

19. Use the *REC* transformation to graph $y = \dfrac{1}{\ln x}$.

Explorations

Identify each statement in questions 20-24 as true or false. Justify your conclusion.

20. The domain of $y_1 = \log|x|$ is the same as the domain of $y_2 = \log x^2$ true, in both instances negative inputs are turned positive making them okay

21. The domain of $y_1 = \log x^{2n}$ is the same as the domain of $y_2 = 2\log x^n$, $\forall n \in \mathbb{Z}$.

22. For $0 < c < 1$, $\lim_{x \to 0^+} \log_c x = \infty$

23. The graphs of $y = (\log x)^2$ and $y = e^x$ intersect at only two points.

24. $\log_c x = -\log_{\frac{1}{c}} x$ $\forall x > 0$. True

25. The change of base law for logarithms suggests any base for a logarithm is as good as any other. From another point of view, this suggests that every logarithm function is a simple transformation (constant translation or scale change) of any other. Find a simple transformation that maps $y = \log x$ onto $y = \ln x$.

26. Two curves are considered tangent if they share a tangent line. Figure 3.5g shows the only exponential function of the form $y = a^x$ that is tangent to its inverse.

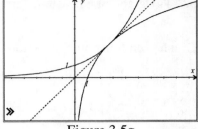

Figure 3.5g

    A. What is the equation of the mutual tangent line to $y = a^x$ and its inverse? Why?

    B. If the formula for the slope of a tangent line to $y = a^x$ at any point x is $m = \ln a \cdot a^x$, then for what value of a is $y = a^x$ tangent to its inverse?

27. Find functions f and g such that the domains of $y = \ln(f(x))$, $y = \ln(g(x))$ and $y = \ln(f(x) \cdot g(x))$ are all distinct, non-overlapping sets.

28. Let $L(x) \equiv \ln(x)$, $f(x) = x + 5$, $g(x) = -x^2 + 2$, $h(x) = e^x$, and $k(x)$ be the piecewise linear function in Figure 3.5h.

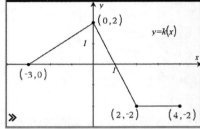

Figure 3.5h

    A. Graph $L(f(x))$, $L(g(x))$, and $L(h(x))$.

    B. Graph $L(k(x))$. State all intercepts.

    C. **[Challenge]** Explore $f(L(x))$, $g(L(x))$, and $h(L(x))$; then hypothesize what happens when any function $y = t(x)$ is composed with $L(x) = \ln(x)$ to obtain $t(L(x))$?

29. Let $f(x) = \log_a(x)$ and $g(x) = \log_c x$ for some values of $a, c \in \mathbb{R}$. Graphs of f and g are shown in Figure 3.5i.

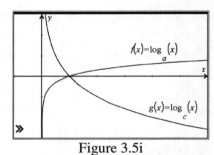

Figure 3.5i

    A. Are the following statements true or false? Explain.

        i. $\log_a c < 0$

        ii. $a \leq 1$

    B. State one or two basic transformations which can map f onto g.

3.6: Logistic Functions

Out yonder there is this huge world, which exists independently of us human beings, and which stands before us like a great, eternal riddle.

– Albert Einstein

Exponential functions often provide good models for real-life growth and decay situations but they are limited. If a population doubles every year, it is reasonable to initially model it by an exponential function. However, populations cannot grow to infinity, so the exponential model fails at some point. A more realistic model for long-term population growth is needed—one that incorporates the exponential nature of the growth or decay while providing a limit to growth. The logistic function satisfies these requirements.

The basic form of a **logistic function** is $y = \dfrac{a}{1+b^x}$ with $b > 0$. Figure 3.6a shows logistic graphs for $a = 1$ and $b \in \{3,\ 10,\ 0.1,\ 0.5\}$. There are two significant characteristics of all of these logistic graphs.

- All contain $\left(0, \dfrac{1}{2}\right)$, an apparent inflection point.

- All have two horizontal asymptotes: the x-axis and $y = 1$.

Also notice that $y = \dfrac{a}{1+b^x}$, requires only three transformations on $y = b^x$ to create the logistic graph.

1. Assuming $b > 0$, applying $T_{0,1}$ to $y = b^x$ gives $y = b^x + 1$ (Figure 3.6b).

2. Applying the reciprocal transformation to this result gives $y = \dfrac{1}{1+b^x}$ (Figure 3.6c).

3. Applying $S_{1,a}$ to step 2 creates the general logistic function, $y = \dfrac{a}{1+b^x}$.

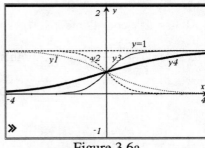

Figure 3.6a

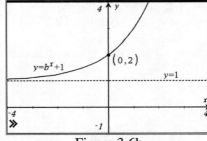

Figure 3.6b

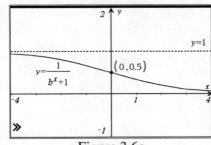

Figure 3.6c

General logistic models are usually presented as $y = \dfrac{a}{1+d\cdot b^x}$ or $y = \dfrac{a}{1+b^{x-c}}$. Both use $T_{0,1}$, the reciprocal transformation, and $S_{1,a}$ as their final steps, but the initial transformations vary. The former applies $S_{1,d}$ first while the latter uses $T_{c,0}$. While they look different, the two are equivalent.

The reciprocal transformation and $S_{1,a}$ combine to create the upper horizontal asymptote at $y = a$ for logistic graphs. Because it creates an upper bound on these functions, a is often called the **carrying capacity** or "growth limit" for logistic functions.

Example 1:

Graph $y = \dfrac{5}{1 + 3^{x-1}}$.

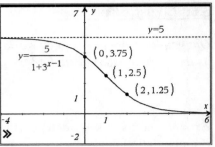

Figure 3.6d

In this logistic equation, $a = 5$, so the end behavior asymptotes are $y = 5$ and $y = 0$. The horizontal translation puts the inflection point at $(1, 2.5)$. Because $b = 3 > 1$, the original exponential is a growth function, so this logistic decays as a result of the reciprocal transformation. Another point on the graph is the y-intercept, $(0, 3.75)$. Its point symmetric image with respect to the inflection point is $(2, 1.25)$ (Figure 3.6d).

Example 2:

For each of the following situations, find an equation for a logistic function that satisfies the given criteria.

A. Carrying capacity 10, inflection point $(2, 5)$, graph passes through $(4, 8)$.

B. Carrying capacity 20, initial value 5, passes through $(-1, 8)$.

C. Passes through the points $(1, 120)$, $\left(-3, \dfrac{40}{11}\right)$ and $\left(-1, \dfrac{200}{7}\right)$.

A. The carrying capacity means $a = 10$, and the inflection point makes the $y = \dfrac{a}{1 + b^{x-c}}$ form easier with $c = 2$. Plug in $(4, 8)$ and the known values of a, c, and solve for b.

$$8 = \frac{10}{1 + b^{4-2}} \quad\Rightarrow\quad 1 + b^2 = \frac{10}{8} \quad\Rightarrow\quad b^2 = \frac{1}{4} \quad\Rightarrow\quad b = \pm\frac{1}{2}$$

Because $b > 0$, the logistic equation is $y = \dfrac{10}{1 + 0.5^{x-2}}$.

B. From the carrying capacity, $a = 20$. Neither $(0, 5)$ nor $(-1, 8)$ has a y-coordinate of 10, so neither is the inflection point. The most efficient approach uses $y = \dfrac{20}{1 + d \cdot b^x}$ with the points plugged in to form a system of equations.

$$\begin{cases} 5 = \dfrac{20}{1 + d \cdot b^0} \\[2mm] 8 = \dfrac{20}{1 + d \cdot b^{-1}} \end{cases}$$

Solving the first equation for d gives $5 = \dfrac{20}{1 + d} \;\rightarrow\; 1 + d = 4 \;\rightarrow\; d = 3$.

Substitution into the latter equation leads to

$$8 = \frac{20}{1 + 3 \cdot b^{-1}} \;\rightarrow\; 1 + 3 \cdot b^{-1} = \frac{5}{2} \;\rightarrow\; \frac{3}{b} = \frac{3}{2} \;\rightarrow\; b = 2$$

The desired equation is: $y = \dfrac{20}{1 + 3 \cdot 2^x}$.

C. Nothing is known about the end behavior, so all that can be attempted is a system of three equations and three unknown parameters. Either logistic form is fine since neither gives easier algebra. Figure 3.6e shows a CAS solution to the system resulting from $y = \dfrac{a}{1 + d \cdot b^x}$.

NOTE: The values for a, b and d, and not the algebraic manipulations are important, so a CAS is appropriate.

The CAS returns two solutions with $b = \pm\frac{1}{3}$. For exponential and logistic functions, $b > 0$, so the final equation is

$$y = \frac{200}{1 + 2 \cdot \left(\frac{1}{3}\right)^x}.$$

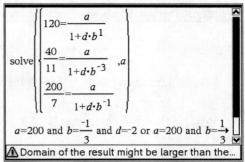

Figure 3.6e

Example 3:

In a storm, a box of 30 bunnies fell off a cargo ship and landed on an island that had enough continuous space and food to support 1000 bunnies. After four months, the population had grown to 100 bunnies. Assuming the population follows a logistic pattern,

    A. How many bunnies are predicted to be on the island after a year?

    B. After how many months will the 800th bunny be born?

    A. A form of the logistic model for this is $f(x) = \dfrac{1000}{1 + d \cdot b^x}$.

$$f(0) = 30 \Rightarrow 30 = \frac{1000}{1 + d \cdot b^0} \Rightarrow d = \frac{97}{3}.$$

$$f(4) = 100 \Rightarrow 100 = \frac{1000}{1 + \frac{97}{3}b^x} \Rightarrow b = \sqrt[4]{\frac{27}{97}}.$$

$$f(x) = \frac{1000}{1 + \frac{97}{3}\left(\frac{27}{97}\right)^{x/4}}$$

$$f(12) \approx 589.167$$

There will be approximately 589 bunnies after one year.

    B. $\quad 800 = \dfrac{1000}{1 + \frac{97}{3}\left(\frac{27}{97}\right)^{x/4}} \Rightarrow x \approx 15.21$

The 800th bunny will be born after about 15 months.

Problems for section 3.6:

Exercises

1. Figure 3.6a shows four logistic functions for $a = 1$ and $b \in \{3, 10, 0.1, 0.5\}$. Which is which?

2. What is the carrying capacity of a logistic function with inflection point at $(12, 48)$? *(handwritten: (12, 96))*

3. Determine another ordered pair point on the logistic function which has an inflection point at $(12, 48)$ and contains the point $(10, 60)$.

4. What is the carrying capacity of the logistic function with equation $y = \dfrac{11}{1 + 5^x}$?

5. [NC] What are the coordinates of the inflection point on the logistic function $y = \dfrac{24}{1 + 4 \cdot 2^x}$?

For questions 6-9, graph the given logistic function.

6. $y = \dfrac{10}{1 + 3\left(\dfrac{1}{2}\right)^x}$

7. $y = \dfrac{12}{1 + \left(\dfrac{2}{5}\right)^{x-3}}$

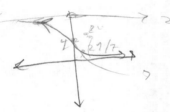

8. $y = -\dfrac{15}{1 + 2(3)^x}$

9. $y = \dfrac{24}{5 + 2^x}$

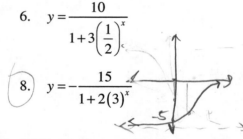

10. [NC] At what point(s) do the graphs of $y = 2^x$ and $y = \dfrac{24}{5 + 2^x}$ intersect?

For questions 11-14, determine an equation for the logistic function described by the given criteria.

11. Carrying capacity 15, initial value 6, and passes through the point $(1, 5)$. *(handwritten: $\dfrac{15}{1 + \frac{3}{2}(4)^x}$)*

12. Carrying capacity 20, inflection point $(-2, 10)$, and passes through $(1, 18)$.

13. Carrying capacity 30 and containing the points $(1, 20)$ and $(3, 10)$. *(handwritten: $\dfrac{30}{1 + d \cdot b^x}$ $\dfrac{30}{1 + \frac{1}{2} \cdot 2^{x-1}}$)*

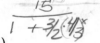

14. Passes through $(2, 75)$, $(4, 120)$, and $\left(-1, \dfrac{50}{3}\right)$.

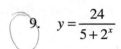

(handwritten work:)

$6 = \dfrac{15}{1 + d}$

$6 + 6d = 15$

$6d = 9$

$d = 3/2$

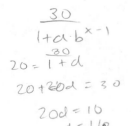

$\dfrac{30}{1 + a \cdot b^{x-1}}$

$20 = \dfrac{30}{1 + d}$

$20 + 20d = 30$

$20d = 10$

$d = 1/2$

Explorations

Identify each statement in questions 15-19 as true or false. Justify your conclusion.

15. The inflection point of a logistic function is always located along the horizontal line midway between the two horizontal asymptotes.

16. One end behavior asymptote of every non-transformed logistic function is $y = 0$.

17. $x \in \mathbb{R}$ is the domain of every logistic function.

18. The inflection point in question 13 is at $(0, 15)$.

19. The inflection point in question 11 is at approximately $(-1.409, 7.5)$. *True*

20. What is the carrying capacity of the logistic function with equation $y = \dfrac{17}{3 + 13 \cdot 5^x}$?

21. What are the *exact* coordinates of the inflection point on the logistic function $y = \dfrac{24}{5 + 2^x}$? $= 12/5$

$\dfrac{12}{5} = \dfrac{24}{5 + 2^x}$

$\left(\log_2 5, \; 12/5 \right)$

$12 + \dfrac{12}{5} \cdot 2^x = 24$

For questions 22-24, determine an equation for the logistic function described by the given criteria.

$\dfrac{12}{5} \cdot 2^x = 12$

22. Has inflection point at $(12, 48)$ and contains the point $(10, 60)$.

$\dfrac{96}{1 + d \cdot b^x}$

$2^x = 5$

$x = \log_2 5$

23. Contains the point $(3, 4)$ and has a y-intercept of 2.

24. Has a y-intercept at 35 and a carrying capacity of 60.

25. The text claimed $y = \dfrac{a}{1 + d \cdot b^x}$ and $y = \dfrac{a}{1 + b^{x-c}}$ are equivalent. Find an equation showing the relationship between b, c, and d.

$y = \dfrac{a}{1 + b^x} - \dfrac{a}{2}$

$y = a \cdot \left(\dfrac{1}{1 + b} - \dfrac{1}{2} \right)$

26. A. Apply $T_{0, -\frac{a}{2}}$ to $y = \dfrac{a}{1 + b^x}$ and prove that the resulting function is odd.

   B. Why do the results of part A prove that $\left(0, \dfrac{a}{2} \right)$ is an inflection point of $y = \dfrac{a}{1 + b^x}$?   *its where it shifts concavity, from growth to decay*

27. Use limits to verify the end behavior asymptotes of logistic functions in both general forms, $y = \dfrac{a}{1 + d \cdot b^x}$ and $y = \dfrac{a}{1 + b^{x-c}}$.

28. For the logistic function $y = \dfrac{a}{1 + d \cdot b^x}$, $y = a$ is one end behavior asymptote. Use this information to

   sketch a graph of $y = \dfrac{x^2}{1 + 3^x}$.

UNIT 4 – DATA ANALYSIS & RATIONAL FUNCTIONS

Mathematics is the Science of Patterns.

– Anonymous

This is a very short unit, but the presentation of the material here is significantly different from other approaches. Ultimately, our goal is for you to realize that there are a scant handful of central ideas driving the mathematics behind rational functions and that each section in this chapter is a creative application and extension of the implications of the prior chapters, particularly the transformations from Chapter 2.

If necessary, take some additional time to digest the deep implications of the effects of transformations in sections 4.2 and 4.3. It will be time well spent.

Enduring Understandings:

- Modeling discrete scatter plots and graphing rational functions are fundamentally the same mathematical idea.

- Mathematics tells you the behavior of a function even when a graph cannot or does not show that behavior.

- A deep understanding of polynomial behavior and transformations dramatically simplifies the graphing of any rational function (and beyond).

This section assumes you have:

- A deep understanding of polynomials, especially including their end behavior and behavior around roots.

- Exposure to basic fits of a function to data.

4.1: A Review of Regressions

> _If I have seen further it is by standing on the shoulders of Giants._
>
> – Sir Isaac Newton

Differences and Ratios

There are two reasons to find functions to model data. First, functions are more efficient and less cumbersome representations of variable relationships, especially in comparison to larger data sets. Second, functions are efficient predictors of y-values. While it is possible to predict without a function, models provide simple, rational justifications for predicting output values.

A kth-degree polynomial models a data set with a fixed Δx and Δy values which first become constant at the kth level. Regardless of Δy outcomes, a data set of n points can always be modeled by an $(n-1)^{st}$-degree polynomial. Because the base form of exponential functions, $y = a \cdot b^x$, has only two parameters, perfectly exponential data requires only two ordered pairs to determine an equation. The same is true for perfect power function data, $y = a \cdot x^b$. The common assumption for all of these approaches is that a derived equation fits all of the data points _exactly_.

So what happens if the pattern is not perfect (for example, data collected in a science experiment)? Considering the effort involved, finding a 19th-degree polynomial to perfectly fit 20 given data points is probably not a worthwhile effort. And even if such a polynomial was found, what would a 19th degree polynomial mean practically? Just because a complicated formula can be found does not mean the added complexity adds any utility or insight into the data. More complex models are almost always harder to compute, are more difficult to interpret, and may disguise the fact that data typically contains variability.

Simpler models often surrender accuracy, and vice versa. Finding models for data is always a balancing act between finding an equation that
1) fits the data reasonably well, and
2) is as simple as possible in explaining the underlying relationship in the data set.

But how is the function used to model a data set determined? How is error measured for the model?

Defining Error

Error is the difference between the actual and predicted outputs (y-values) of a data set. If $y = f(x)$ predicts a set of n points and if (x_i, y_i), $i \in [1, n]$, is one of these, then, $\text{Residual}_i = y_i - f(x_i)$ (Figure 4.1a). A residual measures error, so this equation is equivalent to $Data = Model + Error$.

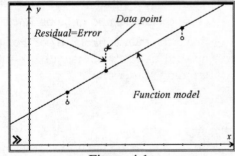

Figure 4.1a

From their definition, residuals are positive when the data value exceeds the predicted value. So each residual gives two pieces of information: the direction and the magnitude of the error for each estimate. Figure 4.1b shows the computation of the residuals using $f(x) = 3x - 2$ to model $\{(1,3),(2,4),(5,10)\}$.

x_i	1	2	5
y_i	3	4	10
$f(x_i) = 3x_i - 2$	1	4	13
$\text{Residual}_i = y_i - f(x_i)$	2	0	-3

Figure 4.1b

Better models usually result in smaller residual values, but that is not enough. Just because a relatively simple model produces small residuals, this does not mean the model is a good fit.

Linear Regressions & Analyzing Residuals

While there are many ways to find an equation for a line, each makes slightly different assumptions and not all arrive at the same conclusion. There is no such thing as a universal "best fit" line, only lines that *best* fit a particular set of assumptions or goals for the fit. The two examples that follow are by no means exhaustive or representative of all possibilities for fitting lines to data.

Example 1:

Model the Fahrenheit (F) vs. Celsius (C) temperature data below

Fahrenheit	42	51	75	88	90
Celsius	6	11	24	31	32

Before assuming a fourth-degree polynomial for the five data points are given, Figure 4.1c shows a scatter plot of the data,[2] providing strong visual evidence that the relationship might be linear.

A low-tech approach finds an equation using two data points representing the entire data set. A benefit of this is that it provides a simple method to determine an equation. The clear downside is that it ignores the majority of the data set. Since the data looks very straight in the $[0,100] \times [0,40]$ window, you can minimize error by

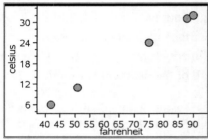

Figure 4.1c

choosing the points with the most extreme x-coordinates, $(42,6)$ and $(90,32)$, give $C = 6 + \dfrac{26}{48}(F - 42)$.

Example 2:

Compute a **linear regression**[3] on the Fahrenheit-Celsius data.

The linear regression is $C \approx 0.542F - 16.684$.[4] If computed from a CAS calculator screen, the regression statistics include an r-value. This value is called the **correlation coefficient** and is measures how close the data came to being linear. The closer $|r|$ is to 1, the more linear the data is likely to be. Positive values of r indicate a general upward trend, while negative r-values indicate general downward-trending scatter plots.

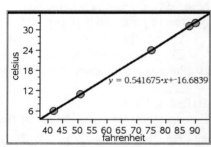

Figure 4.1d

---

[2] To get the graph, first define the data lists on a Calculator screen or enter and name the data in a Lists & Spreadsheets screen. Then press 🏠 → ⑤:**Data & Statistics**. Move the cursor to the bottom center of the screen, click, and choose the Fahrenheit data's name. Then move to the center far left of the screen, click, and choose the Celsius data's name. Press **enter** and the scatter plot appears (Figure 4.1c).

[3] See Appendix A for an explanation of linear regressions.

[4] From Data & Statistics screen in Figure 4.1c, press (menu) → ④:**Analyze** → ⑥:**Regression** → ①:**Show Linear(mx+b)**. The line can be cleared by pressing (menu) → ④:**Analyze** → ⑥:**Regression** → ①:**Hide Linear**.

The models for the Fahrenheit-Celsius data are similar, but not identical. Good data models satisfy the following three measures of their residuals.

1) All of the residuals should be relatively small.[5]

2) About ½ of the residuals should be positive and ½ should be negative.

3) Ideally, there should be no discernable graphical pattern in the residuals.

The first criterion ensures that the model will be close to as many data points as possible. The second suggests that the function should go through the middle of the data. These two criteria apply to all data models, regardless of the type of function chosen to model. The final point evaluates whether the chosen function is from the best possible family. For example, if a linear function was chosen as a model for exponential data, a clear pattern would show in the residual graph. All three criteria are flexible, but any good fitting model will satisfy all three.

The equations from these examples can be compared by analyzing their residuals. Figure 4.1e gives a numerical representation and Figures 4.1f and 4.1g graphical representations of the residuals.

F	C	$C = 6 + \dfrac{26}{48}(F - 42)$	Example 1 Residual	$C \approx 0.542F - 16.684$	Example 2 Residual
42	6	6	0	6.066	-0.0664
51	11	10.875	0.125	10.942	0.0585
75	24	23.875	0.125	23.942	0.0583
88	31	30.917	0.0833	30.983	0.0165
90	32	32	0	32.067	-0.0669

Figure 4.1e

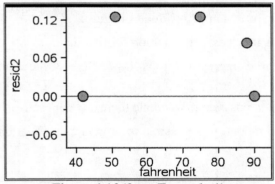

Figure 4.1f (from Example 1)

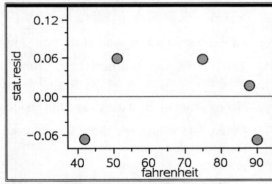

Figure 4.1g (from Example 2)

Following the criteria above,
- The Example 1 residuals are larger in magnitude . Its zero residuals are due to its derivation.
- Example 1's residuals are all non-negative. Example 2's residuals are nearly evenly split between positive and negative values.
- Both residual plots have roughly parabolic patterns.

With respect to these models,
1) Both sets have very small residuals, but Example 2's are generally smaller.
2) The Example 1 residuals fail the second criteria because they contain no negative values.
3) Both residual graphs have patterns, but with only five data points, that is difficult to avoid.

---

[5] There is no universally accepted standard for the meaning of "relatively small" residuals. Being 100 units off would be terrible for the Fahrenheit-Celsius data, but would be considered a near-perfect fit with respect to the National Budget. The key here is to be small relative to the data set. Residuals within 5% of their data points is a good (but not firm) goal.

Overall, the criteria suggest that the linear regression is slightly better. By increasing the *y*-intercept of the line from Example 1, even this criterion could be neutralized. The residual criteria not only help identify when a function is a good fit, but they sometimes offer suggestions for how to improve a function that doesn't fit as well as it could.

Problems for 4.1:

Exercises

1. A model for this (x, y) data gave the following residuals. Find an equation for the model.

x	1	3	5	7
y	8	5	2	-1
Residual	-1	-2	-3	-4

2. A model for this (x, y) data gave the following residuals. Find an equation for the model.

x	-3	-1	1	3	5
y	50	8	6	44	122
Residual	-12	0	4	0	-12

Explorations

Identify each statement in questions 3-7 as true or false. Justify your conclusion.

3. It is always possible to find a model for a set of data for which all residuals are zero.

4. Any equation that hits all of the points in a data set is the "best" fit equation for that data.

5. The correlation coefficient, *r*, from an exponential regression tells how closely the derived exponential equation fit the given data points.

6. Given a model for data set and the residual values, it is possible to re-create the exact data set.

7. Given data values and the residuals from a function model, it is possible to re-create an equation for the function model.

8. A model for a set of data is $y = 2x - 5$, and a scatter plot of the residuals for the data is shown in Figure 4.1h. Recreate the ordered pairs, (x, y), of the original data.

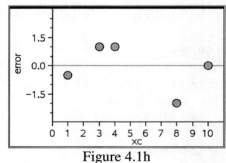

Figure 4.1h

4.2: Rational Functions I: Domain Restrictions & Bouncing Off Infinity

It is a glorious feeling to discover the unity of a set of phenomena that seem at first to be completely separate.

– Albert Einstein

To infinity and beyond!

– Buzz Lightyear

From the Reciprocal Transformation to Vertical Asymptotes
A **rational function** is any function which can be written as a ratio of two polynomials for which the denominator is not zero.

Example 1:

Graph the rational function $r(x) = \dfrac{-1}{x+2}$.

Method 1: Think of $y = \dfrac{-1}{x+2}$ as $T_{-2,0}\left(S_{1,-1}\left(f(x)\right)\right)$ where $f(x) = \dfrac{1}{x}$. Figures 4.2a and 4.2b show this two-step constant transformation.

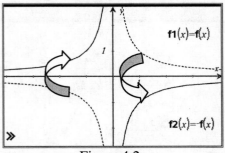

Figure 4.2a

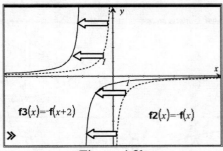

Figure 4.2b

Method 2: $y = \dfrac{-1}{x+2}$ can also be seen as the reciprocal of $y = x+2$ reflected over the *x*-axis, or as the reciprocal of $g(x) = -(x+2)$. The latter involves fewer steps, so it is described next.

Under the reciprocal transformation, the points where $y = 1$ on g (at $x = -3$ and $x = -1$) are invariant, and the *x*-intercept of g becomes a vertical asymptote. These details are marked in Figure 4.2c. Applying the reciprocal transformation to g gives Figure 4.2d, an image identical to that found by Method 1 (Figures 4.2b).

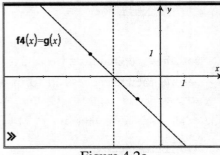

Figure 4.2c

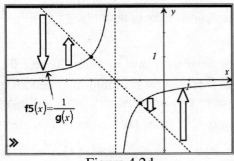

Figure 4.2d

Recall from Section 1.3 that all polynomials pass through the x-axis at their odd powered x-intercepts and bounce off the x-axis at all even-powered x-intercepts. In fact, this behavior is true for any x-intercept *for any continuous function*: at an x-intercept, the graph of every continuous function either bounces off or passes through the axis. Because values do not change sign under the reciprocal transformation, the **vertical asymptotes** derived from those roots also have only two types, even and odd.

The graph of $y = x$ passes through the x-axis at its intercept, while $y = \dfrac{1}{x}$ rises to infinity on one side of the corresponding asymptote, **"passes through infinity"**, and emerges on the opposite side of the vertical asymptote (Figure 4.2e). When graphs of polynomials pass through the x-axis at roots with odd multiplicity, rational functions derived reciprocating polynomials at those roots pass through infinity. Such asymptotes are called **odd vertical asymptotes**.

Similarly, the graph of $y = x^2$ bounces off the x-axis at its intercept, and $y = \dfrac{1}{x^2}$ is positive on both sides of its corresponding asymptote (Figure 4.2f). Because $\lim\limits_{x \to 0} \dfrac{1}{x^2} = +\infty$, the graph of $y = \dfrac{1}{x^2}$ can be described as **"bouncing off infinity"** around its vertical asymptote. Where polynomials bounce off the x-axis at roots with even multiplicity, rational functions bounce off infinity at vertical asymptotes derived from applying the *REC* transformation to polynomials at those roots. Such asymptotes are called **even vertical asymptotes**.

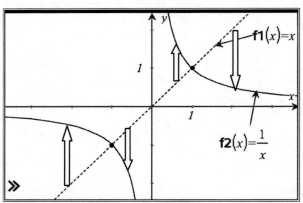

From an odd root to an **odd vertical asymptote**
Figure 4.2e

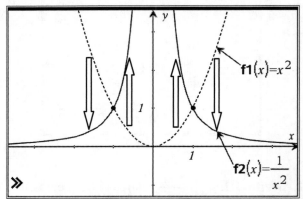

From an even root to an **even vertical asymptote**
Figure 4.2f

Finally because the factors which are used to identify vertical asymptotes cannot be completely simplified out of their equations, they are called **non-removable discontinuities**.

Example 2:

Describe the differences between $f(x) = \dfrac{x^2 + 12x + 20}{x + 2}$ and $g(x) = \dfrac{x^2 + 12x + 19}{x + 2}$.

Both have the same form and nearly the same expression. Both have domain $x \in (-\infty, -2) \cup (-2, \infty)$. Figure 4.2g shows a table of function values near $x = -2$. Despite their closeness in form, the table suggests something significantly different happens to f and g at $x = -2$. f appears to approach $y = 8$ numerically from both sides, while the values of g explode. This suggests that $\lim\limits_{x \to -2} f(x) = 8$, while $\lim\limits_{x \to -2^-} g(x) = -\infty$ and $\lim\limits_{x \to -2^+} g(x) = \infty$. Figures 4.3h and 4.3i show this graphically.

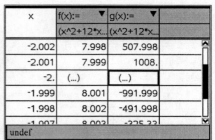

Figure 4.2g[6]

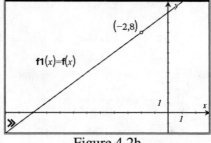

Figure 4.2h

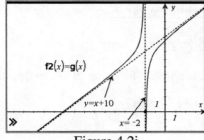

Figure 4.2i

These differences can be supported algebraically. The numerator of f factors and simplifies; g does not.

$$f(x) = \frac{x^2 + 12x + 20}{x+2} = \frac{(x+10)\,(x+2)}{(x+2)} = x+10$$

While these two forms of f look different, other than at $x = -2$, they are algebraically equivalent. The simplified version of f also verifies the earlier numerical limits.

$$\lim_{x \to -2} f(x) = \lim_{x \to -2} (x+10) = 8$$

Substituting $x+10$ for $f(x)$ is permissible because the limit does not ask what happens *at* $x = -2$, which is the only place where the original definition of f and $y = x+10$ differ. A point on a graph with this property is called a **hole**. When a rational function can be algebraically simplified by canceling a common factor in its numerator and denominator, it is *possible* for a hole to appear. Because rational functions algebraically simplify at holes, these points are also called **removable discontinuities**.

Because the numerator of g does not factor over the integers, it cannot simplify, and its discontinuity is non-removable —a vertical asymptote. But g can be rewritten using polynomial division (Figure 4.2j). Notice how the division splits g into two different types of functions: a linear expression ($x+10$) and a residual rational expression $\left(\dfrac{-1}{x+2}\right)$. Interestingly, the linear term is the *same* expression to which f reduced. The rational term's values are very close to 0 everywhere except near $x = -2$, confirming the numerical and graphical findings.

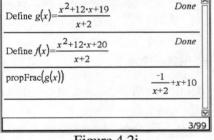

Figure 4.2j

Functions can behave very differently near a point of discontinuity. In algebraic representations, a function can sometimes simplify from a given fractional form, while others retain a rational form even after division. The interaction in rational functions between the quotient and the fractional, remainder term resulting from polynomial division is the focus of the next section.

---

[6] To get a table of values for functions, first define the functions in a Graphs & Geometry or Calculator window. Then open a Lists & Spreadsheet window, and press (menu) → (5): **Function Table** → (1):**Switch to Function Table**. Choose a function name from the pull-down menus and press **enter**. The right arrow creates new columns for more functions. To change the table's settings, press (menu) → (5):**Function Table** → (3):**Edit Function Table Settings**.

Problems for 4.2:

Exercises

Graph each function given in questions 1-3.

1. $f(x) = \dfrac{1}{x+7}$

2. $g(x) = \dfrac{1}{(x+7)^2}$

3. $h(x) = \dfrac{1}{(x+7)^3}$

4. [NC] Use polynomial division to rewrite $y = \dfrac{3x-2}{x+12}$. Graph the result.

5. [NC] Use polynomial division to rewrite $y = \dfrac{-3x^2+6x}{(x-1)^2}$. Graph the result.

6. If $h(x) = \dfrac{x+5}{x-1}$ and $g(x) = \dfrac{x+2}{x+3}$, for what values of x are each of the following defined?

    A. $y = h(x) - g(x)$

    B. $y = h(x) \cdot g(x)$

    C. $y = \dfrac{1}{h(x)+g(x)}$

7. Graph $y = \dfrac{x+2}{x^3+6x^2+12x+8}$. Give reasons for the existence of any holes or vertical asymptotes.

8. Graph $y = \dfrac{x^3+7x^2-25x-14}{x^3+7x^2-25x-15}$. Give reasons for the existence of any holes or vertical asymptotes.

Explorations

Identify each statement in questions 9-13 as true or false. Justify your conclusion.

9. For any three distinct real values of a, b, and c, the graph of $y = \dfrac{(x-a)(x+b)}{(x+b)(x-c)^3}$ has exactly one odd vertical asymptote and exactly one hole.

10. It is impossible for a rational function to have both a vertical asymptote and a hole at the same x-value.

11. The graph of every rational function must contain at least one hole or one vertical asymptote.

12. If all of the vertical asymptotes of a rational function are odd, then the function cannot be even.

13. If the expression $(x-a)$ can cancel from the numerator and denominator of a rational function, then $x = a$ is the location of a hole in the graph of that function.

14. Consider the function defined by $f(x) = \dfrac{(x+3)(x-2)}{(x-2)^2}$. Without graphing the function, determine if f has a hole or a vertical asymptote at $x = 2$. Justify your conclusion.

15. [NC] Use partial fractions to decompose $f(x) = \dfrac{3x-2}{x^2+x-12}$. Without graphing, what do the results say about the graph of f?

16. Given $y_1 = \dfrac{x}{x+3}$.

   A. Use polynomial division to rewrite y_1, and graph the result.

   B. Explain why $y_2 = 1 + \dfrac{3}{x}$ is the reciprocal of y_1.

   C. Graph y_2 followed by its reciprocal.

   D. Are the graphs of parts A and C identical? Why?

17. Under what conditions will the graph of a rational function contain a hole? Under what conditions will the graph of a rational function have a vertical asymptote? Be as precise as possible.

18. Write an equation of a rational function which has three non-removable discontinuities (exactly two of which are even) and two removable discontinuities.

19. Write an equation of a rational function that has a hole at an x-intercept.

20. The arithmetic mean of the test scores for a given class with m students is 83. A second class with n students had a mean score of 91. When combined, the average score for both classes was 88. What is the value of $\dfrac{m}{n}$? Which class was larger and by what factor?

21. The arithmetic mean of 10 test scores is 85. Eliminating the highest and lowest, the average of the remaining tests is 86.5. What is the average of the highest and lowest scores?

22. [NC] If $a + b = 3$ and $ab = 4$, then without solving for a or b, determine the value of $\dfrac{1}{a} + \dfrac{1}{b}$.

23. A man went for a jog. Leaving his house on flat ground, he jogged at 6 mph. After some time, he began running up a hill at 4 mph. Then he turned around and retraced his path back home. He ran to the bottom of the hill at 12 mph, again 6 mph on the flat ground, and his total trip was 2 hours. How far did he jog?

24. A runner sprints for half of his workout at 25 kph and the second half at x kph.

   A. Determine an expression giving his average speed if he ran the same distance in both halves of his workout.

   B. Graph the expression determined in part A.

   C. Use the graph from part B to determine a realistic domain and range for the situation.

   D. Explain why there is a limit to the runner's average speed.

4.3: Rational Functions II: From Data Analysis to Bending Asymptotes

Praised be the human mind which sees more sharply than does the human eye.

– Aaron Bernstein

If you take care of the math, the math takes care of you.

– Kate Plumblee

Polynomial division can be used to change any rational function from the standard form $y = \dfrac{n(x)}{d(x)}$ (with

$d(x) \neq 0$) to $\dfrac{n(x)}{d(x)} = q(x) + \dfrac{r(x)}{d(x)}$, where $q(x)$ is the quotient and $r(x)$ is the remainder from the division.

These different algebraic forms give different clues about the function's behavior.

The $y = \dfrac{n(x)}{d(x)}$ form reveals many characteristics.

- Because division by zero is results in an undefined output, **domain restrictions** occur when $d(x) = 0$.

- A domain restriction is a vertical asymptote if the restriction is non-removable and a hole if it is removable.

- A **y-intercept** occurs when $x = 0$, so $y(0) = \dfrac{n(0)}{d(0)}$ is always a y-intercept if $d(0) \neq 0$.

- An x-intercept occurs when $y = 0$. Because fractions equal zero only when their numerators are zero, solving $n(x) = 0$ for x will determine any x-intercepts, assuming the results are not domain restrictions.

To understand $y = q(x) + \dfrac{r(x)}{d(x)}$, first recall the Section 4.1 equation: $Data = Model + Error$. Here, *Model* approximates the information represented by *Data*. Graphically, the *Model* function runs through the center of the *Data* scatter plot. In a sense, *Model* represents the global data behavior, and *Error* measures the local amount by which *Model* misses each *Data* point. Note the strong similarity between the equation for data analysis, $Data = Model + Error$, and the $y = q(x) + \dfrac{r(x)}{d(x)}$ equation for rational functions. To help identify the relationships between the terms in these two expressions, Figure 4.3a provides multiple representations of a sample data set and a rational function, $g(x) = \dfrac{x^2 + 12x + 19}{x + 2}$.

First, *Data* represents the actual data values just like the y term represents the entire rational function. The left side of each equation represents the overall picture of the respective relationships.

Data	Rational Function
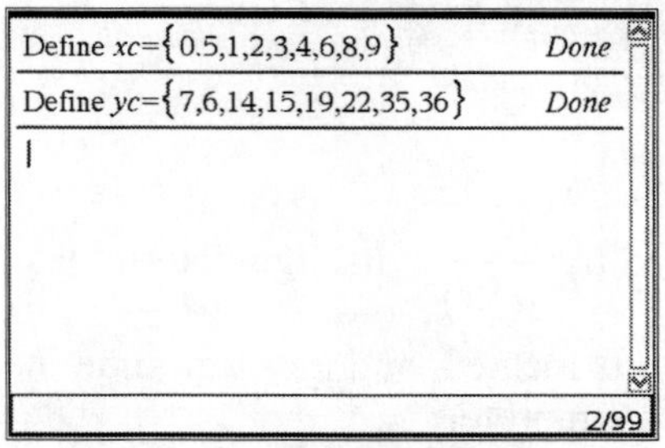 Define $xc=\{0.5,1,2,3,4,6,8,9\}$ *Done* Define $yc=\{7,6,14,15,19,22,35,36\}$ *Done* I 2/99 (the scatter plot)	$$g(x)=\frac{x^2+12x+19}{x+2}=x+10+\frac{-1}{x+2}$$ (the bold curve)
$Model=3.52x+4.52$ (the line)	$q(x)=x+10$ (the dashed line)
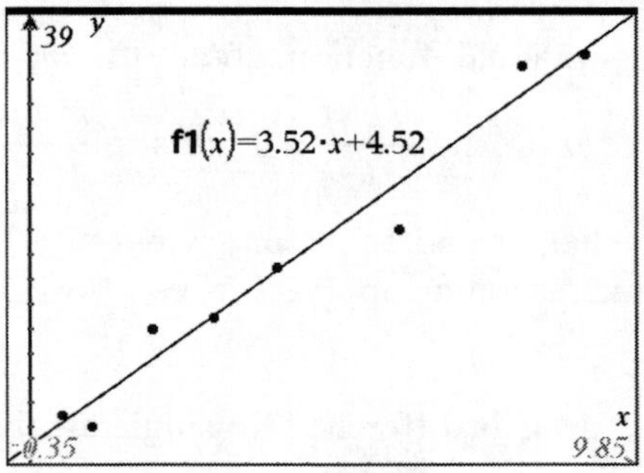 $f1(x)=3.52\cdot x+4.52$	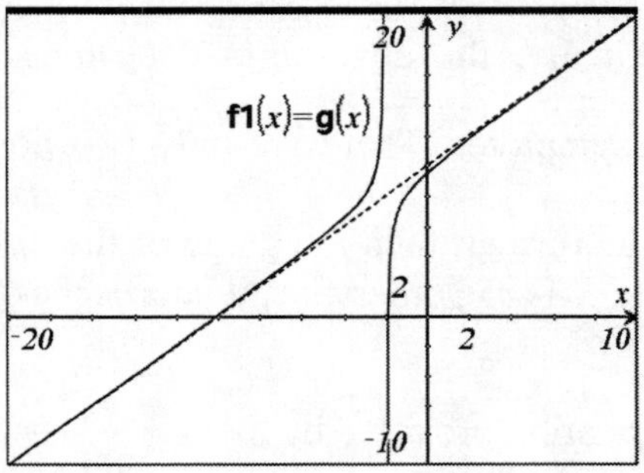 $f1(x)=g(x)$

Figure 4.3a

The graphs of both *Model* for the scatter plot and $y=q(x)$ for the rational function run through the center of their respective graphs. Attempts to zoom out on either would result in both graphs becoming indistinguishable from *Model* or $y=q(x)$, respectively. While *Model* was specifically crafted to run through the center of the data (half of the data on one side for a good model), the reason $y=q(x)$ does the same thing for the graph of a rational function is more subtle.

For rational functions in the $y=q(x)+\dfrac{r(x)}{d(x)}$ form, polynomial division guarantees the degree of r is less than the degree of d, so d dominates r as $|x|\to\infty$ and the $\dfrac{r(x)}{d(x)}$ term *must* tend toward zero as $|x|\to\infty$ making the graph of the rational function approach $y=q(x)$ as $|x|\to\infty$. Because $g(x)\to q(x)$ as $|x|\to\infty$, $y=q(x)$ is called the **end behavior asymptote** of the rational function.

This leaves the $\dfrac{r(x)}{d(x)}$ and *Error* terms. Both are added to their respective models and quantify the distance between the graph and its global behavior model. For the rational function in Figure 4.3a, the error term, $\dfrac{-1}{x+2}$, is small except for values of x close to -2. But as $x\to-2$, this error term rapidly expands in magnitude and the rational function has no limits to its distance from the end behavior asymptote. This is another reason for the existence of vertical asymptotes.

In summary, a function written in the form $f(x) = q(x) + \dfrac{r(x)}{d(x)}$ has many properties.

- The end behavior asymptote for f is $y = q(x)$.

- The error of f from its end behavior asymptote is quantified by $\dfrac{r(x)}{d(x)}$. Its sign determines the side of the end behavior asymptote on which the rational function is located, and its magnitude measures the distance from the rational function to its the end behavior asymptote.

- If $\dfrac{r(x)}{d(x)} = 0$ for any value of x, then there is no error at that point, so f and its end behavior asymptote intersect at that point. While domain restrictions prevent a function from crossing vertical asymptotes, they could intersect an end behavior asymptote any number of times.

- Finally, the error term determines the behavior of the rational function about its end behavior asymptote. There are only two possibilities for the error term, $\lim\limits_{x \to \infty} \dfrac{r(x)}{d(x)} = 0^+$ or $\lim\limits_{x \to \infty} \dfrac{r(x)}{d(x)} = 0^-$. If the former, then the graph of the rational function approaches its end behavior asymptote from above as $x \to \infty$; otherwise, it approaches from below. The same reasoning applies for $x \to -\infty$.

The information provided by the $y = \dfrac{n(x)}{d(x)}$ and $y = q(x) + \dfrac{r(x)}{d(x)}$ forms is different. No single algebraic form ever contains all of the desired information about a function.

Example 1:

Sketch a complete graph of $f(x) = \dfrac{x-2}{(x-5)^2}$.

- The denominator's degree exceeds the numerator's, so $y = q(x) = 0$ is the end behavior asymptote.

- That makes $\dfrac{x-2}{(x-5)^2}$ the error term. For extreme values of x, the constants subtracted do not affect the relative size of the error term, so $\dfrac{x-2}{(x-5)^2}$ behaves like $\dfrac{x}{x^2} = \dfrac{1}{x}$ as $|x| \to \infty$.

- Therefore f approaches its end behavior asymptote just like $y = \dfrac{1}{x}$ approaches its end behavior asymptote, from above as $x \to \infty$ and from below as $x \to -\infty$.

- The error has a multiplicity one zero at $x = 2$, so f linearly crosses its end behavior asymptote at $x = 2$.

- f has an even vertical asymptote at $x = 5$.

- The intercepts of f are $(2, 0)$ and $\left(0, \dfrac{-2}{25}\right)$.

To graph $y = f(x)$,

- plot the end behavior asymptote, vertical asymptote, intercepts and the approach of f to its end behavior asymptote (Figure 4.3b),

- connect the left end-behavior branch of f to its y-intercept, linearly pass through the x-intercept and climb to infinity as $x \to 5^-$ (Figure 4.3c),

- Because $x = 5$ is an even vertical asymptote, f bounces off infinity and drops to connect with the awaiting portion of the graph identified as the right end behavior (Figure 4.3d).

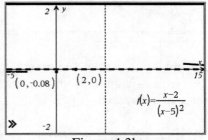

Figure 4.3b

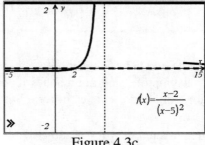

Figure 4.3c

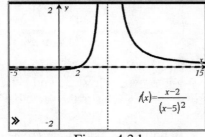

Figure 4.3d

Notice also that when moving from right to left from its vertical asymptote, f must pass through the x-axis at $x = 2$, change directions at a minimum somewhere to the left of $x = 2$, and return to the x-axis. The location of the minimum is not important here, but the behavior of f guarantees its existence.

Example 2:

Compare the graphs of $g(x) = \dfrac{-x^3 + 11x^2 - 34x + 23}{x^2 - 10x + 25}$ and $h(x) = \dfrac{x^4 - 10x^3 + 21x^2 + 41x - 102}{x^2 - 10x + 25}$.

Polynomial division is a good first step (Figure 4.3e). The error term for each is the rational function from Example 1, so the major differences between g, h, and function f from Example 1 are their end behavior asymptotes.

Just as the x-axis is malleable in variable vertical scale changes (Section 2.1), the end behavior asymptote for a rational function can be similarly manipulated. To derive the graphs of g and h from the graph of f, just **bend the asymptote**! Figure 4.3f does this for g and Figure 4.3g shows the result for h. NOTE: *Both transformations are entirely vertical*; any other change would incorrectly alter the vertical asymptote at $x = 5$.

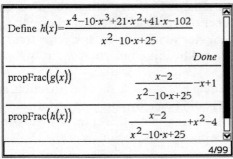

Figure 4.3e

The graph of the error term maintains its relationship to the end behavior asymptote throughout any transformations of the end behavior asymptote. Figure 4.3h zooms in on g to show

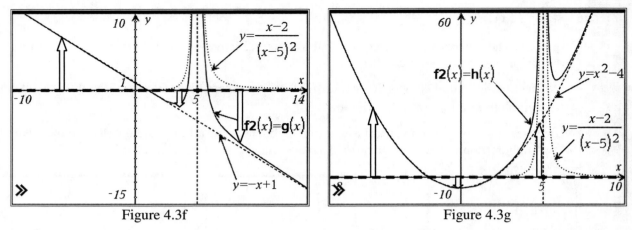

| Figure 4.3f | Figure 4.3g |

that from *right to left*, g still crosses its end behavior asymptote at $x = 2$ from top to bottom and passes its y-intercept below the end behavior asymptote, and approaches its end behavior asymptote from below as $x \rightarrow -\infty$. This is the same behavior exhibited by the error term in the original error term (Figure 4.3d).

Bending the asymptote creates three x-intercepts for g and two for h. The graphs suggest $g(0) > 0$ and $h(0) < 0$. A CAS is ideal for locating these points (Figure 4.3i).

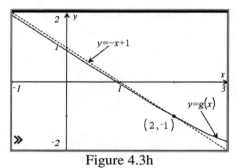

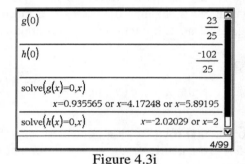

| Figure 4.3h | Figure 4.3i |

As a stand-alone function, the error term approaches the x-axis *exactly the same way* as the final rational function approaches its end behavior asymptote. From this perspective, *any function* can serve as an end behavior asymptote to any error term and the graphical relationships would be the same. The possibilities opened by this connection are truly limitless!

Example 2 drew a rational function starting from the error term and bending the x-axis. This is akin to recreating data by applying a vertical bend to the x-axis of a residual graph to transform it back to the graph of the original data. Section 4.1 also recreated data by adding residuals back to the data model. Whether starting with the residuals or the model, the data is recreated.

Example 3:

Obtain the graph of $b(x) = \dfrac{x^2 - 2x + 1}{(x+1)^3} - x$ twice: once by starting with the error term, and the second time by

starting with the end behavior asymptote.

Note several properties about b.

- The end behavior asymptote is $y = -x$.

- The error term is $\dfrac{x^2-2x+1}{(x+1)^3}$, which shows

  ➤ an odd vertical asymptote at $x=-1$, and

  ➤ the numerator equivalent to $(x-1)^2$ giving b zero error at a multiplicity 2 factor, $x=1$.
    Therefore, b bounces off its end behavior asymptote at $x=1$.

  ➤ For very large values of $|x|$, the error term behaves like $y=\dfrac{(x-1)^2}{(x+1)^3}\xrightarrow{|x|\to\infty}\dfrac{x^2}{x^3}=\dfrac{1}{x}$.

- The y-intercept is $(0,1)$.

Method 1: Start with the error and bend the asymptote.

Figure 4.3j shows the graph of $y=\dfrac{(x-1)^2}{(x+1)^3}$. Figure 4.3k zooms in at $x=1$ to show the bounce off the x-axis. Finally, Figure 4.3l vertically bends the x-axis into $y=-x$ to create a final graph of b. The bounce off the end behavior asymptote does persist even though it is difficult (and sometimes impossible) to obtain a graphing window to show it (Figure 4.3m).

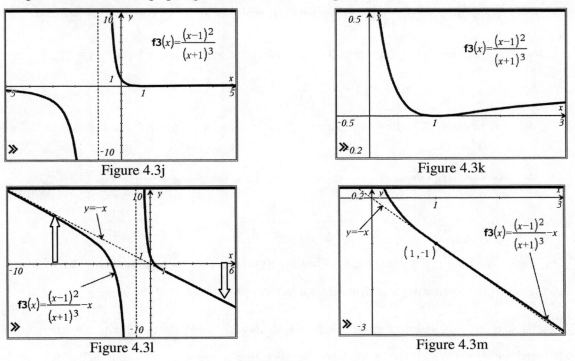

Figure 4.3j   Figure 4.3k

Figure 4.3l   Figure 4.3m

Method 2: Start with the end behavior asymptote and recreate the function.

Plot the end behavior asymptote, vertical asymptotes, y-intercept, and the point where b bounces off its end behavior asymptote. Because the error term behaves like $y=\dfrac{1}{x}$ for large $|x|$, b must approach its end behavior asymptote from below as $x\to-\infty$ and from above as $x\to\infty$ (Figure 4.3n).

To the left of $x=-1$, there are no end behavior asymptote intersections, so b rises and then passes through infinity at the odd vertical asymptote (Figure 4.3o). Finally, the curve passes through its y-intercept, bounces off the end behavior asymptote at $x=1$, and connects to the predicted right side of the graph (Figure 4.3p).

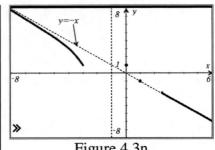

Figure 4.3n

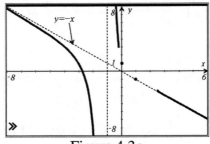

Figure 4.3o

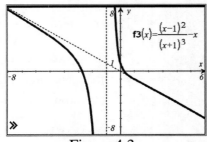

Figure 4.3p

Example 4:

Graph $c(x) = 0.2x^2 - \dfrac{x \cdot (x-3)^3}{x \cdot (x+1)^2 (x-5)^3}$.

First, note several properties about c from its equation.

- The end behavior asymptote is $y = 0.2x^2$.

- The error term is $\dfrac{-x \cdot (x-3)^3}{x \cdot (x+1)^2 (x-5)^3}$ which reveals

  ➤ an even vertical asymptote at $x = -1$ and an odd vertical asymptote at $x = 5$,

  ➤ a hole at $x = 0 \;\Rightarrow\; (x, y) = \left(0, -\dfrac{27}{125}\right)$, and

  ➤ a zero error of multiplicity 3 at $x = 3$. c will cross its end behavior asymptote at $x = 3$ in a "wiggle."

  ➤ For very large values of $|x|$, the error term behaves like $y = -\dfrac{1}{x^2}$:

  $$y = \dfrac{-x \cdot (x-3)^3}{x \cdot (x+1)^2 (x-5)^3} \xrightarrow{|x| \to \infty} -\dfrac{x^4}{x^6} = -\dfrac{1}{x^2}.$$

- There is a hole at the y-intercept, so finding its coordinates requires reducing the x factor from the error term's numerator and denominator before substituting $x = 0$: $\left(0, -\dfrac{27}{125}\right)$.

The end behavior wiggle intercept at $x = 3$ is just another point on the curve, so its y-coordinate is determined by $c(3) = 1.8$. The error term behaves like $y = -\dfrac{1}{x^2}$, so c approaches its end behavior asymptote from below as $|x| \to \infty$. This end behavior, the end behavior asymptote, the vertical asymptotes, the y-intercept/hole, and the point at which c wiggles through its end behavior asymptote are all shown in Figure 4.3q.

Left of $x = -1$, there are no end behavior asymptote intersections, so c falls and bounces off infinity at the even vertical asymptote (Figure 4.3r). Then, c passes through its y-intercept, wiggles through its end behavior asymptote at $x = 3$, rises to infinity on the left of its second vertical asymptote at $x = 5$, passes through infinity (because the $x = 5$ asymptote is odd), and connects to the predicted right side of the graph (Figure 4.3s).

The final graph of $y = c(x)$ in Figure 4.3s suggests there ought to be three singleton x-intercepts. This can be confirmed by solving $c(x) = 0$ for x on a CAS.

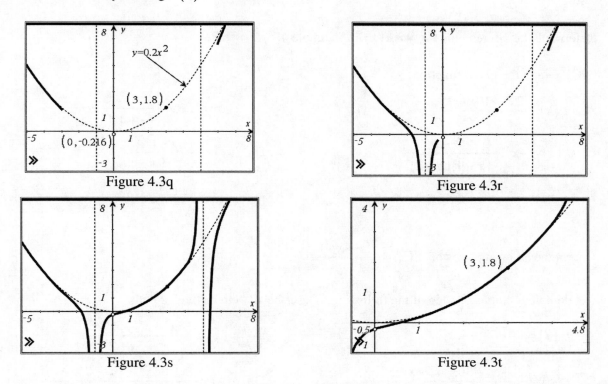

<table>
<tr><td>Figure 4.3q</td><td>Figure 4.3r</td></tr>
<tr><td>Figure 4.3s</td><td>Figure 4.3t</td></tr>
</table>

Images of intersections between rational functions and their bent end behavior asymptotes are often difficult or impossible to obtain due to the close proximity of the two curves and the often rapidly changing function output values at the intersections. Large changes in y-values mask the small changes between the functions at those intersections. Even so, Figure 4.3t shows the wiggle of $y = c(x)$ through its end behavior asymptote at $x = 3$. Just as $y = x^3$ briefly runs parallel to the x-axis as it wiggles through its triple root at $x = 0$, $y = c(x)$ briefly runs parallel to $y = 0.2x^2$ as it wiggles through its triple end behavior asymptote intersection.

There were two approaches to graphing rational functions presented in this section: 1) graphing the error term and then bending the x-axis into the end behavior asymptote, and 2) graphing the end behavior asymptote and building the function by adding the error back onto that asymptote. While both approaches can be used for any rational function, more complicated functions like $y = c(x)$ in Example 4 tend to be easier to handle by adding error back onto the end behavior asymptote.

Above all, be patient. Rational functions have the potential to include a tremendous amount of information including end behavior asymptotes, vertical asymptotes, holes, x- and y-intercepts, end behavior asymptote intercepts, and the pattern of approach of a rational function to its end behavior asymptote for large $|x|$.

Problems for 4.3:

Exercises

1. Determine the *x*-intercepts for $y = c(x)$ in Example 4.

2. [NC] Graph the given function.

    A. $f(x) = \dfrac{2x^2 + 10x - 25}{x+7}$?

    B. $g(x) = 2 - x + \dfrac{x}{(x+4)(x+1)^2}$

    C. $h(x) = -x^2 + \dfrac{2-x}{(x-9)(x-2)^3}$

    D. $i(x) = \dfrac{x-1}{2x^2 - 9x - 5}$

    E. $j(x) = \dfrac{2}{x^2 + 6x + 9}$
    $(x+3)^2$

    F. $k(x) = \dfrac{(x+1)^2}{x^2 - 1}$

3. Determine all asymptotes of $f(x) = \dfrac{x^2 - 4x + 1}{2x - 3}$.
    VA: 3/2

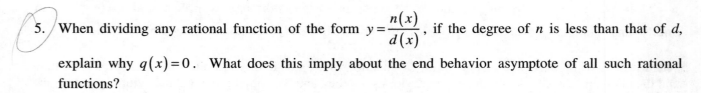

4. Get complete graphs of each of the following. Describe all "interesting" details.

    A. $m(x) = \dfrac{x^3 - 3x - 2}{x^2 - 5x - 6}$

    B. $m(x) = \dfrac{-2x^3 - x^2 + 5x - 2}{x^2 + 3x + 2}$

    C. $n(x) = \dfrac{x^4 - x^3 - 30x^2 - 76x - 56}{x^2 + 1}$

    D. $o(x) = \dfrac{-x^4 + 6x^3 - 5x^2 - 25x + 40}{x^2 - 6x + 9}$

    E. $p(x) = \dfrac{-x^7 + 5x^6 - 5x^5 - 9x^4 + 14x^3 + 5x^2 + 2x + 25}{(x-2)^3(x+1)}$

5. When dividing any rational function of the form $y = \dfrac{n(x)}{d(x)}$, if the degree of *n* is less than that of *d*, explain why $q(x) = 0$. What does this imply about the end behavior asymptote of all such rational functions?

Explorations

Identify each statement in questions 6-10 as true or false. Justify your conclusion.

6. A rational function moves around its end behavior asymptote the same way its error term moves with respect to the *x*-axis.

7. It is possible for a rational function to bounce off its end-behavior asymptote at a hole.

8. [NC] The graph of $y = x^3 + \dfrac{1}{(x-5)^8}$ has two *x*-intercepts. False

9. A rational function could have a third degree denominator and still not have any domain restrictions.

10. A rational function could cross its end behavior asymptote exactly 23 times.

11. Determine the coordinates of the extremum mentioned in the last paragraph of Example 1.

12. Determine an equation of a rational function that crosses its end behavior asymptote, $y = 3x^3 - 3x + 3$, exactly three times.

13. Determine a possible equation for the rational function graphed in Figure 4.3u.

$y = (x-3)^2 + x^2 - 3$

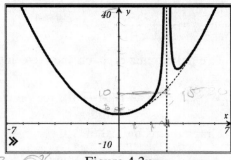

Figure 4.3u

14. Get complete graphs of each of the following. Describe all "interesting" details.

A. $q(x) = \dfrac{(x-3)(x-4)(x+6)}{x^2}$　　　　B. $r(x) = \dfrac{x^3 - 3x^2 - 9x - 5}{x^2 - x - 12}$

C. $s(x) = \dfrac{x^6 + 6x^5 + 5x^4 - 24x^3 - 35x^2 - x}{x^3 + 6x^2 + 9x}$　　　　D. $t(x) = \dfrac{x^6 + 6x^5 + 5x^4 - 24x^3 - 35x^2 - x}{x^3 + 6x^2 - 5}$

15. [NC] In words, describe the graphs of each function. Be sure to note if anything "interesting" happens on each graph at $x = -\pi$.

A. $a(x) = \dfrac{x + \pi}{(x + \pi)^2}$? acts like $\frac{1}{x}$　B. $b(x) = \dfrac{(x + \pi)^2}{x + \pi}$ acts like x　C. $c(x) = \dfrac{(x^2 + \pi^2)^2}{(x^2 + \pi^2)}$ acts like $y = x^2 + \pi^2$

vertical assymptope at $-\pi$　　　hole at $-\pi, 0$　　　no hole or assymtope

16. Determine an equation of a rational function whose graph has the following characteristics .

- y-intercept at $(0, 7)$

- End behavior asymptote at $y = 2x^3 - 3$

- Even vertical asymptotes at $x = -6$ and $x = 2$.

17. Determine an equation of a rational function whose graph has the following characteristics .

- x-intercepts at $x = 2$ and $x = -3$

- Vertical asymptote at $x = 4$

- Hole at $(-5, 0)$

$ax^2 + bx + c$

- Horizontal asymptote at $y = 3$

18. Given $y_1 = \dfrac{x}{x^2 + x - 12}$

   A. Use polynomial division and factoring to rewrite y_1. Graph the result.

   B. Why is $y_2 = x + 1 + \dfrac{-12}{x}$ the reciprocal of y_1?

   C. Graph the reciprocal of y_2.

   D. Are the graphs of parts A and C identical? Why?

19. Determine an equation for a rational function which has a graph like Figure 4.3v.

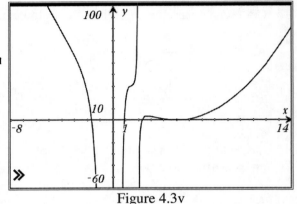

20. A. What is the numerator when a CAS common denominator is applied to $y = c(x)$ (Example 4)?

   B. What does the Fundamental Theorem of Algebra guarantee about the x-intercepts of c?

   C. The graph of $y = c(x)$ had three singleton x-intercepts. Use this to determine the nature of the remaining zeros of $y = c(x)$.

Figure 4.3v

   D. Find all zeros of $y = c(x)$ and show how this confirms your answer to part C.

21. Graph $f(x) = \dfrac{e^x + e^{-x}}{2} = \dfrac{e^x}{2} + \dfrac{e^{-x}}{2}$ by starting with $y = \dfrac{e^x}{2}$ and bending its end behavior asymptote into $y = \dfrac{e^{-x}}{2}$.

22. Reverse the direction of the last question by first graphing $y = \dfrac{e^{-x}}{2}$ and bending its end behavior asymptote into $y = \dfrac{e^x}{2}$. Verify that the intercepts and final graphs are identical.

23. [NC] Graph $y = \log x + \dfrac{\log x - 1.5}{(\log x - 1)(\log x - 2)^2}$.

<u>UNIT 5: TRIGONOMETRY I</u>

It is important that students bring a certain ragamuffin, barefoot irreverence to their studies; they are not here to worship what is known, but to question it.

– Jacob Bronowski

Enduring Understandings:

- There are multiple units of rotation.

- Trigonometric ratios are functions of an angle.

- The symmetry of the circle links triangle side ratios to angles of any magnitude.

- Transformed trigonometric functions have real-life application.

This section assumes you have:

- Right triangle trigonometry

- Exposure to the sine, cosine, and tangent functions and their non-transformed graphs

- The meaning of arc length and sector area for circles and how to compute them.

- The relationships between the sides length measurements of $30° - 60° - 90°$ and $45° - 45° - 90°$ triangles

5.1: Angle Measure, Arc Length, & Sector Area

I advise my students to listen carefully the moment they decide to take no more mathematics courses.
They might be able to hear the sound of closing doors.

– James Caballero

Definition of the radian
While there are 360° (degrees) in a full circle, there is another unit of angle measure—the **radian**. In any circle, a radian is the rotation required to traverse the length of one radius along the circumference of the circle. The circumference of any circle is $C = 2\pi r$, and radian measure is the number of radii on the circumference, so the radian measure of any full circle is 2π. To find the radian measure of an angle, compute the ratio of the corresponding arc length to the radius of the circle: radian measure $= \dfrac{\text{arc length}}{\text{radius length}}$.

Because the units on the right are equivalent, angles measured in radians are unit-less. While acceptable to write "radians" after such an angle measure, it is unnecessary. The converse is also true. Unless other units are explicitly provided for an angle's measure, the angle is *always* assumed to be measured in radians.

Since the radian measure for a complete circle is 2π, the degree (D) and radian (R) measures of any angle are related by the proportion $\dfrac{D}{360°} = \dfrac{R}{2\pi}$.

Example 1:
    A. Convert 135° to radians.
    B. Convert -1.5 radians to degrees.

A. $\dfrac{135°}{360°} = \dfrac{R}{2\pi} \Rightarrow R = \dfrac{3\pi}{4}$

B. $\dfrac{D}{360°} = \dfrac{-1.5}{2\pi} \Rightarrow D = -\dfrac{540°}{2\pi} \approx -85.944°$

Angles are generally measured as rotations starting from the positive *x*-axis. If one of the rays that form an angle is placed on the positive *x*-axis, then the ray at which the angle ends is called its **terminal side**. The direction in which an angle is measured determines its sign. Angles measured in the counter-clockwise direction are positive, while clockwise rotations give negative angles. Angle rotations that have the same terminal side are called **coterminal angles**. Adding any number of full rotations in either direction to an angle gives coterminal angles.

Example 2:
Find two positive and two negative coterminal angles to 392°.

Any angle θ has coterminal angles of the form $\theta + 360n°$, where $n \in \mathbb{Z}$. Therefore, $392° \pm 360°$ gives two positive coterminal angles, 752° and 32°. Two possible negative angles are $392° - 2 \cdot 360° = -328°$ and $392° - 3 \cdot 360° = -688°$.

Arc Length and Sector Area
An arc length is the length of the perimeter traversed along the circumference of a circle within a given central angle. The area of a sector is the portion of a circle's total area contained by a central angle. Both quantities are directly proportional to the central angle defining the length and sector.

In a circle of radius **r** with central angle θ (Figure 5.1a), let l be the arc length and A be the corresponding sector area defined by the angle θ measuring B degrees, or its equivalent R radians. The values of l, A, D, and R are related by the following equivalent proportions, defined by the fraction of the circle each represents.

$$\frac{D}{360°} = \frac{R}{2\pi} = \frac{l}{2\pi r} = \frac{A}{\pi r^2}$$

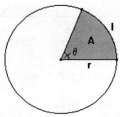

Figure 5.1a

Example 3:
Find the length of an arc in a circle of diameter 10 cm that corresponds to a central angle of $42°$.

Use the two appropriate ratios to set up a proportion.
$$\frac{D}{360°} = \frac{l}{2\pi r} \quad \Rightarrow \quad \frac{42°}{360°} = \frac{l}{10\pi\, cm} \quad \Rightarrow \quad l = \frac{7\pi}{6}cm \approx 3.665\ cm$$

Example 4:
A circle has an arc of length 2 units corresponding to a sector with area 3 square units. What is the radian measure of the central angle of the sector?

$$\frac{2}{2\pi r} = \frac{3}{\pi r^2} \quad \Rightarrow \quad r^2 = 3r \quad \Rightarrow \quad r(r-3) = 0 \quad \Rightarrow \quad r = \{0, 3\} \quad \Rightarrow \quad r = 3$$

$$\text{So, } \frac{R}{2\pi} = \frac{l}{2\pi r} \quad \Rightarrow \quad \frac{R}{2\pi} = \frac{2}{6\pi} \quad \Rightarrow \quad R = \frac{2}{3}\ \text{radians}.$$

Linear and Angular Velocity
There are two common ways to describe the velocity of a rotating object – angular velocity (degrees or radians per unit time) and linear velocity (distance per unit time). When talking about how quickly an object rotates, a convenient starting point is often in revolutions per time unit. Once established, it is easy to transform into other units of angular or linear velocity.

Example 5:
A car with tires of circumference 6 feet is traveling 60 mph. Find the angular velocity of the tires in:

   A. Revolutions per hour

   B. Degrees per hour

   C. Radians per second

   A. There are 5280 ft/mile, so $\quad \frac{60\ mi}{1\ hr} \cdot \frac{5280\ ft}{1\ mi} = \frac{60 \cdot 5280\ ft}{1\ hr} \cdot \frac{1\ rev}{6\ ft} = 52800\ \frac{rev}{hr}$

   B. $\dfrac{52,800\ rev}{1\ hr} \cdot \dfrac{360\ deg}{1\ rev} = 19,008,000\ \dfrac{deg}{hr}$.

   C. $\dfrac{52,800\ rev}{1\ hr} \cdot \dfrac{2\pi\ rad}{1\ rev} \cdot \dfrac{1\ hr}{3600\ sec} = 92.153\ \dfrac{rad}{sec}$.

Problems for Section 5.1:

Exercises

Convert each angle in questions 1-4 to radian measure.

1. 300°   2. −110°   3. 630°   4. $\dfrac{800°}{\pi}$

Convert each angle in questions 5-8 to degree measure.

5. $\dfrac{7\pi}{3}$ 6. $\dfrac{4\pi}{5}$ 7. 4.5   8. $-\pi^2$

For questions 9-11, find one positive and one negative coterminal angle to each given angle. Then write a single expression to represent all of its coterminal angles.

9. 150°   10. $\dfrac{3\pi}{7}$ 11. $-\dfrac{14\pi}{5}$

For questions 12-14, state the quadrant of the terminal side of the given angle.

12. 3 radians   13. 7 radians   14. -2 radians

15. Find the length of the arc corresponding to an angle of 40° in a circle of radius 5.

16. An arc of length 13 units corresponds to a central angle of 2.5 radians. What is the circle's radius?

17. Find the area of a sector that has a central angle of $\dfrac{\pi}{8}$ in a circle of radius 7 cm.

18. If the area of a sector in a circle of radius 8 cm is $25\,\text{cm}^2$, find the measure of the central angle of the sector in degrees.

19. A sector of area $40\,\text{cm}^2$ cuts out an arc of length $16\,\text{cm}$. Find the circle's central angle and radius.

Explorations

Identify each statement in questions 20-24 as true or false. Justify your conclusion.

20. All angle measurements can be expressed in both degrees and radians.

21. Every angle is coterminal to an infinite number of angles.

22. If α and β are two coterminal angles measured in degrees, then $\dfrac{\alpha-\beta}{4}\in\mathbb{Z}$.

23. The sector area varies directly with the arc length within a given circle.

24. One radian is equivalent to a rotation of a quarter circle.

25. A circle has an arc whose length is 2 less than its radius. The corresponding sector's area is numerically 3 more than the radius. What is the radius of the circle?

26. Example 4 used $\dfrac{R}{2\pi} = \dfrac{l}{2\pi r}$ in its last step. Prove $\dfrac{R}{2\pi} = \dfrac{A}{\pi r^2}$ also works and explain why it does.

27. A spinning top has a black dot on its side. After spinning through a 30° angle from its initial position, the dot faces the door.

    A. Through what total degree measure will the top have spun the next time the dot faces the door?

    B. If the top makes 2 rotations every second, how long will it take from its initial position for the dot to pass the door for the second time?

28. A sector of a non-trivial circle has a corresponding radius, arc length, and area all of equivalent numerical measure.

    A. What is the radian measure of the sector's central angle?

    B. What is the area of the circle?

29. The radius, arc length, and area of a sector of a unit circle, in that order, form an arithmetic sequence. What is the area of the sector?

$$\frac{16\,cm}{2\pi r^2} = \frac{40\,cm^2}{\pi r^2}$$

$$80\pi r^2 = 16\pi r^2$$

$$5 = r$$

$$\frac{\theta°}{360} = \frac{16}{10\pi}$$

$$5760 = 10\pi\theta°$$

$$\frac{576}{\pi} = \theta°$$

5.2: The Trigonometric Functions

A new idea comes suddenly and in a rather intuitive way.
But intuition is nothing but the outcome of earlier intellectual experience.

– Albert Einstein

Right-triangle trigonometry

A triangle's side lengths are closely related to the measures of its angles. In right-angled triangles (Figure 5.2a), the three common trigonometric ratios are defined as follows.

Given any acute angle θ in a right-angled triangle:

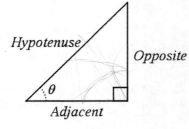

$$\sin\theta = \frac{\text{length of the side opposite } \theta}{\text{length of the hypotenuse}} \left(\text{or} \frac{\text{Opposite}}{\text{Hypotenuse}} \right)$$

$$\cos\theta = \frac{\text{Adjacent}}{\text{Hypotenuse}}$$

Figure 5.2a

$$\tan\theta = \frac{\text{Opposite}}{\text{Adjacent}}$$

Tangent can also be defined in terms of sine and cosine.

$$\tan\theta = \frac{\text{Opposite}}{\text{Adjacent}} = \frac{\text{Opposite/Hypotenuse}}{\text{Adjacent/Hypotenuse}} = \frac{\sin\theta}{\cos\theta}$$

Three more ratios are historically defined as the reciprocals of the sine, cosine and tangent ratios where the denominators do not equal zero.

$$\csc\theta = \frac{\text{hypotenuse}}{\text{opposite}} = \frac{1}{\sin\theta} \qquad \sec\theta = \frac{\text{hypotenuse}}{\text{adjacent}} = \frac{1}{\cos\theta} \qquad \cot\theta = \frac{\text{adjacent}}{\text{opposite}} = \frac{1}{\tan\theta}$$

Because the last four can be expressed in terms of sine and cosine, it seems that none are necessary for any computational purposes. Remember, though, the definition of each function is its unique side ratio, not its algebraic equivalence to other trigonometric functions.

Example 1:

If $\cos\theta = \dfrac{3}{5}$ for some acute angle θ, find the values of the other five trigonometric ratios.

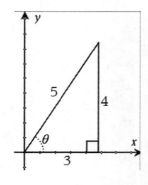

While there are many right triangles for which $\cos\theta = 3/5$, all are similar, guaranteeing equivalent ratios for corresponding sides. If the adjacent side has length 3 and the length of the hypotenuse is 5, by the Pythagorean Theorem the side opposite the angle is $b = 4$ (Figure 5.2b). Therefore,

$$\sec\theta = \frac{5}{3}, \ \sin\theta = \frac{4}{5}, \ \csc\theta = \frac{5}{4}, \ \tan\theta = \frac{4}{3}, \text{ and } \cot\theta = \frac{3}{4}.$$

Figure 5.2b

Trigonometric ratios for special triangles

Some trigonometric ratios can be derived from special triangular relationships. Figures 5.2c and 5.2d show the $45°-45°-90°$ and $30°-60°-90°$ triangle relationships.

1. The $45°-45°-90°$ triangle is an isosceles right triangle, so the legs are congruent. The hypotenuse is found using the Pythagorean theorem.

2. The $30°-60°-90°$ triangle. The side-length relationships are a problem set exercise.

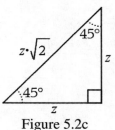

Figure 5.2c

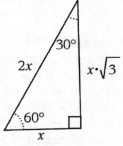

Figure 5.2d

Example 2:

Determine the exact value of $\sin 45°$.

From the $45°-45°-90°$ triangle, $\sin 45° = \dfrac{\text{opposite}}{\text{hypotenuse}} = \dfrac{x}{\sqrt{2} \cdot x} = \dfrac{1}{\sqrt{2}}$

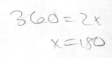

Figure 5.2e gives sine, cosine, and tangent ratios for angles of $30°$, $45°$, and $90°$. Secant, cosecant and cotangent ratios can be found using the reciprocal relationships. Circular trigonometry extends these right triangle trigonometric ratios to all real angles.

x in degrees	$0°$	$30°$	$45°$	$60°$	$90°$
x in radians	0	$\dfrac{\pi}{6}$	$\dfrac{\pi}{4}$	$\dfrac{\pi}{3}$	$\dfrac{\pi}{2}$
$\sin x$	0	$\dfrac{1}{2}$	$\dfrac{1}{\sqrt{2}}$	$\dfrac{\sqrt{3}}{2}$	1
$\cos x$	1	$\dfrac{\sqrt{3}}{2}$	$\dfrac{1}{\sqrt{2}}$	$\dfrac{1}{2}$	0
$\tan x$	0	$\dfrac{1}{\sqrt{3}}$	1	$\sqrt{3}$	undefined

Figure 5.2e

The Unit Circle

A unit circle centered at the origin, ***the*** unit circle, redefines trigonometric ratios in terms of coordinates. If $P(x, y)$ is any point on the unit circle, O is the origin and θ is the counterclockwise angle between the x-axis and point P, then P, O, and the point $(x,0)$ form a right triangle with legs of length x and y and a hypotenuse of 1 (Figure 5.2f). Therefore,

$$\cos \theta = \frac{x}{1} = x \text{ and } \sin \theta = \frac{y}{1} = y.$$

This means the components of P can be redefined as functions of θ. So, $P = (x, y) = (\cos\theta, \sin\theta)$ and the other four trigonometric functions become $\tan\theta = \dfrac{y}{x}$, $\csc\theta = \dfrac{1}{y}$, $\sec\theta = \dfrac{1}{x}$, and $\cot\theta = \dfrac{x}{y}$.

Figure 5.2f

Extending the unit circle to a circle of any radius r produces triangles similar to the one formed with the unit circle.

$$\sin\theta = \frac{y}{r} \qquad \cos\theta = \frac{x}{r} \qquad \tan\theta = \frac{y}{x}$$

$$\sec\theta = \frac{r}{x} \qquad \csc\theta = \frac{r}{y} \qquad \cot\theta = \frac{x}{y}$$

The symmetry of the circle allows this definition to hold for angles of any magnitude, bearing in mind the fact that the signs of the x and y values change (and hence the signs of the ratios) depending on the quadrant in which the terminal side of the angle falls (Figure 5.2g). The radius of the circle, r, is always positive. For this reason, every trigonometric function of any angle can be re-expressed as the positive or negative of the same function of a specific Quadrant I angle.

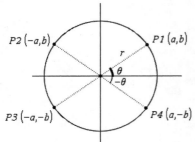

With θ defined by a point on a circle and the trigonometric functions defined for any radius, there are no more size restrictions for angle

Figure 5.2g

measures. Trigonometry can now be applied to circles of all sizes and to positive and negative angles of any magnitude.

On the unit circle, rotating $(1,0)$ by θ and $-\theta$ creates points $P1$ and $P4$, respectively. Because $P1$ and $P4$ are reflection images over the x-axis, their x-coordinates are equal and their y-coordinates are opposite (see Figure 5.2g). The same holds for angles in Quadrants II and III. It follows from this fact that $\forall\theta\in\mathbb{R}$, $\cos(-\theta)=\cos\theta$ making cosine an *even function*. Similarly, $\forall\theta\in\mathbb{R}$, $\sin(-\theta)=-\sin\theta$, makes sine an *odd function*.

Example 3:
Find the exact value of $\sin 210°$.

$210°$ is the Quadrant III image of $30°$ under $r_x \circ r_y$. Because the sine function corresponds to the y-coordinate which is negative, $\sin 210° = -\sin 30° = -\dfrac{1}{2}$.

Example 4:
Rewrite $\cos 462°$ in terms of a first quadrant angle.

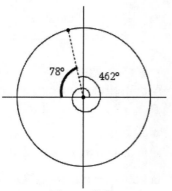

$462°$ terminates in Quadrant II (Figure 5.2h). Its corresponding acute angle from the x-axis is $78°$. Because the x-value is negative in Quadrant II, $\cos 462° = -\cos 78°$.

Cofunctions
The acute angles in any right triangle are **complementary angles**. Using the right triangle definitions of sine and cosine, the cosine of any acute angle equals the sine of the complementary angle. Such pairs of functions are said to be **cofunctions**. By the same reasoning, cosecant and secant are cofunctions. So are cotangent and tangent.

Figure 5.2h

$$\cos\theta = \sin\left(\frac{\pi}{2}-\theta\right) \qquad \csc\theta = \sec\left(\frac{\pi}{2}-\theta\right) \qquad \cot\theta = \tan\left(\frac{\pi}{2}-\theta\right)$$

The symmetry of the circle allows these relationships to be extended to all real angles, and not simply angles in Quadrant I, and so these statements hold $\forall \theta \in \mathbb{R}$.

Example 5:

If $\cos \theta = a$, find, the value of $\sin\left(\theta - \frac{\pi}{2}\right)$ in terms of a.

$$\sin\left(\theta - \frac{\pi}{2}\right) = -\sin\left(\frac{\pi}{2} - \theta\right) \quad \text{odd function}$$
$$= -\cos \theta \quad \text{cofunctions}$$
$$= -a$$

Other Relationships

An equation for a circle of radius r centered at the origin is $x^2 + y^2 = r^2$. $\forall r > 0$, dividing by r^2 gives $\frac{x^2}{r^2} + \frac{y^2}{r^2} = 1$ or $\left(\frac{x}{r}\right)^2 + \left(\frac{y}{r}\right)^2 = 1$. Substituting the definitions of the sine and cosine proves the Pythagorean Identity, $\forall \theta \in \mathbb{R}$, $\sin^2 \theta + \cos^2 \theta = 1$. The three most common variations of the Pythagorean Identity are listed below. Versions 2 and 3 can be derived quickly from version 1.

1. $\sin^2 \theta + \cos^2 \theta = 1$ 2. $\tan^2 \theta + 1 = \sec^2 \theta$ 3. $\cot^2 \theta + 1 = \csc^2 \theta$

Example 6:

If $\tan \theta = -\frac{3}{2}$ and $\frac{\pi}{2} < \theta < \pi$, find the value of $\cos \theta$.

Method 1: A triangle in Quadrant II gives a picture of this situation. From Figure 5.2i,
$$(-2)^2 + 3^2 = r^r \implies r = \pm\sqrt{13}$$

The radius of a circle is positive, so $r = \sqrt{13}$ and $\cos \theta = \frac{x}{r} = \frac{-2}{\sqrt{13}}$.

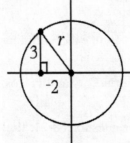

Method 2:

$$\tan^2 \theta + 1 = \sec^2 \theta \implies \left(-\frac{3}{2}\right)^2 + 1 = \sec^2 \theta \implies \sec^2 \theta = \frac{13}{4}$$

So, $\sec \theta = \pm\frac{\sqrt{13}}{2} \implies \sec \theta = -\frac{\sqrt{13}}{2} \implies \cos \theta = -\frac{2}{\sqrt{13}}$.

Figure 5.2i

Example 7:
Kate is standing 15 feet away from a tree. There is a mirror lying on the ground 4 feet from her in which she can see a reflection of the top of the tree. If Kate's line of view to the mirror makes an angle of 52° with the ground, find Kate's height and the height of the tree.

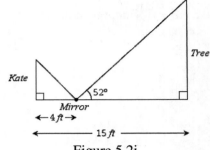

If Kate's height is k, then $\tan 52° = \dfrac{k}{4} \Rightarrow k = 5.12\,ft$. Physics says the angle of incidence of light from the top of the tree equals the angle of reflection of the light, (Figure 5.2j) so the two triangles are similar allowing proportions. Letting t be the height of the tree,

$$\frac{5.12}{4} = \frac{t}{11} \Rightarrow t = 14.08\,ft.$$

Figure 5.2j

Problems for Section 5.2:

Exercises

1. If $\sin\theta = \dfrac{3}{4}$ and $\theta \in \left[0, \dfrac{\pi}{2}\right]$, find an exact value for $\tan\theta$.

For questions 2-5, use the given information to find exact values for all six trigonometric functions.

2. $\tan\theta = \dfrac{12}{5}$, $\theta \in \left[\pi, \dfrac{3\pi}{2}\right]$

3. $\sec\theta = -\dfrac{\sqrt{13}}{3}$, $\theta \in \left[\dfrac{\pi}{2}, \pi\right]$

4. $\cos\theta = \dfrac{1}{5}$

5. $\cos\theta = \sin\theta$, $\theta \in \left[0, \dfrac{\pi}{2}\right]$

6. Find exact coordinates of the point where a 120° ray intersects a radius 5 circle centered at the origin.

7. [NC] Simplify $\cos^2 \dfrac{11\pi}{3} + \sin^2 \dfrac{11\pi}{3}$.

For questions 8-21, find exact values for the given expression.

8. $\cos 510°$

9. $\csc(-225°)$

10. $\tan\dfrac{29\pi}{3}$

11. $\sin\left(-\dfrac{17\pi}{6}\right)$

12. $\sin\dfrac{\pi}{2}$

13. $\cos\dfrac{5\pi}{6}$

14. $\tan\left(\dfrac{-\pi}{4}\right)$

15. $\cot\pi$

16. $\sec\dfrac{\pi}{3}$

17. $\csc\left(-\dfrac{\pi}{4}\right)$

18. $\sin\dfrac{7\pi}{3}$

19. $\cos\left(\dfrac{205\pi}{2}\right)$

20. $\tan 390°$

21. $\cot\left(\dfrac{-21\pi}{3}\right)$

For questions 22-25, solve for all possible values of θ within the given restrictions.

22. $\sin\theta = -\dfrac{\sqrt{3}}{2}$, $\theta \in [0, 2\pi)$

23. $\tan\theta = -1$, $\theta \in \left[-\dfrac{\pi}{2}, \dfrac{\pi}{2}\right)$

24. $\csc\theta = 2$, $\theta \in [0, 4\pi)$

25. $\sec\theta = \sqrt{2}$, $\theta \in \mathbb{R}$

For questions 26-34, assume $\cos\theta = a$ and $\theta \in \left[0, \dfrac{\pi}{2}\right]$. Then restate each expression in terms of a only.

26. $\cos(180° - \theta)$ *a+*

27. $\cos(180° + \theta)$ *−a*

28. $\cos(360° + \theta)$

29. $\sec(720° - \theta)$

30. $\cos(90° + \theta)$

31. $\sin\theta$ *1/2 a*

32. $\sin(\theta - 90°)$ *= a*

33. $\tan(\theta)$

34. $\csc(360° - \theta)$

35. [NC] If $\sin 238° \approx -0.848$, determine an approximation for $\sin 842°$.

$Tan 1.8 = \dfrac{15000}{a}$

$a = 477307.747$

36. If $\cos 2 = a$, what is $\csc(2(\pi - 1))$ in terms of a?

37. [NC] Arrange the following quantities in ascending order.

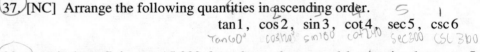

$\tan 1$, $\cos 2$, $\sin 3$, $\cot 4$, $\sec 5$, $\csc 6$

Tan60° cos90° sin180 cot240 sec300 csc360

38. An airplane flying at 15,000 feet above the ground begins its descent. Its angle of depression to the airport is $1.8°$. What is the plane's horizontal distance to the airport?

39. A flagpole of height 18 ft casts a shadow of 25 ft along the ground. Find the angle of elevation from the tip of the shadow to the top of the flagpole.

40. A tent with a 9 ft tall center pole is to be anchored with ropes that make an angle of $28°$ with the ground. Find the length of one of the ropes.

Explorations

Identify each statement in questions 41-45 as true or false. Justify your conclusion.

41. $\forall x \in \mathbb{R}$, $\sin x = \cos\left(x - \dfrac{\pi}{2}\right)$ *True*

42. $\forall x \in \mathbb{R}$ and $n \in \mathbb{Z}$, $\sin x = \sin(x + 2\pi n)$

43. $\sqrt{2}$ is the x-coordinate of the point on a circle of radius 4 with terminal side at angle $\dfrac{3\pi}{4}$.

44. $\forall x \in \mathbb{R}$, $\sec^2 x - \tan^2 x = \sin^2 x + \cos^2 x$

45. $\sin\left(\dfrac{1}{x}\right) = \csc x$

46. Verify the general side length relationships in the triangles in Figures 5.2c and 5.2d using the properties of squares and equilateral triangles.

47. In Figure 5.2k, circle O is the unit circle. If the $m\angle AOB = \theta$, then determine the lengths of $\overline{OA}$, $\overline{AB}$, $\overline{OD}$, $\overline{CD}$, $\overline{OF}$, and $\overline{EF}$ in terms of θ only.

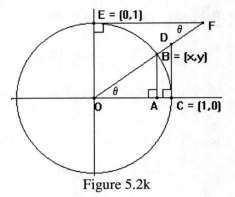

Figure 5.2k

48. In this section, the circle equation, $x^2 + y^2 = r^2$, was used to prove that $\forall \theta \in \mathbb{R}$, $\sin^2 \theta + \cos^2 \theta = 1$. Use $\sin^2 \theta + \cos^2 \theta = 1$ to prove $\tan^2 \theta + 1 = \sec^2 \theta$ and $1 + \cot^2 \theta = \csc^2 \theta$.

5.3: Graphs of Trigonometric Functions.

Every mathematician worthy of the name has experienced ... the state of lucid exaltation in which one thought succeeds another as if miraculously ... this feeling may last for hours at a time, even for days. Once you have experienced it, you are eager to repeat it but unable to do it at will, unless perhaps by dogged work...

– Andre Weil

Graphing $y = \sin x$

By extending the sine function to the y-coordinates on the unit circle, sine became a function with domain $x \in \mathbb{R}$. Figure 5.3b is a scatter plot of the sine values from Figure 5.3a with a smooth, connected plot (Figure 5.3c) representing the fact that $y = \sin x$ is now defined and continuous $\forall x \in \mathbb{R}$. The y-coordinates for angles $x \notin [0, 2\pi]$ repeat the original values. Repetitious functions like $y = \sin x$ are called **periodic** functions.

While the domain of $y = \sin x$ is $x \in \mathbb{R}$, its range is $y \in [-1,1]$. Because it is periodic, the sine function has an infinite number of x-intercepts located at $x = n\pi \ \forall \ n \in \mathbb{Z}$. The **period** of the sine function (the horizontal distance required to complete one cycle) is 2π.

x	0	$\dfrac{\pi}{6}$	$\dfrac{\pi}{4}$	$\dfrac{\pi}{3}$	$\dfrac{\pi}{2}$	$\dfrac{2\pi}{3}$	$\dfrac{3\pi}{4}$	$\dfrac{5\pi}{6}$	π
$\sin x$	0	$\dfrac{1}{2}$	$\dfrac{\sqrt{2}}{2}$	$\dfrac{\sqrt{3}}{2}$	1	$\dfrac{\sqrt{3}}{2}$	$\dfrac{\sqrt{2}}{2}$	$\dfrac{1}{2}$	0
x	π	$\dfrac{7\pi}{6}$	$\dfrac{5\pi}{4}$	$\dfrac{4\pi}{3}$	$\dfrac{3\pi}{2}$	$\dfrac{5\pi}{3}$	$\dfrac{7\pi}{4}$	$\dfrac{11\pi}{6}$	2π
$\sin x$	0	$-\dfrac{1}{2}$	$-\dfrac{\sqrt{2}}{2}$	$-\dfrac{\sqrt{3}}{2}$	-1	$-\dfrac{\sqrt{3}}{2}$	$-\dfrac{\sqrt{2}}{2}$	$-\dfrac{1}{2}$	0

Figure 5.3a

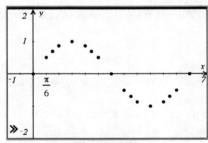

Figure 5.3b

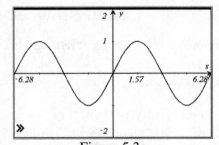

Figure 5.3c

The continuous graph in Figure 5.3c reconfirms the odd function relationship for sine ($\sin(-x) = -\sin x$) stated at just prior to Example 3 in Section 5.2. This is even further reinforced by the graphical symmetry of $y = \sin x$ with respect to the origin.

Graphing $y = \cos x$

By similar reasoning, cosine is a periodic function and can be graphed by plotting points to show the pattern of one cycle. Alternatively, cosine could be graphed via transformations. Because sine and cosine are cofunctions, $\cos x = \sin\left(\dfrac{\pi}{2} - x\right)$. Because sine is an odd function,

$$\cos x = \sin\left(-\left(x - \dfrac{\pi}{2}\right)\right) = -\sin\left(x - \dfrac{\pi}{2}\right) \qquad . \qquad \text{Therefore,}$$

$$\cos x = T_{0,\frac{\pi}{2}} \circ r_{x-axis}\left(\sin(x)\right). \qquad \text{Figures 5.3d an d5.3e show these}$$

transformations to the final graph of cosine.

As a result of the transformational relationship between cosine and sine, the two functions must have the same domain, range, and period. Therefore, the x-intercepts of cosine are the image of sine's intercepts

under $T_{\frac{\pi}{2},0}$: $x = (2n+1)\dfrac{\pi}{2} \quad \forall\, n \in \mathbb{Z}$.

The graphs of the remaining trigonometric functions can be derived using transformations on the sine and/or cosine.

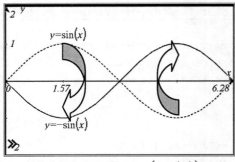

Figure 5.3d: $r_{x-axis}\left(\sin(x)\right)$

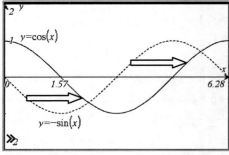

Figure 5.3e: $\cos x = T_{0,\frac{\pi}{2}} \circ r_{x-axis}\left(\sin(x)\right)$

Example 1:
Graph $y = \tan x$.

- From before, $\forall x = \pi n,\, n \in \mathbb{Z},\ \sin x = 0$. $\tan x = \dfrac{\sin x}{\cos x}$, so $y = \tan x = 0$ at the same x-values as $y = \sin x$. That is, $y = \tan x$ has x-intercepts at all integer multiples of π.

- For $x \in \left(0, \dfrac{\pi}{2}\right)$, $\sin x$ increases and $\cos x$ decreases. With an increasing numerator and decreasing denominator, the tangent ratio increases.

- As $x \to \left(\dfrac{\pi}{2}\right)^{-}$, $\cos x \to 0^{+}$ and $\tan x \to \infty$. Therefore, $x = \dfrac{\pi}{2}$ is a vertical asymptote for the graph of tangent. Every x-intercept of $y = \cos x$ is odd, so the graph of $y = \tan x$ has odd vertical asymptotes at all of cosine's x-intercepts. Figure 5.3f shows the vertical asymptotes, x-intercepts, and the graph of $y = \tan x$ for $x \in \left[0, \dfrac{\pi}{2}\right]$.

- Because $x = \dfrac{\pi}{2}$ is an odd vertical asymptote for tangent, the graph of $y = \tan x$ passes through infinity as x increases. Moving right over $x \in \left(\dfrac{\pi}{2}, \pi\right)$, tangent's numerator is positive and decreasing, while its denominator is negative and increasing in magnitude. Overall, tangent is negative and decreases in magnitude until $x = \pi$. Periodicity completes the graph (Figure 5.3g).

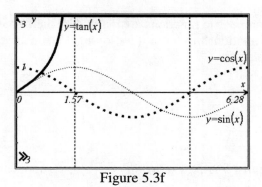

Figure 5.3f

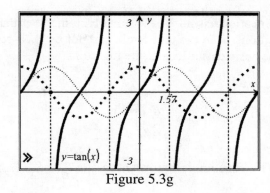

Figure 5.3g

The graph of $y = \tan x$ has vertical asymptotes at $x = (2n+1)\dfrac{\pi}{2} \ \forall \ n \in \mathbb{Z}$, making its domain $x \in \mathbb{R}$ except at these values. Due to the odd vertical asymptotes, the range of tangent is $y \in \mathbb{R}$.

Example 2:
Graph $y = \csc x$.

Because $y = \csc x = \dfrac{1}{\sin x}$, the graph of cosecant can be found using the *REC* transformation on $y = \sin x$.

As noted in the Section 2.5 introduction of the *REC* transformation, this means that

    i)   points where $\sin x = \pm 1$ remain the same,

    ii)   points where $\sin x = 0$ are eliminated, and all other points not mentioned in i are changed.

    iii) $\forall x \in \mathbb{R}$, $|\sin x| \le 1$, so it follows that $\forall x$, $|\csc x| \ge 1$.

Figure 5.3h shows the transformation. Thus, $y = \csc x$ has odd vertical asymptotes at $x = n\pi$, $\forall n \in \mathbb{Z}$. Its domain is $x \in \mathbb{R}$, except for the points where vertical asymptotes exist. Its range is $y \in (-\infty, -1] \cup [1, \infty)$ and there are no x-intercepts.

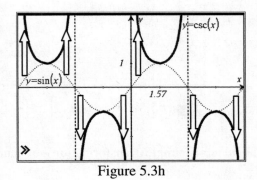

Graphs for $y = \sec x$ and $y = \cot x$ also can be found using the *REC* transformation.

Figure 5.3h

Domain considerations

Notice that $\csc x = \dfrac{1}{\sin x}$ has odd vertical asymptotes wherever $\sin x = 0$. When $\sin x$ is defined as $y = REC(\csc x)$, the vertical asymptotes are transformed back to zeros, giving the sine graph. Given the non-complex fraction requirement for the *REC* transformation, $REC(\csc x) = \sin x \ \forall x \in \mathbb{R}$. This is equivalent to saying that $(\sin x)(\csc x) = 1$ so long as neither function is zero. Without this requirement, the expression $\dfrac{1}{\csc 0} = \dfrac{1}{\left(\frac{1}{\sin 0}\right)}$ is undefined because the overall denominator is undefined at $x = 0$. Some CAS use the reversible interpretation of the *REC* transformation and return $\dfrac{1}{\csc 0} = 0$. The same holds for secant with cosine and tangent with cotangent. There are domain issues with reciprocal definitions of trigonometric functions, but as long as both functions are being evaluated only over their *common* domain, mutual reciprocation has no issues. Finally, remember that all trigonometric functions are *defined* by their coordinate definitions from the unit circle and not by their reciprocal relationships. If a question of domain arises, rely on the basic definitions of the functions.

Example 3:

Graph $y = 3\sin\left(2\left(x - \dfrac{\pi}{2}\right)\right) + 1$.

This function is equivalent to $y = T_{\frac{\pi}{2},1} \circ S_{\frac{1}{2},3}(\sin x)$. Figures 5.3i-5.3l show the step-by-step application of these transformations.

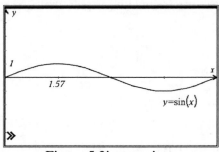

Figure 5.3i: $y = \sin x$

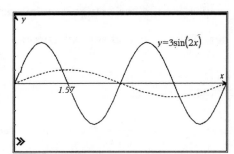

Figure 5.3j: $y = 3\sin 2x$

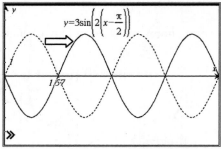

Figure 5.3k: $y = 3\sin\left(2\left(x - \dfrac{\pi}{2}\right)\right)$

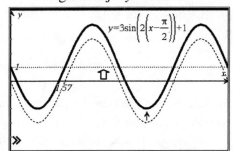

Figure 5.3l: $y = 3\sin\left(2\left(x - \dfrac{\pi}{2}\right)\right) + 1$

In general, simple transformations of sine graphs have equations of the form $y = A\sin(b(x-h)) + k$ or $\dfrac{y-k}{A} = \sin\left(\dfrac{x-h}{B}\right)$, where $S_{A,B}$ and $T_{h,k}$ are the respective scale changes and translations from the parent function $y = \sin x$.

- $|A|$ is called the **amplitude** of the graph.
- $|B|$ identifies the number of cycles the function goes through in a normal period for the function (that is, in a horizontal distance of 2π). In other words, the period of the transformed function is given by $\dfrac{2\pi}{|b|}$. Note that $B = \dfrac{1}{b}$ in the sine equations above.
- h gives the horizontal shift (sometimes called "phase shift") of the function.
- k gives the vertical shift of the function, which is also the midline of the graph.

Problems for Section 5.3:

Exercises

For questions 1-4, graph at least one complete cycle of the given function.

1. $y = 3\sin\left(4\left(x - \frac{\pi}{4}\right)\right) + 2$

2. $y = -\frac{1}{4}\cos\left(\frac{x}{2} + \frac{\pi}{4}\right) - 3$

3. $y = 2\tan\left(3x + \frac{\pi}{6}\right) - 1$

4. $y = 3 + 2\csc\left(\frac{\pi}{6}(x - 1)\right)$

Determine a possible equation for each graph in questions 5-8.

5.

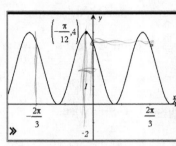

$y = 2\cos\left(3\left(x + \frac{\pi}{12}\right)\right) + 2$

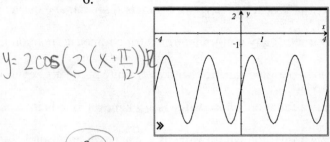

6.

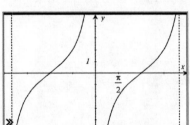

7.

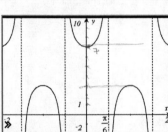

8.

For questions 9-17, graph the given function.

9. $y = |\sin x|$

10. $y = \cos^2 x$

11. $y = \sqrt{\tan x} + 1$

12. $y = \sqrt{\cos x}$

13. $y = \sin(|x|)$

14. $y = \sec x$

15. $y = \cot x$

16. $y = \sec^2 x - 1$

17. $y = \ln(\sin x)$

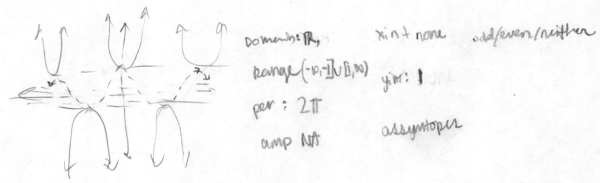

Explorations

Identify each statement in questions 18-22 as true or false. Justify your conclusion.

18. The amplitude of $y = -3\cos x + 1$ is -3.

19. Sine and cosine are the only trigonometric functions with domain $x \in \mathbb{R}$.

20. $y = \pi \sin\left(x + \dfrac{\pi}{2}\right) + 5$ is an even function.

21. There is a constant translation of the tangent function whose image is an even function.

22. There is a constant translation of the secant function whose image is an even function and another translation whose image is an odd function.

23. Determine if each of tangent, cotangent, secant, and cosecant are even, odd, or neither. Justify.

24. Why doesn't any non-transformed trigonometric function have an even vertical asymptote?

25. Create a periodic function with an infinite number of even vertical asymptotes.

26. Create a function whose value is 1 whenever sine is increasing and -1 otherwise.

27. Create a function whose value is 1 whenever cosecant is increasing and 0 otherwise.

5.4: Solving Equations and Sinusoidal Applications

Use for yourself little, but give to others much.

– Albert Einstein

Solving equations
Example 1:
Solve $1.3^x \cos x = 1.3^x \ \forall x \in [0, 2\pi)$.

$$1.3^x \cos x - 1.3^x = 0 \ \Rightarrow \ 1.3^x (\cos x - 1) = 0$$

Since $1.3^x \neq 0$, $\cos x = 1$, therefore $x = 0$ is the only solution.

Example 2:
Solve $\tan x = \sec x \ \forall x \in \mathbb{R}$.

Both functions are undefined when $\cos x = 0$. Rewriting the functions in terms of sine and cosine gives
$\dfrac{\sin x}{\cos x} = \dfrac{1}{\cos x} \ \Rightarrow \ \sin x = 1$. Because $\sin x = 1$ when $\cos x = 0$, *there is no solution*. A careless disregard
for the domain of the given equation would have produced a wrong solution.

Example 3:
Solve $\cos^4 x - 5\cos^2 x + 4 = 0 \ \forall x \in \mathbb{R}$.

This pattern is quadratic. If $A = \cos^2 x$, the equation can be rewritten as:
$$A^2 - 5A + 4 = 0 \ \Rightarrow \ (A-1)(A-4) = 0$$
Substituting back for A gives $\cos^2 x = 1$ or $\cos^2 x = 4 \ \Rightarrow \ \cos x = \pm 1$ or $\cos x = \pm 2$. Because ± 2 are
outside the range of cosine, $\cos x = \pm 1 \ \Rightarrow \ x = n\pi \ \forall n \in \mathbb{Z}$.

Sinusoidal applications
Section 5.3 introduced graphs and transformations of the trigonometric functions. Any function which can be
re-written as a combination of basic transformations on a sine function is called a **sinusoidal function**. Just
as exponential functions can be used to model growth and decay, sinusoidal functions can be used to model
quantities that have an oscillating pattern.

Example 4:
As of 2009, the Singapore Flyer was the largest Ferris wheel in the world with a diameter of 150 meters and a
total height of 165 meters. One complete rotation of the wheel takes approximately 30 minutes. If Faiz is
sitting in the highest capsule when the wheel begins to rotate, write an equation to express his height above
the ground as a function of time.

It is helpful to plot the given information as a first step (Figure 5.4a). The
graph shows one complete cycle of the Singapore Flyer, allowing visual
determination of the amplitude, horizontal stretch, phase shift and vertical
shift for the corresponding sinusoidal function.

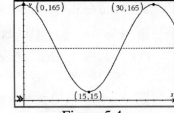

Figure 5.4a

There are multiple options for equations. In this case, the graph begins

at a peak at $t = 0$. A cosine equation $\left(\dfrac{y-k}{A} = \cos\left(\dfrac{x-h}{B} \right) \right)$ seems to be a good choice.

- There are no reflections, so A and B are both positive.

- $A = \dfrac{150}{2} = 75$

- The period is 30 minutes, so $B = \dfrac{30}{2\pi} = \dfrac{15}{\pi}$.

- Because the peak is at $t = 0$, no phase shift is *required*.

- The midline has been translated up 15+75, so the total vertical shift is 90.

All together, this gives $h(t) = 75\cos\left(\dfrac{\pi}{15}t\right) + 90$.

Problems for Section 5.4:

Exercises

Solve each equation for $x \in \mathbb{R}$.

1. $2\cos^2 x + \cos x - 1 = 0$

2. $\tan^2 x - 3\tan x + 1 = 0$

3. $2\sin^4 x + 5\sin^2 x = 3$

4. $2\cos^2 x + 7\sin x + 2 = 0$

5. $\sec x + 2\tan x - \csc x = 2$

6. $e^{-x}(3\sin x + 1) = 0$

Solve for $x \in [0, 2\pi)$.

7. $\sin 2x = \dfrac{1}{2}$

8. $2\sin^3 x = 3\sin x$

9. The phases of the moon are periodic and have a cycle of approximately 30 days. The percentage illumination *I* of the moon is a sinusoidal function of the number of days passed. A new moon has 0% illumination and a full moon has 100% illumination. Data from the U.S. Naval Observatory states that a full moon was recorded on June 7, 2009, and the next new moon was on June 22, 2009.

  A. Determine a sinusoidal function that models the given data.

  B. Use your model to estimate the percentage illumination of the moon on June 27, 2009.

10. The London Eye was the third largest Ferris wheel in the world in 2009. It has a radius of 61 meters and its outer edge rotates at 26 cm/sec. The highest point one of its passenger capsules can go is 135 meters above the ground.

  A. Find the angular velocity of the wheel in revolutions per minute.

  B. Find the time it takes to complete one revolution.

  C. Assuming that when the ride begins, Susanna was at the two o'clock position and ascending, determine a sinusoidal equation for Susanna's height with respect to time in minutes.

  D. Find the first positive time when Susanna was at her lowest point.

  E. If a ride is more than one complete revolution but less than two, how long was Susanna on the wheel if she got off at the lowest point?

11. Refer to the Singapore Flyer information from Example 4.

    A. What is Faiz's height above the ground after 10 minutes?

    B. What is the first positive time at which Faiz is 75 m above the ground? Is he ascending or descending?

The following questions pertain to the data presented in the tables below regarding the number of daylight hours on the first day of each month in three different cities around the world.[7]

Set 1: Nome, Alaska

Month	Jan	Feb	Mar	Apr	May	Jun	Jul	Aug	Sept	Oct	Nov	Dec
Daylight hours	4:13	7:05	10:12	13:38	17:01	20:22	21:09	18:05	14:33	11:16	7:51	4:48

Set 2: Stuttgart, Germany

Month	Jan	Feb	Mar	Apr	May	Jun	Jul	Aug	Sept	Oct	Nov	Dec
Daylight hours	8:21	9:28	11:02	12:53	14:35	15:54	16:06	15:05	13:29	11:39	9:53	8:34

Set 3: Dar-Es-Salaam, Tanzania

Month	Jan	Feb	Mar	Apr	May	Jun	Jul	Aug	Sept	Oct	Nov	Dec
Daylight hours	12:31	12:24	12:14	12:02	11:52	11:45	11:44	11:49	11:59	12:10	12:21	12:30

12. Plot the data for each city on a scatter plot.

13. Determine and graph a sinusoidal model for the daylight pattern in each city.

14. Analyze the similarities and differences in the patterns among the cities.

15. Is Dar-Es-Salaam North or South of the Equator? How can you tell?

16. Use your model to predict the approximate length of the day in Stuttgart on May 15.

17. According to your model, if the sun rises at 4:00 am on the longest day of the year in Nome, at what time will it set?

18. One student chose to use degrees for this problem because she said it would essentially allow her to avoid using a horizontal stretch. Was she right? Explain.

Explorations

Identify each statement in questions 19-23 as true or false. Justify your conclusion.

19. The equation $a^x \cos x = 0$ for any $a \in \mathbb{R}$ has exactly two solutions in $[0, 2\pi)$.

20. $f(x) = a \sin(bx)$ has b zeros in $x \in [0, 2\pi)$.

21. $g(x) = \sin x + \cos x$ is a **sinusoidal function**.[8]

22. The period of $h(x) = \sin\left(\dfrac{x}{2}\right) + \cos\left(\dfrac{x}{3}\right)$ is 6π.

23. $j(x) = \sin^2 x$ is a sinusoidal function.

24. In terms of $a \in \mathbb{Z} > 0$, how many zeros does the graph of $y = e^{ax} \sin(ax)$ have for $x \in [0, 2\pi)$?

---

[7] Data accessed at http://www.timeanddate.com on June 24, 2009.

[8] A sinusoidal function is any function that can be written as a single sinusoid under $T_{h,k}$ and/or $S_{a,b}$ only.

5.5: Advanced Transformations on Trigonometric Functions

Mathematics has beauties of its own—a symmetry and proportion in its results, a lack of superfluity, an exact adaptation of means to ends, which is exceedingly remarkable and to be found only in the works of the greatest beauty. When this subject is properly ... presented, the mental emotion should be that of enjoyment of beauty.   *– J. W. A. Young*

Variable stretches

Consider the graph of $y = 2\sin x$ (Figure 5.5a). As shown, the sine function can be pictured as a function bouncing between a "ceiling" at its amplitude ($y = 2$) and a "floor" at the opposite end of its amplitude ($y = -2$).

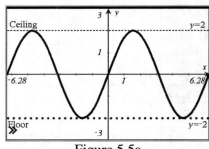

Figure 5.5a

Example 1:

Use variable stretches to develop the graph of $y = x\sin x$.

The difference between the graphs of $y = 2\sin x$ and $y = x \cdot \sin x$ is that the amplitude of the latter is variable. As in Figure 5.5a, graphs of trigonometric functions bounce between the graphs of their ceiling and floor functions. In the case of $y = x \cdot \sin x$, the ceiling is $y = x$ and the floor is $y = -x$. Figure 5.5b shows these boundaries and $y = x\sin x$. In general, the ceiling and floor functions are called **envelope curves**, as suggested by the graphs in Figures 5.5a and 5.5b.

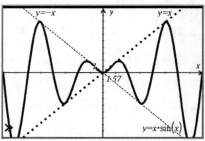

Figure 5.5b

In Figure 5.5b, $y = -x$ and $y = x$ intersect, exchanging relative positions. The fact that the "floor" is actually above the "ceiling" on the left of the *y*-axis is irrelevant. Like all products of two odd functions, $y = x \cdot \sin x$ is an even function.

Example 2:

Graph $y = 2^x \cos x$.

First, graph the envelope curves for this function, $y = 2^x$ and $y = -2^x$. The cosine curve touches its ceiling at inputs which are even multiples of π and its floor at odd multiples of π (Figure 5.5c). Due to the speed of its exponential envelope curves, it is difficult to observe graphically what happens for negative *x*-values. Even so, algebraic analysis confirms that the graph oscillates forever between its ceiling and floor, and has an infinite number of zeros.

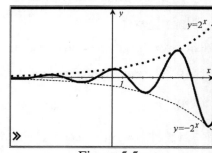

Figure 5.5c

Variable slides

The graph of $y = \sin x$ can also be pictured as an oscillating curve about the x-axis, its "midline." Applying $T_{0,2}$ shifts every point and the midline up two units. $y = \sin x + 0$ has a midline of $y = 0$ and $y = \sin x + 2$ has a midline of $y = 2$ (Figure 5.5d).

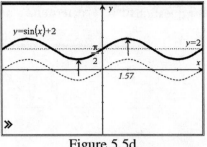

Figure 5.5d

Example 3:

Graph $y = \sin x + x$.

This is the same graphical transformation referred to in Section 2.1 as "bending the midline" and as "bending the asymptote" of rational functions in Unit 4. The end behavior asymptote of a rational function and the midline of a trigonometric function both define the centers of their respective graphs. The functions maintain their relationships with their respective midlines, regardless of how the midline changes or what function defines the midline. Trigonometric functions oscillate infinitely around their midlines, maintaining the same vertical distance from the midline as the original function did from the x-axis. Therefore $y = \sin x$ and $y = \sin x + x$ intersect their respective midlines at every integer multiple of π input. Because the parent sine function has an amplitude of 1, $y = \sin x + x$ is at its greatest vertical distance of 1 from its midline at all odd integer multiples of $\dfrac{\pi}{2}$, the points where the parent sine function is also at its peak.

Just as $y = \sin x$ has a ceiling and floor at $y = 0 \pm 1$, it is often helpful to think of $y = \sin x + x$ as having a ceiling and floor one unit above and below its midline of $y = x$. In other words, $y = \sin x + x$ oscillates between $y = x + 1$ and $y = x - 1$. The graph of $y = \sin x + x$ (Figure 5.5e) shows these relationships with $y = x$ as the midline and a sine curve oscillating around it.

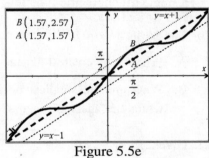

Figure 5.5e

Problems for Section 5.5:

Exercises

[NC] Graph each given function in questions 1-12.

1. $y = 2x \sin x$

2. $y = |x| \sin x$

3. $y = \cos x + x$

4. $y = \dfrac{\cos x}{x^2 + 1}$

5. $y = x \left(x^2 - \pi^2 \right) \sin x$

6. $y = \sin x + \left(x - \pi \right)^2$

7. $y = x^2 \cos x$

8. $y = e^x + \cos x$

9. $y = \sqrt{|x|} + \tan x$

10. $y = \left(4 - x^2 \right) \cos x + x$

11. $y = \dfrac{\sin x + 5}{1 + 2^x}$

12. $y = \sqrt{x} \cos x$

For questions 13-14, determine a possible equation for the given graph.

13.

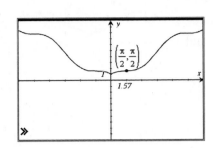

14.

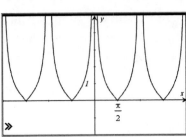

$y = |\tan x|$

Explorations

Identify each statement in questions 15-18 as true or false. Justify your conclusion.

15. $f(x) = x \cdot \cos x$ is an even function. ~~True~~ False

16. $g(x) = x + \cos x$ is an even function. False

17. The graph of $y = x + \csc x$ is identical to the graph of $y = \dfrac{1}{x + \sin x}$

18. The domain of a sum of two functions is the union of the domains of the individual functions.

19. If a line crosses the x-axis at an angle of θ from the positive x-axis, write the slope of that line in terms of a single trigonometric function.

20. The greatest *vertical* distance between the graph of $y = \sin x + x$ and its midline is 1.

   A. What is the greatest distance between the graph of $y = \sin x + x$ and its midline?

   B. What are the smallest positive coordinates of the points on $y = \sin x + x$ and its midline between which that distance occurs?

21. Given $f(x) = 2\sin x \cos x$.

   A. Graph $y = f(x)$ using $y = 2\sin x$ as the variable amplitude.

   B. Graph $y = f(x)$ using $y = 2\cos x$ as the variable amplitude.

   C. Determine an equation for $y = f(x)$ using only one trigonometric function.

22. The end of Example 1 claimed that $y = x\sin x$ is an even function. Prove this claim.

23. A. Graph $y = \dfrac{\tan x}{x}$.

   B. Is the function even, odd, or neither? Prove your claim.

   C. Use the graph to determine $\displaystyle\lim_{x \to \infty} \dfrac{\tan x}{x}$.

24. Sketch a graph of $y = R(x + \cos x) = \dfrac{1}{x + \cos x}$.

5.6: Inverse Trigonometric Functions

I'm sorry to say that the subject I most disliked was mathematics. I have thought about it. I think the reason was that mathematics leaves no room for argument. If you made a mistake, that was all there was to it.

– Malcolm X

A helicopter $5000\,ft$ west of a building is told to hover at the same altitude over a point $9000\,ft$ north of the building. Finding the angle through which the pilot would need to turn in order to fly to his new location in a straight line requires finding an angle as opposed to a trigonometric ratio. The answer is the acute angle for which $\tan\theta = 9/5$.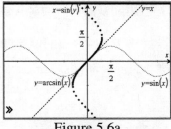

Solutions to questions like this require inverse *functions*. This is complicated by the fact that basic trigonometric functions are infinitely periodic, which means there are infinitely many solutions to such equations. Trigonometric functions are *not* one-to-one, so their inverse relations are not functions.

To create graphs of inverse trigonometric functions, as with all inverses, first apply $r_{y=x}$ to the graphs of the parent functions.   To create inverse trigonometric functions, the ranges of the images must be restricted. Figures 5.6a and 5.6b show the graphs of sine and cosine, their inverse relations as dotted curves, and their final restricted inverse *functions* as bolded solid curves.   These inverse *functions* of sine and cosine are denoted $y = \arcsin x = \sin^{-1} x$ and $y = \arccos x = \cos^{-1} x$, respectively.

Figure 5.6a

Similar transformations create the graph of $y = \arctan x = \tan^{-1}x$ (Figure 5.6c).

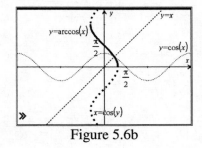

Figure 5.6b

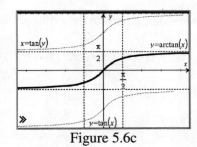

Figure 5.6c

Defining inverse trigonometric functions
Due to the range restrictions, the basic inverse trigonometric functions are defined as follows.

$$y = \arcsin x = \sin^{-1} x \text{ iff } x = \sin y \text{ and } -\frac{\pi}{2} \le y \le \frac{\pi}{2},$$

$$y = \arccos x = \cos^{-1} x \text{ iff } x = \cos y \text{ and } 0 \le y \le \pi$$

$$y = \arctan x = \tan^{-1} x \text{ iff } x = \tan y \text{ and } -\frac{\pi}{2} < y < \frac{\pi}{2}$$

$y = \arcsin x$ and $y = \arccos x$ have domain $x \in [-1,1]$, while $y = \arctan x$ has domain $x \in \mathbb{R}$. The output values for inverse trigonometric functions are the **principal values** of the angles and are always located in Quadrant I and either Quadrant II or IV, depending on the function.

Example 1:

Graph $f(x) = \sin^{-1}(2x)$.

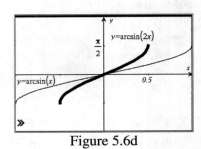

This is equivalent to $y = S_{1,\frac{1}{2}}(y = \arcsin x)$. Figure 5.6d gives the

original and transformed curves.

Figure 5.6d

To see the differences between inverse trigonometric functions and relations, consider $\arccos\left(\cos\dfrac{\pi}{6}\right) = \dfrac{\pi}{6}$.

This was straightforward, but solving the equation $\cos y = \cos\dfrac{\pi}{6}$ $\forall y \in \mathbb{R}$, gives $y = 2n\pi \pm \dfrac{\pi}{6}$, $\forall n \in \mathbb{R}$. There are infinitely many solutions to the second equation because the relation has no range restrictions while range restrictions limit inverse trigonometric functions to a single solution.

Finally, range restrictions on inverse trigonometric functions sometimes alter outputs depending on the value of the input angle. For instance, $\arccos\left(\cos\dfrac{\pi}{6}\right) = \dfrac{\pi}{6}$, but $\arccos(\cos 300°) = 60°$, not $300°$.

Problems for Section 5.6:

Exercises

1. If $\tan x = -\sqrt{3}$, compute the principal value of x.

2. Explain why $\arccos(\cos 300°) = 60°$, not $300°$.

3. Solve for $0 \le x \le 2\pi$.

   A. $\cos x = \dfrac{1}{2}$ B. $\sin 3x = -\dfrac{1}{2}$

4. Solve $\forall x \in \mathbb{R}$: $\tan x = -\sqrt{3}$.

[NC] For questions 5-18, evaluate the given expression.

5. $\cos^{-1}\left(\dfrac{1}{2}\right)$

6. $\operatorname{arccsc}\left(\sqrt{2}\right)$

7. $\operatorname{arccot}\left(-\sqrt{3}\right)$

8. $\sin\left(\arccos\dfrac{\sqrt{3}}{2}\right)$

9. $\sec\left(\sin^{-1}\left(-\dfrac{7}{25}\right)\right)$

10. $\operatorname{arcsec}\left(\sec 255°\right)$

11. $\arcsin\left(\tan\dfrac{\pi}{4}\right)$

12. $\arccos\left(-\dfrac{\sqrt{3}}{2}\right)$

13. $\tan^2\left(\operatorname{arcsec}\left(-\dfrac{13}{12}\right)\right)$

14. $\cos\left(\arctan(-2)\right)$

15. $\operatorname{arccsc}\left(\csc\left(\dfrac{7\pi}{6}\right)\right)$

16. $\sin\left(\arcsin(0.654)\right)$

17. $\sin\left(2\arccos\left(\dfrac{1}{2}\right)\right)$

18. $\sin\left(\arctan\left(-\sqrt{3}\right)-\arcsin\left(-\dfrac{1}{2}\right)\right)$

For questions 19-24, state the domain and range of the given relation.

19. $y = 3\arccos x$

20. $y = -4\sin^{-1}(3x)$

21. $x = 2\sin y$

22. $y = \dfrac{\arctan(-3x)}{2} - \dfrac{\pi}{4}$

23. $3x = \cos(4y)$

24. $4y = 2\arccos\left(\dfrac{x}{3}\right) + 1$

For questions 25-29, graph the given function and describe the transformations that have taken place on the basic inverse trigonometric function. State the domain and range of the function.

25. $f(x) = 3\arccos\left(\dfrac{x}{2}\right)$

26. $g(x) = -5\arcsin(x-1)$

27. $h(x) = 5\tan^{-1}\left(\dfrac{x}{3}\right)$

28. $i(x) = x\arctan x$

29. $j(x) = x^2\arctan x$

Explorations

Identify each statement in questions 30-34 as true or false. Justify your conclusion.

30. $\arcsin x + \arccos x = \dfrac{\pi}{2}$

31. $\forall \alpha \in \mathbb{R},\ \tan\left(\arctan\alpha\right) = \alpha$

32. $\forall x \in \mathbb{R},\ \sin^{-1}\left(\sin x\right) = x$

33. $\arcsin x = \arcsin(-x)$

34. $\operatorname{arcsec} x = \dfrac{1}{\arccos x}$

35. Give the domain and range of $y = 3x\arcsin x$.

36. Find, in terms of $k \in \mathbb{R}$, an expression for all the y-intercepts of the graph of $x = \sin\left(3(y-k)\right)$.

UNIT 6: TRIGONOMETRY II

A positive attitude may not solve all your problems,
but it will annoy enough people to make it worthwhile.

– Herm Albright

Enduring Understandings:

- Equivalence is at the core of all mathematical proof.

- Recognizing and converting between different forms of a mathematical expression can reveal previously obscured characteristics about the expression.

- There are just a few basic identities; all others can be derived from these.

- Trigonometry has useful applications, and with appropriate technology, right-triangle trigonometry can address a wide variety of real-world situations.

This section assumes you have:

- Basic knowledge of trigonometric relationships, graphs and their transformations.

- Familiarity with variable dilations and translations

- Algebraic and graphical understanding of inverses

6.1: Basic Trigonometric Identities

> *Sometimes the questions are complicated and the answers are simple.*
>
> – Theodor "Dr. Seuss" Geisel

Unit 5 introduced three Pythagorean identities for trigonometric functions.

$$\sin^2 x + \cos^2 x = 1 \qquad\qquad \tan^2 x + 1 = \sec^2 x \qquad\qquad 1 + \cot^2 x = \csc^2 x$$

An **identity** is a statement of equivalence between two expressions that is true for all values in the domain *common* to all the functions involved.

Some examples of identities from previous units are $\tan x = \dfrac{\sin x}{\cos x}$, $\csc x = \dfrac{1}{\sin x}$ and $\sin\left(\dfrac{\pi}{2} - x\right) = \cos x$.

Following are other types of identities and strategies for handling them.

Example 1:

If $\cos x = a$, find $\cot\left(x - \dfrac{\pi}{2}\right)$ in terms of a.

$$\cot\left(x - \frac{\pi}{2}\right) = -\cot\left(\frac{\pi}{2} - x\right) \qquad \text{(odd function)}$$

$$= -\tan x \qquad \text{(cofunctions)}$$

$$= -\left(\sec^2 x - 1\right) \qquad \text{(Pythagorean Identity)}$$

$$= -\left(\frac{1}{a^2} - 1\right) \qquad \left(\sec x = \frac{1}{\cos x}\right)$$

Therefore, $\cot\left(x - \dfrac{\pi}{2}\right) = 1 - \dfrac{1}{a^2}$.

Identities are used to change the form of mathematical expressions, typically reducing complicated relationships into simpler, more manageable forms. As with other algebraic operations, rewriting an expression involving trigonometric functions may reveal other information about the function.

Example 2:

Simplify $\dfrac{(\sec x - 1)(\sec x + 1)}{\sin^2 x}$ to a trigonometric expression involving as few terms as possible.

$$\frac{(\sec x - 1)(\sec x + 1)}{\sin^2 x} = \frac{\sec^2 x - 1}{\sin^2 x} \qquad \text{(multiplication)}$$

$$= \frac{\tan^2 x}{\sin^2 x} \qquad \text{(Pythagorean identity)}$$

$$= \frac{\sin^2 x}{\cos^2 x} \cdot \frac{1}{\sin^2 x} \qquad \text{(definition of tangent \& cancel common factors)}$$

$$= \sec^2 x \qquad \text{(reciprocal identity)}$$

The Logic Behind Verifying Identities

An important point to remember when proving identities is that *the stated equality has not yet been established*. While identities are typically presented as equations, their point is to establish the equality. Therefore, when proving an identity, all algebraic manipulations that assume equality are forbidden. In particular, strategies involving both sides of an equation simultaneously (cross-multiplication, adding equivalent values to both sides, etc.) typically cannot be used because they assume the very equality the proof is attempting to establish—a critical circular reasoning flaw.

A proof of an identity is logically consistent only if one side is can be completely manipulated into the form of the other side. However, if work on one side reaches a seemingly unworkable point, insights can often be gained by shifting attention to the other side, essentially working from the both ends of the problem to the middle. *At no point* can both sides be manipulated simultaneously. In the end, though, a *proof* must be presented uni-directionally, completely transforming one side into the other. These strategies may help.

---

1. Typically work from the more complicated side first.

2. Use the form of the inactive side as a guide for manipulating the active side. If parts of the expressions match, all that remains is to prove the remaining portions are equivalent.

3. When expressions are cumbersome, consider converting everything into sines and cosines. While the expressions may become more complicated initially, the number of possible identities required will have been significantly reduced.

4. Make sure every function operates on the same input. For example, identities involving both $\sin x$ and $\cos(2x)$ usually need to be rewritten to include functions of x only.

5. Algebraic strategies like factoring, multiplying expressions, adding fractions, simplifying complex fractions, etc. are often useful.

Example 3:

Verify the identity $\dfrac{\cot^2 x}{1+\csc x} = \dfrac{1-\sin x}{\sin x}$.

Method 1: Rewrite the more complicated left side.

$$\frac{\cot^2 x}{1+\csc x} = \frac{\csc^2 x - 1}{1+\csc x} \qquad \left(\text{Pythagorean identity}\right)$$

$$= \frac{(\csc x - 1)\,\cancel{(\csc x + 1)}}{\cancel{(1+\csc x)}} \qquad \left(\text{factor and cancel}\right)$$

$$= \frac{1}{\sin x} - 1 \qquad \left(\text{reciprocal identity}\right)$$

$$= \frac{1-\sin x}{\sin x} \qquad \left(\text{common denominator}\right)$$

QED

Method 2: Change all functions into sines and cosines, beginning with the left hand side.

$$\frac{\cot^2 x}{1+\csc x} = \frac{\dfrac{\cos^2 x}{\sin^2 x}}{1+\dfrac{1}{\sin x}}$$

$$= \frac{\sin^2 x}{\sin^2 x} \cdot \frac{\dfrac{\cos^2 x}{\sin^2 x}}{1+\dfrac{1}{\sin x}} \qquad \text{(multiply by a special form of 1)}$$

$$= \frac{\cos^2 x}{\sin x \cdot (\sin x + 1)} \qquad \text{(simplify)}$$

$$= \frac{1-\sin^2 x}{\sin x \cdot (\sin x + 1)} \qquad \text{(Pythagorean identity)}$$

$$= \frac{(1-\sin x)(\sin x + 1)}{\sin x \cdot (\sin x + 1)} \qquad \text{(factor)}$$

$$= \frac{1-\sin x}{\sin x} \qquad \text{(cancel)}$$

QED

Method 2 is more complicated, but employs different strategies. While some approaches are more elegant or efficient than others, almost any approach can succeed with sufficient stamina and attention to detail.

Example 4:
Verify $\sin^2 x - \sin^4 x = \cos^2 x - \cos^4 x$.

    Method 1: Work from both ends to an identical statement and then rewrite the proof.

Thought process		Presentation of proof
Left-hand side	Right-hand side	$\sin^2 x - \sin^4 x = \sin^2 x \cdot \left(1-\sin^2 x\right)$
$\sin^2 x - \sin^4 x$	$\cos^2 x - \cos^4 x$	$= \sin^2 x \cdot \cos^2 x$
$= \sin^2 x \cdot \left(1-\sin^2 x\right)$	$= \left(1-\cos^2 x\right) \cdot \cos^2 x$	$= \left(1-\cos^2 x\right) \cdot \cos^2 x$
$= \sin^2 x \cdot \cos^2 x$	$= \sin^2 x \cdot \cos^2 x$	$= \cos^2 x - \cos^4 x$

    Method 2:

$$\sin^2 x - \sin^4 x = \left(1-\cos^2 x\right) - \left(1-\cos^2 x\right)^2 \qquad \text{(convert to cosines)}$$

$$= \left(1-\cos^2 x\right)\left[1-\left(1-\cos^2 x\right)\right] \qquad \text{(factor)}$$

$$= \left(1-\cos^2 x\right) \cdot \cos^2 x \qquad \text{(simplify)}$$

$$= \cos^2 x - \cos^4 x \qquad \text{(expand)}$$

QED

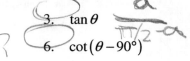

Problems for Section 6.1:

Exercises

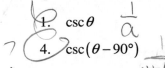

For questions 1-6, assume $\sin\theta = a$ and $\theta \in \left[0, \dfrac{\pi}{2}\right]$. Then rewrite the given expression in terms of a.

1. $\csc\theta$ $\dfrac{1}{a}$

2. $\cos\theta$ $\pi/2 - a$

3. $\tan\theta$ a $\pi/2 - a$

4. $\csc(\theta - 90°)$ $\dfrac{1}{\pi/2 - a}$

5. $\cos(\pi - \theta) = \pi$

6. $\cot(\theta - 90°)$

7. If a given equation is an identity, what is true about the graphs of each side of the equation?

Graphically determine if the equations in questions 8-15 are likely to be identities.

8. $\sqrt{x^2 - 9} = x - 3$

9. $\cos(2x) = 2\cos^2 x - 1$

10. $2\sin\left(x + \dfrac{11\pi}{6}\right) = \sqrt{3}\sin x - \cos x$

11. $\cos^2 x - \sin^2 x = \cos x$

12. $\cos(x + \pi) = \cos x + \cos\pi$

13. $\cos\left(x - \dfrac{\pi}{2}\right) = -\sin x$

14. $\sin(2x) = \sin(2)\cdot\sin x$

15. $\tan\left(x - \dfrac{\pi}{2}\right) = -\cot x$

16. For each equation from questions 8-15 that was *not* an identity, rewrite one side to make it an identity.

Verify each identity in questions 17-26.

17. $\cos\left(x - \dfrac{\pi}{2}\right) = \sin x$

18. $\dfrac{1}{1 - \sin^2 x} = 1 + \tan^2 x$

19. $\tan^2 x - \cot^2 x = \sec^2 x - \csc^2 x$

20. $\dfrac{\tan^2 x + 1}{1 + \cot^2 x} = \tan^2 x$

21. $\tan\theta = \dfrac{\sin\theta\cdot\cos\theta}{1 - \sin^2\theta}$

22. $\dfrac{\sec x + 1}{\csc x + 1} = \dfrac{\csc x - 1}{\sec x - 1}\cdot\tan^4 x$

23. $\dfrac{1}{\sin x + 1} + \dfrac{1}{1 - \sin x} - 2 = 2\tan^2 x$

24. $\dfrac{\sin^2 z - \cos^2 z}{1 - 2\sin z\cdot\cos z} = \dfrac{1 + 2\sin z\cdot\cos z}{\sin^2 z - \cos^2 z}$

25. $\dfrac{\sin A\cdot\cos B + \cos A\cdot\sin B}{\cos A\cdot\cos B - \sin A\cdot\sin B} = \dfrac{\tan A + \tan B}{1 - \tan A\cdot\tan B}$

26. $\sin^3 r - \cos^3 r = \sin r - \cos r + \sin^2 r\cdot\cos r - \sin r\cdot\cos^2 r$

Explorations

Identify each statement in questions 27-32 as true or false. Justify your conclusion.

27. $\sin(30° + 60°) = \sin 30° + \sin 60°$

28. The two sides of a trigonometric identity must be equivalent for *all* values of the input variable.

29. $\forall x \in \mathbb{R}$, $\sec^2 x - \tan^2 x = 1$

30. Multiplying both sides of an identity by the same expression is a valid strategy to verify an identity.

31. $\forall x \in \mathbb{R}$, $\sec^2 x - \tan^2 x = \sin^2 x + \cos^2 x$

32. The graph of $y = \cot^2 x - \csc^2 x + 5$ is symmetric with respect to the *y*-axis.

33. Solve for $x \in [0, 2\pi)$: $\sin\left(x - \dfrac{\pi}{2}\right) = 1$.

34. A. Prove $\tan\theta = \dfrac{\sec\theta}{\csc\theta}$.

   B. While the identity in part A is true for the common domain values of both expressions, state the values of θ for which it does not hold.

   C. How do the graphs of $y = \tan x$ and $y = \dfrac{\sec x}{\csc x}$ compare? Be specific.

35. Prove the exponential identity $\dfrac{e^x - e^{-x}}{e^x + e^{-x}} = \dfrac{e^{2x} - 1}{e^{2x} + 1}$.

6.2: Sum & Difference Identities

I have no special talents. I am only passionately curious.

– Albert Einstein

There were several problems in section 6.1 which proved that sine and cosine do not distribute over addition. An invocation of the unit circle definition of the cosine with the distance formula shows what actually happens when a trigonometric function is applied to a sum.

The Sum of Angles Identity for Cosine

Define four points on the unit circle. Point P is at $(1,0)$. Let point A be the image of P under a counterclockwise rotation of α units, and let point B be the image of A under a counterclockwise rotation of β units. Finally, let point C be the image of P under a clockwise rotation of β units (Figure 6.2a).

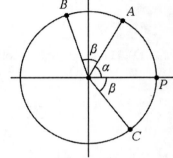

By angle addition, B is also the image of P under a counterclockwise rotation of $\alpha+\beta$ units. Likewise A is the image of C under the same rotation.

Figure 6.2a

The coordinates of points A and B are $A=\left(\cos(\alpha),\sin(\alpha)\right)$ and $B=\left(\cos(\alpha+\beta),\sin(\alpha+\beta)\right)$. Point C can be defined similarly and then simplified using even and odd function properties:
$$C=\left(\cos(-\beta),\sin(-\beta)\right)=\left(\cos(\beta),-\sin(\beta)\right).$$

$\overline{PB}\cong\overline{CA}$ because their central angles are congruent. The distance formula gives
$$PB=CA$$
$$\sqrt{\left[\cos(\alpha+\beta)-1\right]^2+\left[\sin(\alpha+\beta)-0\right]^2}=\sqrt{\left[(\cos\alpha)-(\cos\beta)\right]^2+\left[(\sin\alpha)-(-\sin\beta)\right]^2}\ .$$

Because $y=\sqrt{x}$ is one-to-one, this equation simplifies to
$$\left(\cos(\alpha+\beta)-1\right)^2+\left(\sin(\alpha+\beta)-0\right)^2=(\cos\alpha-\cos\beta)^2+(\sin\alpha+\sin\beta)^2$$
$$\underline{\cos^2(\alpha+\beta)}-2\cos(\alpha+\beta)+1+\underline{\sin^2(\alpha+\beta)}=\boxed{\cos^2\alpha}-2\cos\alpha\cdot\cos\beta+\underline{\cos^2\beta}+\boxed{\sin^2\alpha}+2\sin\alpha\cdot\sin\beta+\underline{\sin^2\beta}$$
$$1\!\!\!/+1-2\cos(\alpha+\beta)=1\!\!\!/+1-2\cos\alpha\cdot\cos\beta+2\sin\alpha\cdot\sin\beta$$
$$\diagdown\!\!\!2\cos(\alpha+\beta)=\diagdown\!\!\!2\left(\cos\alpha\cdot\cos\beta-\sin\alpha\cdot\sin\beta\right)$$
$$\cos(\alpha+\beta)=\cos\alpha\cdot\cos\beta-\sin\alpha\cdot\sin\beta.$$

This proof also can be demonstrated by defining a multi-variable distance function on a CAS. Once the function is defined, CAS commands perform the algebraic manipulations that take the distance equation to the final form of the sum identity for the cosine. Note how the CAS takes care of background algebra such as simplifying the Pythagorean identities and factoring, allowing the focus to remain on the fact that this is essentially a distance problem (Figure 6.2b).

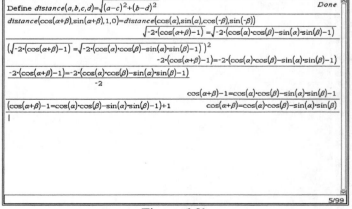

Figure 6.2b

Example 1:
Derive a formula for $\cos(\alpha - \beta)$.

$$\cos(\alpha - \beta) = \cos(\alpha + (-\beta))$$

$$= \cos\alpha \cdot \cos(-\beta) - \sin\alpha \cdot \sin(-\beta) \qquad \text{(sum of angles identity for cosine)}$$

$$= \cos\alpha \cdot \cos\beta + \sin\alpha \cdot \sin\beta \qquad \text{(even and odd functions)}$$

Example 2:
Simplify $\cos\left(x - \dfrac{\pi}{2}\right)$.

Chapter 5 accomplished this using transformations and symmetry. This now can be achieved algebraically with the sum of angles identity for cosine.

$$\cos\left(x - \frac{\pi}{2}\right) \;=\; \cos x \cdot \cos\frac{\pi}{2} + \sin x \cdot \sin\frac{\pi}{2} \;=\; \cos x \cdot (0) + \sin x \cdot (1) \;=\; \sin x$$

The Sum of Angles Identity for Sine

Because $\cos\left(x - \dfrac{\pi}{2}\right) = \sin x$, then

$$\sin(\alpha + \beta) = \cos\left((\alpha + \beta) - \frac{\pi}{2}\right) \qquad \text{(substitution)}$$

$$= \cos\left(\alpha + \left(\beta - \frac{\pi}{2}\right)\right) \qquad \text{(regrouping)}$$

$$= \cos\alpha \cdot \cos\left(\beta - \frac{\pi}{2}\right) - \sin\alpha \cdot \sin\left(\beta - \frac{\pi}{2}\right) \qquad \text{(sum of angles identity for cosine)}$$

$$= \cos\alpha \cdot \sin\beta - \sin\alpha \cdot \left(-\sin\left(\frac{\pi}{2} - \beta\right)\right) \qquad \text{(Example 2 \& odd functions)}$$

$$= \cos\alpha \cdot \sin\beta + \sin\alpha \cdot \cos\beta \qquad \text{(cofunctions)}$$

Example 3:
Given $\cos A = \dfrac{12}{13}$ and $\sin B = \dfrac{4}{5}$ with $A, B \in \left[0, \dfrac{\pi}{2}\right]$, determine a simplified value of $\sin(A + B)$.

Using the Pythagorean identity,

$$\sin^2 A + \cos^2 A = 1 \;\Rightarrow\; \sin^2 A + \left(\frac{12}{13}\right)^2 = 1 \;\Rightarrow\; \sin^2 A = \frac{25}{169} \;\Rightarrow\; \sin A = \pm\frac{5}{13}.$$

The negative answer is extraneous because $A \in \left[0, \dfrac{\pi}{2}\right]$. Similarly, $\cos B = \dfrac{3}{5}$. Therefore,

$$\sin(A + B) \;=\; \sin A \cdot \cos B + \cos A \cdot \sin B \;=\; \frac{5}{13} \cdot \frac{3}{5} + \frac{12}{13} \cdot \frac{4}{5} \;=\; \frac{63}{65}.$$

The Sum of Angles Identity for Tangent

Substituting, $\tan\theta = \dfrac{\sin\theta}{\cos\theta} = \dfrac{\sin(A+B)}{\cos(A+B)} = \dfrac{\sin A \cdot \cos B + \cos A \cdot \sin B}{\cos A \cdot \cos B - \sin A \cdot \sin B}$. An advantage of this formula is that it

expresses $\tan(A+B)$ with nothing more required than the identities for $\sin(A+B)$ and $\cos(A+B)$. The disadvantage is that it is algebraically rather cumbersome. However, it is possible to improve its appearance by dividing by $\cos A \cos B$.

$$\tan(A+B) = \frac{(\sin A \cdot \cos B + \cos A \cdot \sin B)/(\cos A \cdot \cos B)}{(\cos A \cdot \cos B - \sin A \cdot \sin B)/(\cos A \cdot \cos B)} = \frac{\tan A + \tan B}{1 - \tan A \cdot \tan B}$$

Example 4:

Use a sum or difference identity to find an exact value for $\tan 105°$.

$$\tan 105° = \tan(60° + 45°) = \frac{\tan 60° + \tan 45°}{1 - \tan 60° \cdot \tan 45°} = \frac{\sqrt{3}+1}{1-\sqrt{3}}$$

The sum and difference identities for the sine, cosine and tangent are summarized below. The difference identities have not all been proven yet. As shown in Example 1, though, they are simple to derive by replacing the second angle with its opposite and using even and odd function properties.

1.	$\sin(A \pm B) = \sin A \cdot \cos B \pm \cos A \cdot \sin B$
2.	$\cos(A \pm B) = \cos A \cdot \cos B \mp \sin A \cdot \sin B$
3.	$\tan(A \pm B) = \dfrac{\tan A \pm \tan B}{1 \mp \tan A \cdot \tan B}$

Example 5:

Verify the identity, $\cos A + \cos B = 2\cos\left(\dfrac{A+B}{2}\right) \cdot \cos\left(\dfrac{A-B}{2}\right)$.

Add the cosine sum and difference identities:

$$\cos(x+y) = \cos x \cdot \cos y - \sin x \cdot \sin y$$
$$\underline{+\cos(x-y) = \cos x \cdot \cos y + \sin x \cdot \sin y}$$
$$\cos(x+y) + \cos(x-y) = 2\cos x \cdot \cos y$$

Now, let $A = x+y$ and $B = x-y$. Therefore, $2x = A+B \Rightarrow x = \dfrac{A+B}{2}$ and $2y = A-B \Rightarrow y = \dfrac{A-B}{2}$.

Substituting these values into the initial sum gives the desired result:

$$\cos(x+y) + \cos(x-y) = 2\cos x \cdot \cos y \Rightarrow \cos A + \cos B = 2\cos\left(\frac{A+B}{2}\right) \cdot \cos\left(\frac{A-B}{2}\right)$$

Problems for Section 6.2:

Exercises

[NC] For questions 1-8 , determine an exact value that does not involve trigonometric expressions.

1. $\sin 75°$

2. $\csc\left(\dfrac{11\pi}{12}\right)$

3. $\sin\left(-15°\right)$

4. $\tan 195°$

5. $\cos\left(\dfrac{7\pi}{12}\right)$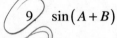

6. $\sec 255°$

7. $2\tan 465°$

8. $\sin 105°$

For questions 9-14, if $\tan A = \dfrac{2}{3}$ and $\cos B = \dfrac{4}{5}$ $\forall A, B \in \left[0, \dfrac{\pi}{2}\right]$, determine exact values for each expression.

9. $\sin\left(A+B\right)$

10. $\sec\left(B-A\right)$

11. $\cot\left(A+B\right)$

12. $\sec^2\left(B+A\right)-1$

13. $\csc\left(A-B\right)$

14. $\cos\left(A+B-\dfrac{\pi}{2}\right)$

For questions 15-18, simplify and, if possible, evaluate without using a calculator.

15. $\cos 65° \cdot \cos 55° - \sin 65° \cdot \sin 55°$

16. $\dfrac{\tan 60° - \tan 40°}{1+\tan 60° \cdot \tan 40°}$

17. $\sin\dfrac{\pi}{7}\cdot\cos\dfrac{\pi}{5}+\sin\dfrac{\pi}{5}\cdot\cos\dfrac{\pi}{7}$

18. $\dfrac{3}{\sin\dfrac{\pi}{3}\cdot\cos\dfrac{\pi}{6}+\cos\dfrac{\pi}{3}\cdot\sin\dfrac{\pi}{6}}$

Verify the identities in questions 19-20.

19. $\sin\left(\dfrac{3\pi}{2}+x\right)=-\cos x$

20. $\sin\theta = \sin\left(\pi-\theta\right)$

[NC] Evaluate the given expressions in questions 21-22.

21. $\sin\left(\arctan\left(\dfrac{4}{3}\right)+\arccos\left(\dfrac{12}{13}\right)\right)$

22. $\sec\left(\arccos\left(-\dfrac{1}{2}\right)+\dfrac{\pi}{4}\right)$

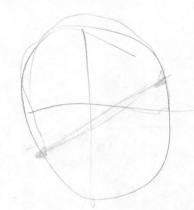

Explorations

Identify each statement in questions 23-28 as true or false. Justify your conclusion.

23. $\csc(A+B) = \csc A + \csc B$

24. If $\forall A, B \in \mathbb{R}$, $\cos(A+B) = 0$, then $\cos A = \sin B$.

25. $\sin^2\left(\dfrac{\pi}{6}\right) = \sin^2\left(-\dfrac{\pi}{6}\right)$ implies that $\sin\theta = \sin(-\theta)$ is an identity.

26. Once the sum and difference identities for sine and cosine have been established, exact values for the sine and cosine of any given angle can be determined.

27. Given the exact value of one trigonometric function at a given angle and the terminal side of that angle, the exact values of all other trigonometric functions can be determined.

28. A proof of any identity allows the manipulation of both sides of the proposed equality so long as that manipulation is not simultaneous.

29. A. If $\cos A = \dfrac{12}{13}$ and $\sin B = \dfrac{4}{5}$ with $A, B \in \left[0, \dfrac{\pi}{2}\right]$, verify that $\sin(A-B) = -\dfrac{33}{65}$.

   B. What does the result from part A say about the relative sizes of A and B? Why?

   C. Without computing either A or B, identify the quadrant in which $A-B$ is located.

30. If $\tan A = \dfrac{4}{3}$ and $0 \le A \le \dfrac{\pi}{2}$, find an exact value for $\sin\left(\dfrac{\pi}{2} - \dfrac{A}{2}\right)$.

31. If $A+B = \cos^{-1}(-1)$ and $A \in \left[-\pi, -\dfrac{\pi}{2}\right]$, then in what quadrant must angle B fall?

32. Using angle sum and difference identities to prove $\cos A - \cos B = -2\sin\left(\dfrac{A+B}{2}\right) \cdot \sin\left(\dfrac{A-B}{2}\right)$.

33. Solve for $x \in [0, 2\pi)$: $\sin\dfrac{\pi}{3} \cdot \cos x + \cos\dfrac{\pi}{3} \cdot \sin x = \dfrac{1}{2}$.

34. Determine an identity for $\sec(A+B)$ in terms of $\sec A$ and $\sec B$ only.

35. If A, B and C are angles in a triangle, find an expression for $\sin(A+B)$ in terms of C only.

6.3: Double Angle & Power-Reducing Identities

> _It takes 20 years of hard work to become an overnight success._
>
> – Diana Rankin

Double Angle Identities

Because $\sin(2A) = \sin(A+A)$, double angle identities are special cases of the angle sum identities.

Example 1:

Find a simplified expression for $\sin(2A)$

$$\sin(2A) \;=\; \sin(A+A) \;=\; \sin A \cdot \cos A + \cos A \cdot \sin A \;=\; 2\sin A \cdot \cos A$$

Simplified versions of the primary double-angle identities are listed below. The three versions of $\cos(2A)$ are derived from Pythagorean substitutions.

$$1. \quad \sin(2A) = 2\sin A \cdot \cos A$$
$$2. \quad \cos(2A) = \cos^2 A - \sin^2 A \qquad (1)$$
$$= 2\cos^2 A - 1 \qquad (2)$$
$$= 1 - 2\sin^2 A \qquad (3)$$
$$3. \quad \tan(2A) = \frac{2\tan A}{1 - \tan^2 A}$$

Example 2:

If $\sin x = \dfrac{6}{7}$, find the value of $\cos(2x)$.

There are three identity options for $\cos(2x)$, but the given information is in terms of $\sin x$ only, suggesting the version of $\cos(2x)$ involving only the sine function.

$$\cos(2x) \;=\; 1 - 2\sin^2 x \;=\; 1 - 2\left(\frac{6}{7}\right)^2 \;=\; -\frac{23}{49}$$

Notice that the result is independent of the quadrant in which x is located.

Example 3:

If $\tan x = \dfrac{4}{3}$ and $\pi < x < \dfrac{3\pi}{2}$, in which quadrant is $2x$ located, and what is the value of $\sin(2x)$?

Because $|\tan x| > 1$ in Quadrant III, it follows that $x \in \left(\dfrac{5\pi}{4}, \dfrac{3\pi}{2}\right)$. Therefore $2x \in \left(\dfrac{5\pi}{2}, 3\pi\right)$, and $2x$ must be a Quadrant II angle.

$\sin(2x) = 2\sin x \cdot \cos x$, so values for $\sin x$ and $\cos x$ are required. A right triangle with legs of lengths 3 and 4 in Quadrant III gives $\sin x = -\dfrac{4}{5}$ and $\cos x = -\dfrac{3}{5}$, so

$$\sin(2x) = 2\sin x \cdot \cos x = 2\left(-\frac{4}{5}\right)\left(-\frac{3}{5}\right) = \frac{24}{25}$$

Power-Reducing Identities

Because $\cos(2x)$ can be expressed solely in terms of $\sin^2 x$ or $\cos^2 x$, conversely, each of these identities can be written in terms of $\cos(2x)$ only.

$$\cos(2x) = 1 - 2\sin^2 x \quad \Rightarrow \quad \sin^2 x = \frac{1 - \cos(2x)}{2}$$

$$\cos(2x) = 2\cos^2 x - 1 \quad \Rightarrow \quad \cos^2 x = \frac{1 + \cos(2x)}{2}$$

From these, it follows that $\tan^2 x = \dfrac{1 - \cos(2x)}{1 + \cos(2x)}$. These three are called **Power-Reducing Identities**.

Ultimately, these are nothing more than restatements of the double angle identities, which in turn, are revisions of the angle sum formulas. They are different algebraic representations of the same relationships, but the new forms offer a more direct way to compute half angle ratios.

Example 4:

Find an exact value for $\cos\left(\dfrac{13\pi}{8}\right)$.

The cosine power reducing formula gives

$$\cos^2\left(\frac{13\pi}{8}\right) = \frac{1 + \cos\left(2 \cdot \dfrac{13\pi}{8}\right)}{2} = \frac{1 + \cos\left(\dfrac{13\pi}{4}\right)}{2} = \frac{1 + \dfrac{-\sqrt{2}}{2}}{2} = \frac{2 - \sqrt{2}}{4}$$

$$\Rightarrow \quad \cos\left(\frac{13\pi}{8}\right) = \pm\sqrt{\frac{2 - \sqrt{2}}{4}}$$

Since $\dfrac{13\pi}{8} \in \left(\dfrac{3\pi}{2}, 2\pi\right)$, $\cos\left(\dfrac{13\pi}{8}\right) > 0 \Rightarrow \cos\left(\dfrac{13\pi}{8}\right) = +\dfrac{\sqrt{2 - \sqrt{2}}}{2}$.

Half-Angle Identities

Generalizing Example 4, the cosine ratio for a half angle can be expressed as $\cos\left(\dfrac{1}{2}\theta\right) = \pm\sqrt{\dfrac{1 + \cos\theta}{2}}$. Algebraically, the plus or minus results from applying the SQR transformation; the actual sign of the ratio is then determined by the quadrant in which $\dfrac{1}{2}\theta$ falls. Half angle identities for the sine and tangent can be derived through the process demonstrated in Example 4.

Problems for Section 6.3:

Exercises

[NC] For questions 1-6, determine an exact value of the given expression using double angle identities.

1. $\cos\left(\dfrac{13\pi}{12}\right)$

2. $\sin(105°)$

3. $\sec\left(-\dfrac{3\pi}{8}\right)$

4. $\tan 22.5°$

5. $\cos\left(\dfrac{5\pi}{8}\right)$

6. $\csc\left(\dfrac{17\pi}{24}\right)$

Find exact values for the given expressions in questions 7-12, if $\tan A = \frac{2}{3}$ and $\cos B = \frac{4}{5}$ $\forall A, B \in \left(0, \frac{\pi}{2}\right)$.

7. $\tan(2A)$

8. $\sin(2A)$

9. $\cos(2B)$

10. $\sin\left(\frac{A}{2}\right)$

11. $\cos\left(\frac{1}{2}B\right)$

12. $\cos(3B)$

[NC] Simplify, and if possible, evaluate without a calculator each expression in questions 13-16.

13. $1 - 2\sin^2\left(\frac{5\pi}{12}\right)$

14. $\sin^2\left(\frac{\pi}{8}\right) - \cos^2\left(\frac{\pi}{8}\right)$

15. $2\sin\left(\frac{\pi}{9}\right) \cdot \cos\left(\frac{\pi}{9}\right)$

16. $\dfrac{2\tan\left(\frac{3\pi}{8}\right)}{1 - \tan^2\left(\frac{3\pi}{8}\right)}$

For questions 17-18, given $\sin\theta = a$, determine a value for the expression in terms of a that does not involve trigonometric expressions.

17. $\sec(2\theta)$

18. $\sin(3\theta)$

Verify each identity in questions 19-25.

19. $1 - \cos(2x) \cdot \sec^2(x) = \tan^2(x)$

20. $1 - \cos(2x) \cdot \sec^2(x) = \sec^2(x) - 1$

21. $\dfrac{\cos(2m)}{1 - \sin(2m)} = \dfrac{1 + \sin(2m)}{\cos(2m)}$

22. $\dfrac{\sec(x) + 2}{\csc(x) + 2} = \dfrac{\sin(x) + \sin(2x)}{\cos(x) + \sin(2x)}$

23. $\forall x \in (0, \pi)$, $\cot\left(\frac{x}{2}\right) = \dfrac{1 + \cos(x)}{\sin(x)}$

24. $\cos(3x) = 4\cos^3(x) - 3\cos(x)$

25. $\sin^3(x) + \cos^3(x) = \frac{1}{2}(\sin(x) + \cos(x)) \cdot (2 - \sin(2x))$

Evaluate the expressions in questions 26-27 without using a calculator:

26. $\cos\left(2\arcsin\left(\frac{7}{25}\right)\right)$

27. $\sin\left(\frac{1}{2}\arccsc\left(-\frac{3}{2}\right)\right)$

Explorations

Identify each statement in questions 28-32 as true or false. Justify your conclusion.

28. If $\tan A = \dfrac{6}{5}$ and $A \in [0, 90°]$, then $\tan\left(\dfrac{1}{2}A\right) = \dfrac{3}{5}$.

29. $\sin(2\pi - 2x) = \sin(2\pi) - \sin(2x)$

30. $\forall x \in \mathbb{R}$, $\dfrac{4\sin(x) \cdot \cos(x)}{1 - 2\sin^2(x)} = 2\tan(2x)$

31. $\forall x \in \mathbb{R}$, $\tan\left(\dfrac{1}{2}x\right) = \dfrac{1}{2}\tan(x)$

32. Double and half-angle identities are just simplified versions of sum identities.

33. If $\sin(x) = \dfrac{5}{12}$, find all possible values for $\cos\left(\dfrac{1}{2}x\right)$.

34. A. Find an exact value for $\sin 75°$ using a **sum of angles identity**.

    B. Find an exact value for $\sin 75°$ using a **double angle identity**.

    C. [NC] Algebraically prove that the answers to parts A and B are equivalent.

35. Prove the double-angle identities for the cosine and tangent.

36. Derive identities for $\sin\left(\dfrac{1}{2}\alpha\right)$ and $\tan\left(\dfrac{1}{2}\alpha\right)$.

37. A.[NC] Evaluate $\sqrt{\dfrac{1 - \cos(120°)}{\cos(120°) + 1}}$. Your answer should not contain trigonometric functions.

    B. Use an identity to find an expression equivalent to $\sqrt{\dfrac{1 - \cos(120°)}{\cos(120°) + 1}}$ that does not involve square roots.

[NC] For questions 38-40, solve the given equation for all possible real values of x.

38. $1 - \sin x = \cos(2x)$ 39. $\cos(2x) = \cos x$ 40. $\dfrac{4\cos^2 x}{\sin(2x)} - 4\cos x - \csc x + 2 = 0$

41. Solve for all possible values of $x \in [0, 2\pi)$: $4\sin x \cdot \cos x = 1$.

42. The last line of Example 2 claims that the solution is independent of the quadrant in which the original angle x is located. Prove this claim.

43. Iman insists that $\forall n \in \mathbb{Z}$, $\left(\cos^2\left(\dfrac{\pi}{2}n\right) - \sin^2\left(\dfrac{\pi}{2}n\right)\right)^2 = 1$. Is she correct? Justify your answer.

44. Determine an expression for $\cos(3x)$ using only sine functions.

6.4: Many Familiar Formulae & A Revolutionary Concept

> *It is not the strongest of the species that survives, nor the most intelligent,*
> *but the one most responsive to change.*
>
>   – Charles Darwin
>
> *Teachers open the door. You enter by yourself.*
>
>   – Chinese Proverb

Trigonometric Form of the Area of a Triangle

The area of any triangle is $\frac{1}{2}(\text{base})(\text{height})$, but in cases where the height is not known, trigonometric ratios can determine these lengths and thereby the triangle's area. From Figure 6.4a, $\sin C = \frac{h}{b} \;\Rightarrow\; h = b \cdot \sin C$, so

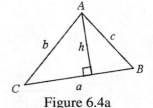

Figure 6.4a

$$Area = \frac{1}{2}a \cdot (b\sin C) = \frac{1}{2}ab\sin C.$$

A trigonometric form of the area of a triangle is therefore, $Area = \frac{1}{2}ab\sin C = \frac{1}{2}ac\sin B = \frac{1}{2}bc\sin A$.

The Law of Sines

In Figure 6.4a, h can be defined differently using the two right triangles: $h = b \cdot \sin C = c \cdot \sin B$. Dividing both height expressions by $b \cdot c$ and generalizing the result leads to the Law of Sines, $\frac{\sin A}{a} = \frac{\sin B}{b} = \frac{\sin C}{c}$.

Example 1:

Solve $\triangle ABC$ given $a = 9$, $m\angle B = 70°$, and $m\angle C = 85°$.

Remember from geometry that **solving a triangle** requires finding all side and angle measures. Given two angles (Figure 6.4b), $m\angle A = 25°$. Then, the Law of Sines gives $\frac{\sin 25°}{9} = \frac{\sin 70°}{b}$ which leads to $b \approx 20.0115$. A similar set-up gives $c \approx 21.2148$.

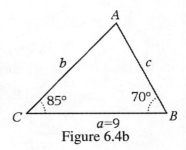

Figure 6.4b

The Law of Cosines

While the Law of Sines is simple in appearance, it cannot be used unless an angle-opposite-side pair is given. These cases require the Law of Cosines.

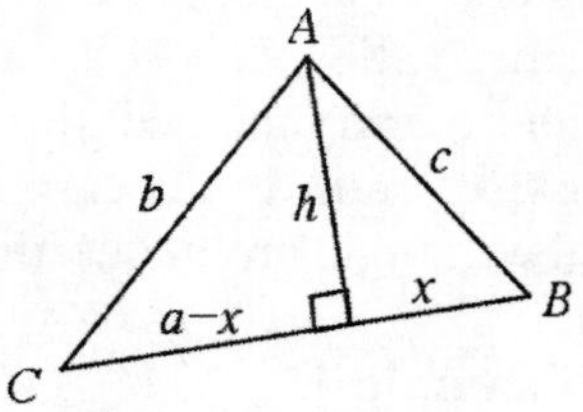

Figure 6.4c uses the altitude from Figure 6.4a to divide $\triangle ABC$ into two right triangles yielding the system of equations below.

$$\begin{cases} x^2 + h^2 = c^2 \\ (a-x)^2 + h^2 = b^2 \\ x = c \cdot \cos B \end{cases}$$

Figure 6.4c

By substitution and elimination, this system yields the Law of Cosines:
$a^2 = b^2 + c^2 - 2b \cdot c \cdot \cos A$. As with the Law of Sines, the altitude could have been drawn to any side yielding the following variations on the Law of Cosines: $b^2 = a^2 + c^2 - 2a \cdot c \cdot \cos B$ and $c^2 = a^2 + b^2 - 2a \cdot b \cdot \cos C$.
The Law of Cosines conveniently works best on exactly the triangles the Law of Sines cannot solve.

Example 2:
Solve $\triangle ABC$ given $a = 3$, $b = 2.4$, and $m\angle C = 85°$.

This time, there is no angle-opposite side pair information given (Figure 6.4d), so the more complicated-looking Law of Cosines must be used. This leads to $c^2 = 2.4^2 + 3^2 - 2.4 \cdot 3 \cdot \cos 85° \approx 13.505$, so $c \approx +3.6749$. The Law of Sines can then be used to find $m\angle A \approx 54.4137°$, giving $m\angle B \approx 40.5863°$ (Figure 6.4e).

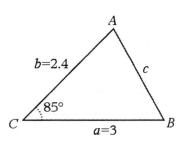

Figure 6.4d

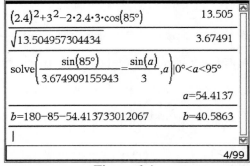

Figure 6.4e

Example 3:
Find the area of $\triangle ABC$ where $a = 12$, $b = 7$, and $c = 10$.

    Method I: With no angle measure provided, the Law of Sines is useless, but the Law of Cosines gives an angle measure which can be used in the trigonometric area formula.

$$12^2 = 7^2 + 10^2 - 2 \cdot 7 \cdot 10 \cdot \cos A \quad \Rightarrow \quad A = \cos^{-1}\left(\frac{5}{140}\right) \approx 1.535$$

$$Area = \frac{1}{2}b \cdot c \cdot \sin A \approx \frac{1}{2} \cdot 7 \cdot 10 \cdot \sin 1.535 \approx 34.978$$

    Method II: With all three sides known, **Heron's formula** also yields the area of a triangle. It states that if a triangle has sides a, b and c, then $s = \dfrac{a+b+c}{2}$ and $Area = \sqrt{s \cdot (s-a)(s-b)(s-c)}$.

$$s = \frac{12+7+10}{2} = 14.5 \quad \Rightarrow \quad Area = \sqrt{14.5(14.5-12)(14.5-7)(14.5-10)} \approx 34.978$$

SSA and the Ambiguous Case
Historical strategies for solving triangles involve appropriate uses of the Laws of Sines and Cosines. These are sufficient for most triangles. However, when given the lengths of two sides of a triangle and a non-included angle, there may be some ambiguity. This fact hinges on congruence proofs for triangles. If three sides (SSS), two sides and an included angle (SAS), two angles and an included side (ASA), or two angles and a non-included side (AAS) are known, the triangle either exists in a unique arrangement or it does not exist at all. In the case of two sides of and a non-included angle (SSA), the existence and the uniqueness of the triangle are in question.

Example 4:
Determine all solutions to $\triangle ABC$ given $a = 7$, $b = 9$ and $m\angle A = 40°$.

    Since the side opposite the given angle is the shorter one, it can be "swung" to give two possible arrangements with the given information (Figure 6.4f): $\triangle AB_1C$ and $\triangle AB_2C$.

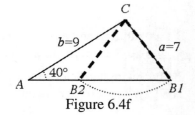

Figure 6.4f

Method 1: The Law of Sines produces two possible values for angle B — the values of B_1 and B_2 (Figure 6.4g). Figure 6.4h provides the remainder of the triangle solution.

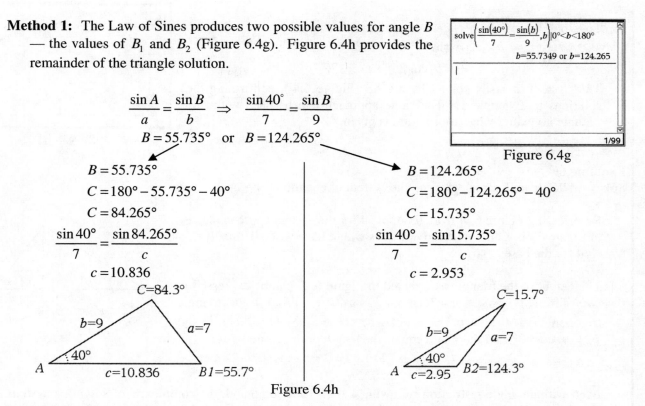

Figure 6.4g

$$\frac{\sin A}{a} = \frac{\sin B}{b} \implies \frac{\sin 40^\circ}{7} = \frac{\sin B}{9}$$

$$B = 55.735^\circ \quad \text{or} \quad B = 124.265^\circ$$

$B = 55.735^\circ$	$B = 124.265^\circ$
$C = 180^\circ - 55.735^\circ - 40^\circ$	$C = 180^\circ - 124.265^\circ - 40^\circ$
$C = 84.265^\circ$	$C = 15.735^\circ$
$\dfrac{\sin 40^\circ}{7} = \dfrac{\sin 84.265^\circ}{c}$	$\dfrac{\sin 40^\circ}{7} = \dfrac{\sin 15.735^\circ}{c}$
$c = 10.836$	$c = 2.953$

Figure 6.4h

Method 2: The Law of Cosines leads to a quadratic equation in terms of the length of side c.
$$7^2 = 9^2 + c^2 - 2 \cdot 9 \cdot c \cdot \cos 40^\circ$$
The solution to this equation confirms the c values found in Method I and verifies that there are two solutions. The Law of Sines can then be used to solve each case for $m\angle B$ and $m\angle C$.

A Revolutionary Concept
While the Laws of Sines and Cosines are certainly adequate for solving triangles, the power of CAS for solving systems of equations arguably makes both obsolete.

While only three pieces of information are required to uniquely define a triangle, every such triangle has at least one side measure specified. *Orient the triangle with the known side as its base and draw an altitude* (Figure 6.4i). If more sides are given, any one of them will work equally well as the base. The result is always a 2x2 system of equations.

Example 5:
Solve $\triangle ABC$ from Example 1 using the system of equations approach.

While only three pieces of information are required to uniquely define a triangle, every such triangle has at least one side measure specified. *Orient the triangle with the known side as its base and draw an altitude* (Figure 6.4i). If more sides are given, any one of them will work equally well as the base.

The height of this triangle then can be expressed in two different ways: $h = c \cdot \sin 70^\circ = b \cdot \sin 85^\circ$. From right triangle trigonometry, the two parts of the base are $y = b \cdot \cos 85^\circ$ and $x = c \cdot \cos 70^\circ$; their sum equals the length of the entire base:
$$x + y = 9 = b \cdot \cos 85^\circ + c \cdot \cos 70^\circ.$$

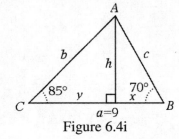

Figure 6.4i

Together, these form a system of equations.

$$\begin{cases} c \cdot \sin 70° = b \cdot \sin 85° \\ 9 = b \cdot \cos 85° + c \cdot \cos 70° \end{cases}$$

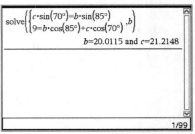

Figure 6.4j

This system is easily solved by a CAS (Figure 6.4j), confirming the solutions to Example 1. The same approach can be used with *any* triangle no matter what information is given.

Example 6:

Solve $\triangle ABC$ from Example 4 using the system of equations approach.

Re-orienting Figure 6.4f from Example 4 for the system approach leads to Figure 6.4i. Either of the given sides would have worked equally well for the base.

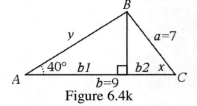

Figure 6.4k

The height of the triangle as oriented in Figure 6.4k can be expressed in two different ways: $h = 7 \cdot \sin x = y \cdot \sin 40°$. From right triangle trigonometry, the two parts of the base are $b1 = 7 \cdot \cos x$ and $b2 = y \cdot \cos 40°$ and their sum equals the length of the entire base.

$$b1 + b2 = 9 = y \cdot \cos 40° + 7 \cdot \cos x$$

Even with an angle restriction following a **such that** command, the complexity of solving a system with simultaneously unknown angles and sides outstrips the Nspire CAS's solve command (Figure 6.4l). Solving each equation for *y* and graphing (Figure 6.4m) simultaneously finds both solutions from Figure 6.4h (this time with the angles given in radians), confirming the results of Example 4.

NOTE: The same approach can be used with *any* triangle no matter what information is given. The system will have 0, 1, or 2 solutions depending on the given information.

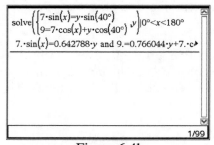

Figure 6.4l

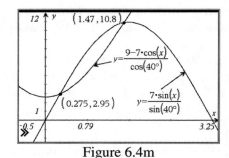

Figure 6.4m

Since the approaches shown in Examples 5 and 6 work for *every* conceivable triangle, including SSA ambiguous cases, the Laws of Sines and Cosines can now be considered historical artifacts.

Problems for Section 6.4:

Exercises

1. In $\triangle ABC$, $m\angle B = 90°$, $a = 7$ and $b = 10$. This data is SSA, so why does $\triangle ABC$ not present an ambiguous case?

For questions 2-3, find the area of $\triangle ABC$ using the given information.

2. $a = 8$ cm, $c = 5$ cm, and $m\angle B = 32°$ 3. $a = 9$, $b = 7$, and $m\angle A = 85°$

For questions 4-12, solve each $\triangle ABC$ using the given information. If there are two solutions, determine both.

4. $b = 4$, $c = 5$, and $m\angle B = 79°$ 5. $m\angle B = 50°$, $a = 6$ and $\angle C$ is a right angle

6. $a = 7$, $b = 8$, and $m\angle A = 35°$ 7. $m\angle C = 0.8$, $b = 250$ and $a = 120$

8. $m\angle A = 1$, $m\angle B = 1.2$ and $a = 10$ 9. $a = 11$, $b = 9$, and $m\angle A = 41°$

10. $m\angle B = 32°$, $m\angle C = 58°$ and $c = 10$ 11. $b = 9$, $c = 18$, and $m\angle B = 30°$

12. $a = 11$, $b = 14$ and $c = 20$

For questions 13-16, use the methods of Examples 5 & 6 to solve $\triangle ABC$.

13. $m\angle B = 50°$, $a = 6$ and $\angle C$ is a right angle   14. $m\angle A = 1$, $m\angle B = 1.2$ and $a = 10$

15. $m\angle C = 0.8$, $b = 250$ and $a = 120$ 16. $a = 11$, $b = 14$ and $c = 20$

Explorations

Identify each statement in questions 17-21 as true or false. Justify your conclusion.

17. The Laws of Sines and Cosines can only be used on non right-angled triangles.

18. When given two sides and a non-included angle in a triangle, it is possible to use the Law of Cosines to solve for the remaining information.

19. Any triangle can be completely solved using just right triangle trigonometry.

20. Every situation in which the Law of Sines can be used can result in two triangles.

21. $\triangle ABC$ exists, if $a = 7$, $b = 11$, and $m\angle A = 100°$.

22. Complete the proof of the Law of Cosines from the system of equations given beside Figure 6.4b.

23. Find the area of a circular segment[9] with a central angle of $52°$ in a circle of radius 12 cm.

24. Find the length of a chord subtending a central angle of 2 radians.

25. Ian measures the angle of elevation to the top of a tower to be $52°$. Moving 30 ft closer to the tower, she finds the angle of elevation to now be $60°$. Determine the height of the tower in feet.

---

[9] A circular segment is the area bounded by the chord and arc defined by a central angle in circle.

26. A lamp-post 16 ft tall is slightly inclined from a vertical position. It casts a shadow of length 20 ft when the angle of elevation of the sun is 55°. Given that the lamp-post is not in danger of falling to the ground, find its angle of inclination from horizontal.

27. The Ngorongoro crater in Tanzania (Figure 6.4h) is home to a large concentration of African wildlife. The caldera's floor diameter is approximately 19 km at a vertical depth of 610 m from the upper rim. If the shortest distance along the rise of the crater (from its rim to its floor) is 670 m, determine the circumference of the upper rim.

Figure 6.4h

28. Two ships leave port together, with the USS Rebecca heading 30° West of North at 13 miles per hour and the USS Michelle at 43° West of South at 10 miles per hour.

    A. How far apart are the ships after four hours?

    B. At the time from part A, the USS Rebecca detours, turning 120° counterclockwise from its original course. After how long will it intersect the path of the USS Michelle?

    C. Are the two ships in danger of colliding? Justify your response.

29. Why does the identity $\sin\theta = \sin(\pi - \theta)$ show that the Law of Sines must be used cautiously in ambiguous triangle cases?

30. Given the triangle with vertices $A(2,1)$, $B(1,5)$ and $C(-3,-1)$.
    A. Find its area.   B. Find $m\angle ABC$.

31. What happens when the Law of Cosines is applied to a right triangle for which neither acute angle is known? Why?

32. How can the Law of Cosines been seen as a generalization of the Pythagorean Theorem?

33. A. Without solving, why must the Figure 6.4f triangle have only one solution?

    B. The system of equations used to solve that triangle was
$$\begin{cases} c\cdot\sin 70° = b\cdot\sin 85° \\ 9 = b\cdot\cos 85° + c\cdot\cos 70° \end{cases}.$$
    Without solving, explain why this system must produce only one solution.

34. Repeat the triangle solution in Example 2, this time using the CAS approach described at the end of the section.

35. Given $\triangle ABC$ where $a = 10$, $b = 12$ and $m\angle B = 75°$.

    A. Why does this information suggest that $\triangle ABC$ may be an ambiguous case?

    B. Use the results of the Law of Sines as an initial step to explain why $\triangle ABC$ is not an ambiguous case.

    C. Use the results of the Law of Cosines as an initial step to explain why this is not an ambiguous case.

    D. Set up a system of equations (as explained at the end of the section) as an initial step. Why does the systems approach prove that this is not an ambiguous case?

6.5: Sums & Differences of Sinusoids

> *A closed mind is a dying mind.*
>
> – Edna Ferber

Sums of Sinusoids with Identical Periods
Example 1:

Graph $y = \sqrt{2}\sin\left(x - \dfrac{\pi}{4}\right)$ using two different approaches.

Method 1. Apply $T_{\frac{\pi}{4},0} \circ S_{1,\sqrt{2}}$ to the parent sine function (Figure 6.5a).

Method 2. Apply the angle sum identity for the sine.

$$y = \sqrt{2}\sin\left(x - \frac{\pi}{4}\right) = \sqrt{2}\left(\sin x\cos\left(\frac{\pi}{4}\right) - \cos x\sin\left(\frac{\pi}{4}\right)\right)$$

$$= \sqrt{2}\left(\frac{1}{\sqrt{2}}\sin x - \frac{1}{\sqrt{2}}\cos x\right) = \sin x - \cos x$$

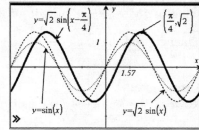

Figure 6.5a

From the perspective of bending asymptotes this suggests one trigonometric function oscillating about another, and it does not matter which is considered the transformed midline. Figure 6.5b shows midline $y = -\cos x$ with $y = \sin x$ wrapped around it, maintaining the same distance from the transformed midline as it ordinarily would from the *x*-axis. In particular, notice how the final graph intersects its midline at $n\pi$, $n \in \mathbb{Z}$, and is at its greatest vertical distance of 1 from the midline at odd integer multiples of $\dfrac{\pi}{2}$.

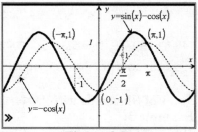

Figure 6.5b

A constant horizontal translation on a sinusoid can always be expressed using an angle sum identity with one of the angles variable and the other constant. By its definition, the angle sum identity used in this way guarantees that its simplified form is always a sum of a sine and a cosine with the same argument, and therefore, the same period. Conversely, the sum or difference of a sine and cosine with the same period can always be re-written as a constant horizontal translation on a single sinusoid with that same period. It also makes intuitive sense that if the addends are periodic and repeat over identical periods, then their sum must repeat over the same period.

There are some other notable relationships in Example 1 in the equation $\sqrt{2}\sin\left(x - \dfrac{\pi}{4}\right) = \sin x - \cos x$.

Comparing the individual amplitudes of the addends (1 and -1) to the amplitude of the final graph $\left(\sqrt{2}\right)$ suggests that the three amplitudes may have a Pythagorean relationship. Less obvious is the fact that tangent of the translation angle is equivalent to the ratio of the addends' coefficients: $\tan\left(-\dfrac{\pi}{4}\right) = \dfrac{-1}{1}$.

Consider a generic sum of the form $y = a\sin x + b\cos x$. This can be expressed as a single, horizontally translated sinusoid, which can in turn be expanded using an identity as follows.

$$a \cdot \sin x + b \cdot \cos x \;=\; c \cdot \sin(x + h) \;=\; c \cdot \sin x \cdot \cos h + c \cdot \cos x \cdot \sin h.$$

This leads to the system of equations $\begin{cases} a = c \cdot \cos h \\ b = c \cdot \sin h \end{cases}$ which can be solved for h by division.

$$\tan h = \frac{b}{a} \implies h = \tan^{-1}\left(\frac{b}{a}\right)$$

Solving for c is not as obvious. Squaring both sides of the systems equations and adding eliminates h.

$$a^2 + b^2 = c^2 \cos^2 h + c^2 \sin^2 h$$

$$a^2 + b^2 = c^2 \left(\cos^2 h + \sin^2 h\right)$$

$$a^2 + b^2 = c^2$$

In general, the sum of any two sinusoids with the same period is a stretched, phase-shifted sinusoid with the same period. The amplitude of this sum always has a Pythagorean relationship with the amplitudes of the addends, and $h = \tan^{-1}\left(\frac{b}{a}\right)$.

Example 2:
Rewrite $y = 3\sin x + 4\cos x$ using a single sinusoid.

$y = 3\sin x + 4\cos x = c \cdot \sin(x + h)$, so

$$c = \sqrt{3^2 + 4^2} = 5 \quad \text{and} \quad h = \tan^{-1}\left(\frac{4}{3}\right) = 0.927$$

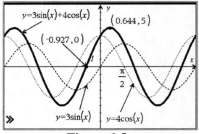

Therefore, $y = 5\sin(x + 0.927)$. The coordinates shown in Figure 6.5c confirm these computations and the period of 2π.

Figure 6.5c

Sums of Sinusoids with Different Periods
Example 3:
Graph $f(x) = \sin(2x) + \cos(3x)$.

As its components are periodic, so is f. Sine's period is π and cosine's is $\frac{2\pi}{3}$, so f has period 2π—the least common multiple of its addends' periods. The graph of f confirms its period. The graph can be pictured as $y = \cos 3x$ with $y = \sin(2x)$ as its bent midline.

Figure 6.5d shows the midline as a dashed curve, but $y = \cos(3x)$ could have been the bent midline with an identical result. Also notice that while f is periodic, it is certainly not sinusoidal.

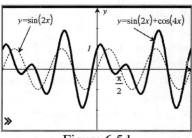

Figure 6.5d

More Envelope Curves
Example 4:
Graph $g(x) = \cos(2x) + \cos(4x)$.

As in Example 1, Figure 6.5e shows g as the graph of $y = \cos(4x)$ with $y = \cos(2x)$ as its variable center line. The graph of g has period π and is not sinusoidal.

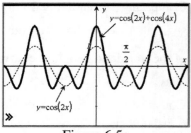

Figure 6.5e

Now picture a sinusoidal graph enveloping the graph of g. Figure 6.5f shows the curve , $y = 2\cos x$. The first zeros of g left and right of the y-axis are $x \approx \pm 0.523599$. Trial and error suggests that these are $x = \pm\dfrac{\pi}{6}$. That suggests that the internal curve has period $2 \cdot \left(\dfrac{\pi}{6} - \dfrac{-\pi}{6}\right) = \dfrac{2\pi}{3}$ making $y = \cos(3x)$ a possible equation for the internal curve. In this way, g can be re-expressed as a sinusoid under a variable stretch.

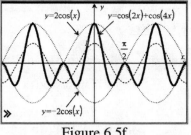

Figure 6.5f

$$y = \cos(2x) + \cos(4x) = 2\cos x \cdot \cos(3x)$$
$$= 2\cos\left(\frac{2x}{2}\right) \cdot \cos\left(\frac{6x}{2}\right)$$

While the right-side internal rational terms could be reduced, their form highlights the algebraic relationship between the components of the sum form and those in the product form of the curve. In general, this is because $\cos(\alpha) + \cos(\beta) = 2\cos\left(\dfrac{\alpha - \beta}{2}\right) \cdot \cos\left(\dfrac{\alpha + \beta}{2}\right)$ (proven in Example 5, Section 6.2).

Example 5:
Rewrite $y = \cos(3x) + \cos(5x)$ using envelope curves, and graph the function with its envelopes.

$$y = \cos(3x) + \cos(5x)$$
$$= 2\cos\left(\frac{5x - 3x}{2}\right) \cdot \cos\left(\frac{5x + 3x}{2}\right)$$
$$= 2\cos\left(\frac{2x}{2}\right) \cdot \cos\left(\frac{8x}{2}\right)$$

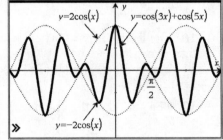

Figure 6.5g

The envelopes are $y = \pm 2\cos x$ (dashed curves in Figure 6.5g).

An application of this property relates to the sound waves created by instruments during the process of tuning. When two instruments produce nearly the same (but not identical) notes, the resulting sound has audible "beats" due to the constructive and destructive patterns of the sound waves from the two instruments. This can be illustrated by the high and low amplitude portions of the graph of $y = \cos x + \cos(2x)$ in Figures 6.5c-6.5e. The beats from the nearly identical sound waves are modeled by the envelope curves. The closer the two instruments are to producing the same sound, the closer their sound waves come to having the same period, making their difference approach zero and increasing the period of the envelope curves. For this reason, musicians know that to get their instruments in tune, they need to increase the time interval between beats until they are no longer noticeable. Ideally, the period reaches infinity at which point there is no more difference in the periods of the sound waves.

Problems for Section 6.5:

Exercises

Rewrite each equation in questions 1-4 as a single sinusoid.

1. $y = \sin x + \cos x$

2. $y = 4\sin(2x) + 4\cos(2x)$

3. $y = 4\sin\left(2x + \dfrac{\pi}{6}\right) + 4\cos\left(2x + \dfrac{\pi}{6}\right)$

4. $y = 7\cos(3x) + 24\sin(3x)$

Graph each function in questions 5-8 and determine any envelope curves.

5. $y = \cos(2x) + \cos(3x)$

6. $y = \cos(x) + \cos(4x)$

7. $y = \cos(3x) - \cos(4x)$

8. $y = 2\cos(2x) + 2\cos(5x)$

Explorations

Identify each statement in questions 9-13 as true or false. Justify your conclusion.

9. The graph of $y = \sin(x) - 3\cos\left(x - \dfrac{\pi}{4}\right)$ is a sinusoid.

10. The graph of $y = \sin(2x) + \cos(3x)$ is a sinusoid.

11. The graph of $y = \sin(2x) + \cos(3x)$ is periodic.

12. The period of the graph of $y = \cos(3x) + \cos(5x)$ is $\dfrac{2\pi}{15}$.

13. For any two functions of the forms $y_1 = \sin(ax)$ and $y_2 = \cos(bx)$, the function $f(x) = y_1 + y_2$ can be re-written as a variable stretch of some sinusoid.

14. If the transformations on the angle of the sinusoids are the same, as in question 13, explain why the angle expression is irrelevant to the determination of an equation for the sum of the sinusoids.

The method of combining sinusoids with the same period requires both original sinusoids to have the same horizontal translation. Questions 15-16 explore how to handle sums of sinusoids with different initial translations.

15. Rewrite $y = 5\sin\left(2x + \dfrac{\pi}{6}\right) - \cos(2x)$ as a single sinusoid.

16. Rewrite $y = 5\sin\left(x + \dfrac{5\pi}{3}\right) + 12\cos\left(x - \dfrac{3\pi}{4}\right)$ as a single sinusoid.

17. Given the results of the last question, explain why there is really no need to discuss cases in which both addends are sines or both are cosines.

Prove the identities given in questions 18-19.

18. $\sin A + \sin B = 2\sin\left(\dfrac{A+B}{2}\right)\cos\left(\dfrac{A-B}{2}\right)$

19. $\sin A - \sin B = 2\cos\left(\dfrac{A+B}{2}\right)\sin\left(\dfrac{A-B}{2}\right)$

For questions 20-26, graph the given function and determine any envelope curves.

20. $y = \sin(2x) + \sin(3x)$

21. $y = 3\sin(4x) - 3\sin x$

22. $y = \sin(2x) + \cos(3x)$ (from Example 3)

23. $y = 5\sin(2x) + 5\cos(3x)$

24. $y = 2\sin(4x) + 2\cos(7x)$

25. $y = x \cdot \sin(2x) + x \cdot \cos(3x)$

26. $y = x \cdot \sin(2x) + x \cdot \cos(3x) + 2x$

27. [NC]   The smallest magnitude zeros of $y = \cos x + \cos(2x)$ in Example 1 were calculated to be $x \approx \pm 1.0472$. The example suggested these were equivalent to $x = \pm \dfrac{\pi}{3}$. Prove this claim.

UNIT 7: PARAMETRIC FUNCTIONS & VECTORS

In times of change, the learners will inherit the earth while the learned find themselves beautifully equipped to deal with a world that no longer exists.

– Eric Hoffer

Enduring Understandings:

- Parametrically defined functions give valuable information about the relationships between variables that cannot be found from their Cartesian forms.

- The choice of parameter allows you to control the graph without changing or restricting the function.

- Vectors give a sense of location and direction from a point, and are often the only means of understanding 3-dimensional space.

- Vectors and parametric functions are equivalent forms of the same idea and have many useful applications.

This section assumes you have:

- Comfort with trigonometric functions.

- Ability to set up and solve systems of equations.

7.1: A Review of Parametric Functions

Perhaps the most surprising thing about mathematics is that it is so surprising. The rules which we make up at the beginning seem ordinary and inevitable, but it is impossible to foresee their consequences.

– E. C. Titchmarsh in N. Rose *Mathematical Maxims and Minims*, 1988

Consider the following set of ordered pairs.

x	2	3.5	5	6.5	8
y	13	12	11	10	9

Because Δx and Δy are both constant, the data is linear. One way to model the data is to use a point-slope form of a line. An alternative is to define each variable independently, as in

$$\begin{cases} x = 2 + 1.5t \\ y = 13 - t \end{cases}, \ t \in \mathbb{R} \ .$$

Here, both x and y are defined starting from the initial ordered pair and changing by the respective Δx and Δy values. The third variable introduced in the equations, t, is the **parameter,** and it controls the values of both x and y. Equations defined this way are called **parametric equations**. When defined only in terms of x and y, they are called **Cartesian**[1], or **rectangular equations**.

Example 1:
An ant and a spider walk across a 3 foot by 6 foot table which is oriented as shown in Figure 7.1a. The ant walks from the lower right corner to the midpoint of the top edge of the table. The spider walks from 1 foot below the top right corner and proceeds to the lower left corner of the table. Assume both creatures travel along linear paths. State Cartesian equations for the insects' paths, determine the point at which their paths cross, and state whether it is possible to know if the spider met the ant.

- The ant's path is easiest in point-slope form, $y_{ant} - 0 = -(x - 6)$.

  The spider is $y_{spider} = \dfrac{1}{3}x$.

- Solving the system created by the linear equations gives their point of intersection: $\left(\dfrac{9}{2}, \dfrac{3}{2} \right)$.

- Even though their paths cross, it is impossible to determine if the ant and spider meet. Without knowing each insect's velocity, it cannot be known if they were in the same position *at the same time*.

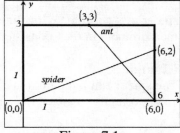

Figure 7.1a

Example 2:
The ant from Example 1 travels at 1 foot per second and the spider at 0.6 feet per second, determine if they meet under the conditions stated in Example 1.

Method 1:
A simple method determines the distance each has to travel and then uses each velocity to find the time required to travel each distance. Using the relationship $distance = rate \times time$, Figure 7.1b computes the distance and then the time for the ant (2.12 sec) and the spider (2.64 sec). Based on this, the ant clears the intersection at $(4.5, 1.5)$ about half a second before the spider arrives at the point.

$distance(6,0,4.5,1.5)$	2.12132
$\dfrac{2.12132 03435596 \cdot \_ft}{1 \cdot \_ft}$	$2.12132 \cdot \_s$
$distance(6,2,4.5,1.5)$	1.58114
$\dfrac{1.5811388300842 \cdot \_ft}{0.6 \cdot \_ft}$	$2.63523 \cdot \_s$
	5/99

Figure 7.1b

---

[1] Named for Rene Descartes, 17th century mathematician/philosopher inventor of the system.

Method 2:

A slightly more complicated (but ultimately far more useful) approach involves rewriting the linear equations from Example 1 in a way that incorporates time.

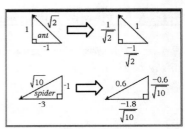

The slope of the ant's path is -1, so every time the ant travels 1 unit each horizontally and vertically, it travels $\sqrt{2}$ units along its linear path. Scaling the triangle down to a hypotenuse of 1 foot/second says the ant travels $\dfrac{1}{\sqrt{2}}$ feet/second horizontally and vertically.

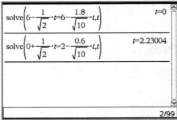

Figure 7.1c

Figure 7.1c shows the similar set-up for the spider. That means the animals' x- and y-coordinates can be written

$$Ant:\begin{cases} x = 6 - \dfrac{1}{\sqrt{2}}t \\ y = 0 + \dfrac{1}{\sqrt{2}}t \end{cases}, t \geq 0 \qquad and \qquad Spider:\begin{cases} x = 6 - \dfrac{1.8}{\sqrt{10}}t \\ y = 2 - \dfrac{0.6}{\sqrt{10}}t \end{cases}, t \geq 0.$$

Figure 7.1d

In these sets of equations, both axis coordinates have been rewritten in terms of a third variable, t. Relations in this form are said to be written in **parametric functions**.

If the ant and the spider are at the same place at the same time, their x- and y-coordinates will be equal at the same value of t. Figure 7.1d shows that the t-values are not equivalent, confirming the result of Method 1 that the creatures do not meet even though their paths intersect.

Also notice the unit consistency for the ant's and spider's parametric equations. Using $y = 2 - \dfrac{0.6}{\sqrt{10}}t$ from the

spider as an example, 2 is the initial "height", measured in feet, $-\dfrac{0.6}{\sqrt{10}}$ is the rate (in feet/second) at which the

spider's y-coordinate declines, and t is measured in seconds. Combining these terms gives a final measurement in feet, consistent with the y-coordinate meaning.

Example 3:

The spider can jump up to a distance of 2 inches in an attack. Does it get an opportunity to attack the ant? If so, at what time(s) does this happen?

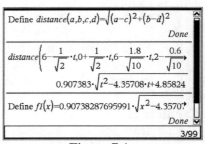

On the surface, this is not an easy question because the distance between the animals (Figure 7.1e) appears to be nothing close to a simple function. Even so, if the distance between the animals is expressed as a function of x, the distance function can be graphed and the minimum distance value pulled from the graph.

Figure 7.1e

CAUTION: Students sometimes get confused when reading the coordinates of the distance function this way. Remember that even though the coordinates are written as (x, y) for graphing purposes, they must be interpreted as (time, distance).

Figure 7.1f shows a graph of the distance function and the line $y = \dfrac{2}{12}$. The fractional form to accommodate for the parametric function being stated in feet. Notice that the minimum point on the

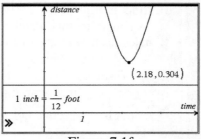

Figure 7.1f

distance curve, $(x, y) = (\text{time,distance}) = (2.18, 0.304)$, does not intersect $y = \dfrac{2}{12}$, so the spider was never close enough to the ant to attack. At $t \approx 2.18 \sec$, the animals were closest at distance $\approx 0.304\, ft \approx 3.65\, in$.

Overall, parametric equations are sometimes more cumbersome because they have several equations for each function, but they have the ability to give additional information and combine different, dependent relationships all related to a single, common, independent variable. As with the distance exploration in Example 3, equations in parametric form can also contain more information than equations in Cartesian form.

Where the first three examples explored the implications of two parametrically-defined functions, there are always multiple ways to parameterize curves. Unlike equations in Cartesian form, the choice of parameterization can influence the appearance and direction of the curve.

Example 4:

Graph the parametrically defined function $\begin{cases} x = t + 5 \\ y = t^2 \end{cases}$, describe the graph, and then rewrite the parametric

equations in Cartesian form to confirm the graphical patterns.

Figure 7.1g shows a graph of the parametric function.[2] The image appears to be that of a parabola with its vertex at $(5, 0)$. To verify, solve the given x equation for t and substitute into the given y equation.

$$t = x - 5 \implies y = (x - 5)^2$$

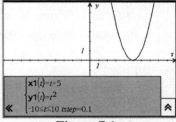

The Cartesian form of the equation confirms that the given parametric equations are equivalent to $T_{5,0}\left(y = x^2\right)$. While the horizontal translation

Figure 7.1g

is written as though "opposite" in the Cartesian form, this is not the case for the parametric form.

Rewriting parametric functions into Cartesian equivalents (Example 4) is called **eliminating the parameter**.

Example 5:
Graph and describe each parametrically defined function. Then eliminate the parameter to confirm the descriptions.

$$f : \begin{cases} x = t + 1 \\ y = 3t^2 - 5 \end{cases}, t \in \mathbb{R} \qquad g : \begin{cases} x = 5 - 2t \\ y = 1 + t \end{cases}, t \in [-1, 4] \qquad h : \begin{cases} x = 3^t \\ y = t \end{cases}, t \geq 0$$

The graph of each graph is shown Figures 7.1h-7.1j, respectively.

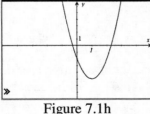

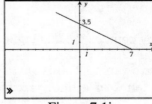

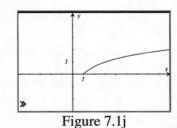

Figure 7.1h   Figure 7.1i   Figure 7.1j

Figure 7.1h shows f for $t \in [-10, 10]$. Like Example 4, the graph is a parabola, this time with its vertex at

---

[2] In a **Graphs & Geometry** window, press (menu) → ③:**Graph Type** → ②:**Parametric** to change the mode to parametric. Then enter the x and y equations in the respective $x1(t)$ and $y1(t)$ fields. The t-range defaults to $[0, 2\pi]$, but this can be manually adjusted on the bottom line as was done in Figure 7.1g.

$(1,-5)$. Again notice that the coordinates of the vertex are the constant terms in the respective parametric equations. To eliminate the parameter, solve the x equation for t and substitute into the y equation:

$$t = x - 1 \implies y = 3(x-1)^2 - 5$$

This is the Cartesian equivalent of $T_{1,-5}\left(y = 3x^2\right)$.

Function g is linear with x-intercept 7, y-intercept 3.5, and slope -0.5, but Figure 7.1i shows the graph as a segment, not a complete line. Sometimes there are multiple ways to eliminate parameters, so this time solve for t in the y equation and substitute into the x equation to get $x = 5 - 2(y-1) = -2y + 7$. Notice that this equation is the same form as the slope-intercept form of a line with the variables reversed. Here, the y-coefficient is the reciprocal of the slope and the constant is the x-intercept, confirming the graph. The segment's endpoints are confirmed through the t-restrictions: $t = -1 \implies (x,y) = (7,0)$ (right) and $t = 4 \implies (x,y) = (-3,5)$ (left).

While the parametric equations initially suggest an exponential function, the graph of function h (Figure 7.1j) appears logarithmic. Because $t \in [0,\infty)$, $y \in [0,\infty)$ as well, but the range of exponential functions restricts the x-values to $x \geq 1$. The final confirmation that this is a logarithmic graph happens when the parameter is eliminated. Solving for t in the x equation and substituting gives $y = \log_3 x$. The t-restrictions cause only the Quadrant I portion of the graph to appear.

Example 6:

Determine parametric equations for the following.

A. The line segment connecting $(3,2)$ and $(1,-5)$

B. The unit circle

    A. There are several parameterizations that work, but an easy one results directly from an equation of the line containing the points. In point-slope form, that line could be written $y = \dfrac{7}{2}(x-3) + 2$, so a

    parametric equivalent could be $\begin{cases} x = t \\ y = \dfrac{7}{2}(t-3) + 2 \end{cases}$. Because $x = t$, the t restrictions are $1 \leq t \leq 3$.

    An alternative approach requires noticing that the segment starts at $(3,2)$ after which the x-values drop 2 and the y-values drop 7, giving equations $\begin{cases} x = 3 - 2t \\ y = 2 - 7t \end{cases}, t \in [0,1]$.

    B. In Chapter 5, the coordinates of any point on the unit circle were defined as $\begin{cases} x = \cos\theta \\ y = \sin\theta \end{cases}$ where θ is the measure of rotation from the positive x-axis. This is already a parametric set of equations for the unit circle as x and y are defined in terms of a third variable. Using this, $\theta \in \mathbb{R}$ is acceptable, but $\theta \in [0, 2\pi]$ is sufficient.

    As an aside, eliminating the parameter for the unit circle parameterization is different than the earlier examples. Remember that an equation for the unit circle is $x^2 + y^2 = 1$. Both variables are squared, so squaring the component functions is a good start.

$$x = \cos\theta \quad \Rightarrow \quad x^2 = \cos^2\theta$$
$$y = \sin\theta \quad \Rightarrow \quad y^2 = \sin^2\theta$$

Adding gives $x^2 + y^2 = \cos^2\theta + \sin^2\theta \quad \Rightarrow \quad x^2 + y^2 = 1$, confirming the suggested parameterization.

This definition of the unit circle is a *parametric* function even though it is not a Cartesian function (its graph fails the vertical line test). For parametric functions, t (or θ) is the only input while both x and y are output variables. Because every value of θ creates a unique ordered pair on the unit circle, each parametric relation in part B of Example 6 is actually a *parametric function*.

Problems for 7.1:

Exercises

For questions 1-8, assume $t \in \mathbb{R}$ and graph each parametric function and describe the function defined.

1. $a : \begin{cases} x = t \\ y = t^2 - 1 \end{cases}$
2. $b : \begin{cases} x = t^2 - 1 \\ y = t \end{cases}$

3. $c : \begin{cases} x = 7t + 1 \\ y = 3 - 2t \end{cases}$
4. $d : \begin{cases} x = t^2 \\ y = -t^2 - 2 \end{cases}$

5. $e : \begin{cases} x = 4 + \cos t \\ y = 2 - \sin t \end{cases}$
6. $f : \begin{cases} x = 4\cos t \\ y = 3 + \sin t \end{cases}$

7. $g : \begin{cases} x = \log_2\left(\sqrt{t}\right) \\ y = t^3 \end{cases}$
8. $h : \begin{cases} x = 2\cos t \\ y = 3\sin t \end{cases}$

9. Eliminate the parameter for each function in questions1-8. Use the resulting Cartesian equation to confirm the form of the graph.

10. In Atlanta, GA US Highway 41 and West Wesley Road intersect each other at a right angle with US 41 running due North-South and West Wesley running perfectly East-West. Assume Car A approaches this intersection northbound on US41 at a constant 35 miles per hour and Car B approaches from the east on West Wesley at a constant 30 miles per hour. There are no other cars near the intersection at that moment and both drivers unfortunately decide to ignore the stop light at the intersection. At 12:31:51, Car A is 1000 feet from the intersection and Car B is 800 feet from the intersection.

   A. To the nearest second, when does Car A reach the intersection?

   B. When does Car B reach the intersection?

   C. Do the cars collide? Explain.

   D. Write parametric equations for the paths of the cars.

   E. When are the *x*-coordinates of the cars equivalent? What is the physical meaning of this time?

   F. At what time are the cars physically closest to each other?

11. Why is the vertical line test not a good universal graphing tool for determining if a relation is a function?

For questions 12-15, write a set of parametric equations to represent the given Cartesian relation.

12. $y = 7x - 3$ 13. $42x - 137y = 9$

14. $x - y^2 = 32$ $X = 32 - t^2$ 15. $y^2 - x^2 = 4$

$y = \sqrt{-32 + 2t}$

Explorations

Identify each statement in questions 16-20 as true or false. Justify your conclusion.

16. Every Cartesian function can be parameterized in more than one way.

17. Every Cartesian function can be expressed as a parametric function.

18. Every parametric function can be expressed as a Cartesian function.

19. When eliminating the parameter from parametric equations, the best approach is to solve for the parameter in one equation and substitute the result into the other equation.

20. If two parametrically defined curves intersect, they do so simultaneously.

21. A function is defined parametrically by $\begin{cases} x = \tan t \\ y = \sec t \end{cases}$. Sketch this function for $t \in [0, 2\pi]$, name the family for which this function is an example, and rewrite the function in a more familiar Cartesian form by eliminating the parameter.

22. Graph $f : \begin{cases} x = \sec t \\ y = \sec t \end{cases}$. Then determine a Cartesian equivalent for f and use the parameterization to explain any unique features of the graph.

Define parametric equations for each graph described in questions 23-27.

23. A circle centered at the origin with radius 3.

24. A *line* through the points $(2,4)$ and $(-7,1)$.

25. A *line segment* with endpoints $(2,4)$ and $(-7,1)$.
                                   $\dfrac{-3}{-9}$ $X = 9t - 7$ $t \in [1, 0]$

$y = 3t + 1$

26. A *ray* starting at the point $(2,4)$ and going through $(-7,1)$.

27. A horizontally-oriented parabola.

28. Another approach to eliminating the parameter from the unit circle parametric equations in part B of Example 6 is to solve for θ in the x equation and substitute the result into the y equation. Determine the final equation using this approach, and explain why the approach used in Example 6 is more elegant.

7.2: Remaining in Control

An expert is someone who knows some of the worst mistakes that can be made in his subject, and how to avoid them.

– Werner Heisenberg in *Physics and Beyond*, 1971

As seen in the solutions for parameterization of a line segment in Example 6 of Section 7.1, it is always possible to write different parametric equations for a function. This section explores the implications of different parameterizations and what can be gained from the additional work required to express a function in parametric form.

Starting points and direction
Example 1:
Describe the differences in the graphing of the line segment connecting $(3,2)$ and $(1,-5)$ under each of the following variations.

A. $\begin{cases} x=t \\ y=3.5t-8.5 \end{cases}, t \in [1,3]$ B. $\begin{cases} x=3+2t \\ y=2+7t \end{cases}, t \in [-1,0]$ C. $\begin{cases} x=2-1\sin t \\ y=-1.5-3.5\sin t \end{cases}, t \in [0,\infty)$

In all three sets of parametric equations, the parameter can be eliminated by solving either equation and substituting into the other to establish that each represents at least a portion of the line through $(3,2)$ and $(1,-5)$, or the line $y=3.5x-8.5$.

A. $\begin{cases} x=t \\ y=3.5t-8.5 \end{cases}$ is the simplest parametric form, substituting $x=t$ in the slope-intercept form of the line. As in Example 6 of Section 7.1, $x=t$ also forces the domain, or t, restrictions to be $1 \le t \le 3$. Because t (and therefore x) moves from 1 to 3, this particular parameterization will graph from left to right.

B. $\begin{cases} x=3+2t \\ y=2+7t \end{cases}$ looks almost like the second parameterization of the segment in Example 6 of Section 7.1, but the t-domain is different. The equations suggest that the starting point is $(3,2)$, but because the domain goes from $t=-1$ to $t=0$, this parameterization also graphs from left to right the same way the part A parameterization did. Because the "width" of the domain for part B is 1 $(0-(-1))$ while the width of the domain for part A is 2 $(3-1)$, this second parameterization actually graphs faster.

C. At first glance, the domain of $\begin{cases} x=2-1\sin t \\ y=-1.5-3.5\sin t \end{cases}$ suggests this parameterization might graph the entire line and not just the desired segment, but t is embedded within a sine function and the range of sine is $[-1,1]$, so this parameterization is restricted. When $t=0$, $(x,y)=(2,-1.5)$, the midpoint of the segment. At $t=\dfrac{\pi}{2}$, $(x,y)=(1,-5)$, so the graph initially moves from right to left. At $t=\pi$, the graph returns to the midpoint and finally reaches the other endpoint at $t=\dfrac{3\pi}{2}$. As $t \to \infty$, the segment is regraphed an infinite number of times. Because it takes π units of t to graph the entire segment, this particular parameterization is the "slowest" of the parameterizations in this example.

Example 2:
Prove that each parameterization describes $y = x^2$. Describe the effects of each parameterization.

A. $\begin{cases} x = t \\ y = t^2 \end{cases}, t \in \mathbb{R}$

B. $\begin{cases} x = \log(t) \\ y = \left(\log(t)\right)^2 \end{cases}, t \in (0, \infty)$

C. $\begin{cases} x = \cos(t) \\ y = \cos^2(t) \end{cases}, t \in \mathbb{R}$

In all three sets of parametric equations, the expression for the y equation is the square of that of the x equation, so $y = x^2$, proving that each parameterization is a representation of $y = x^2$.

A. For $\begin{cases} x = t \\ y = t^2 \end{cases}$, t and x are identical, so there is no substantial difference between $y = x^2$ and this parameterization, and $t \in \mathbb{R}$ generates all x-values from the domain of $y = x^2$.

B. For $\begin{cases} x = \log(t) \\ y = \left(\log(t)\right)^2 \end{cases}$, the logarithm functions require t to be positive, explaining the t-restrictions given with the function. However, since $\log(t)$ has range $\mathbb{R}$, this gives $x \in \mathbb{R}$ and consequently, $y \in \mathbb{R}$, resulting in the complete graph of $y = x^2$.

C. For $\begin{cases} x = \cos(t) \\ y = \cos^2(t) \end{cases}$, the range of the cosine and squaring functions take the infinite t domain to the much more restricted $x \in [-1,1]$ and $y \in [0,1]$ (Figure 7.2a). This parameterization repeatedly graphs the portion of $y = x^2$ between $(-1,1)$ and $(1,1)$ an infinite number of times.

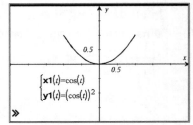

Figure 7.2a

From the analyses in Examples 1 and 2, it should be clear that different parameterizations of a curve offer not only different algebraic forms, but also affect the starting point, direction of graphing, and the speed of the graphing. Many attempt to control the graphing speed of a parametric graph by adjusting the *tstep* only[3]. These examples show that it is also possible to control the speed of a graph within the parametric equations, regardless of the *t*-domain settings.

In neither Example 1 nor 2 is the parameter graphed when the parametric function is shown. The *t*-values determine x and y, but only the ordered pairs (x, y) are actually graphed. There are some circumstances where graphs of (t, x) or (t, y) are useful, but t usually remains out of sight when graphing parametrically. The main purpose of t is to control x and y.

"As Is" Behavior of Transformations
An advantage of working with parametrically-defined relations is that transformations on these curves behave "as written" and not "opposite" as in the case of equations in Cartesian form. For example, applying $T_{1,-5}$ to $y = 3x^2$ yields $y + 5 = 3(x-1)^2$ in Cartesian form and $\begin{cases} x = t+1 \\ y = 3t^2 - 5 \end{cases}$ as a potential parametric form (from Example 5, Section 7.1). The "as written" format of parametric functions happens precisely because the x and

---

[3] The *tstep* is the distance between successive *t*-values at which a calculator computes function values in a parametric function. On an Nspire, the *tstep* is set on the bottom line of a parametric equation graphing command line to the right of the *t* domain values. This can be seen on the bottom line of Figure 7.1g.

y-coordinates are controlled by different equations. If you want to move $y = 3x^2$ one unit right and five units down, all of the x-values increase by one and the y-values decrease by five, precisely as indicated by the parametric equations above.

[handwritten: can you move the function without changing all values]

In general, a generic Cartesian function defined by $y = f(x)$ is written its simplest parametric form by $\begin{cases} x = t \\ y = f(t) \end{cases}$. If $T_{h,k} \circ S_{a,b}$ is applied, the algebraic result will be $\begin{cases} x = a \cdot t + h \\ y = b \cdot f(t) + k \end{cases}$. Solving the first equation

for t and the second for $f(t)$ gives $t = \dfrac{x-h}{a}$ and $f(t) = \dfrac{y-k}{b}$. Substitution produces $\dfrac{y-k}{b} = f\left(\dfrac{x-h}{a}\right)$,

one of the algebraic forms resulting from applying $T_{h,k} \circ S_{a,b}$ to the Cartesian function $y = f(x)$.

Because transformations behave as written for functions in parametric form, rewriting parametric equations into Cartesian form explains the seemingly "reversed" behavior of Cartesian form of equations. While this explains the algebraic form of Cartesian *functions* under transformations, the same is true for the algebraic form of Cartesian relations like the unit circle.

Example 3:

Describe the differences in the graphing each of the following transformations of the unit circle. Assume $t \in [0, 2\pi]$ for each graph.

A. $\begin{cases} x = \cos t + 5 \\ y = \sin t - 1 \end{cases}$ 　　　　B. $\begin{cases} x = 4\cos t \\ y = -2\sin t \end{cases}$ 　　　　C. $\begin{cases} x = -\cos t - 3 \\ y = 3\sin t + 1 \end{cases}$

Example 6 of Section 7.1 parameterized the unit circle with $\begin{cases} x = \cos t \\ y = \sin t \end{cases}$ for

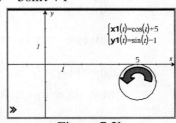

$t \in [0, 2\pi]$, starting the graph at $(1,0)$ and tracing counterclockwise.

A. In parametric mode, adding 5 to the x-equation moves the graph 5 units right. Subtracting 1 from the y moves the graph down one, making parametric function A a unit circle centered at $(5,-1)$ (Figure 7.2b). The graph begins at $(6,-1)$ and graphs counterclockwise.

Figure 7.2b

B. Parametric function B applies $S_{4,-1}$ *[handwritten: 4]* to the unit circle. Any two-dimensional dilation of a circle with unequal magnitudes creates an ellipse, and a negative dilation also creates a reflection. Therefore, parametric function B is an ellipse with a horizontal major axis of 8 and a minor axis of 4, centered at the origin. The vertical reflection keeps the starting point on the right at $(4,0)$, but changes the graphing direction to clockwise (Figure 7.2c).

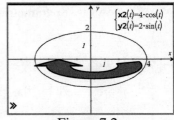

Figure 7.2c

C. Parametric function C applies $T_{-3,1} \circ S_{-1,3}$ to the unit circle resulting in a translated, horizontally reflected, vertically stretched ellipse whose graph begins at $(-4,1)$ and traverses clockwise (Figure 7.2d).

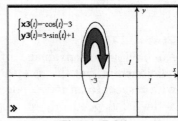

Because transformations of relations in parametric form can be read "as is," Figure 7.2d

dilations and translations are straight-forward and simple in this format. Obviously, the starting point of a parametrically-defined graph is determined by the starting point of the t-domain. As shown by these three examples, the starting point of a transformation of the unit circle is the image of the point $(1,0)$ under the same transformation. For example, the starting point for parametric function C is $T_{-3,1} \circ S_{-1,3}\big[(1,0)\big]=(-4,1)$. The direction arrow of the graphing can also be understood as the image of the direction arrow of the original unit circle under the same transformations, explaining the stretched appearance of the arrows in Figures 7.2b-7.2d.

Inverses

As noted in Section 2.6, the inverse of a function f converts the output of the original relation back to its original input. That is, application of an inverse transformation is algebraically equivalent to switching x and y. Where Section 2.6 focused on inverses as reflection transformations over the line $y=x$, the "as is" nature of parametric equations easily allows us to algebraically define and graph the inverse of any parametrically-defined relation simply by switching the x and y expressions.

Example 4:

Define parametric equations and use them to obtain a graph of the inverse of each of the following relations.

   A. The line $f(x)=-2x+9$

   B. The function $g(x)=x\cdot\cos x$

   C. The circle, h, of radius 3 whose center is at the point $(7,-3)$

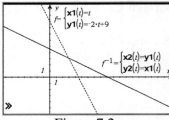

Figure 7.2e

   A. The quickest parameterization of the given line is $f:\begin{cases}x=t\\y=-2t+9\end{cases}$, so its inverse can be written $f^{-1}:\begin{cases}x=-2t+9\\y=t\end{cases}$. Figure 7.2e shows f (dashed) and f^{-1} (solid). Oblique lines are one-to-one and therefore always invertible, making f^{-1} a function.

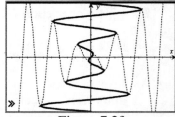

Figure 7.2f

   B. Function g is a cosine curve with variable amplitude. Its inverse is obviously not a function, but Figure 7.2f (g dashed and g^{-1} in bold) shows that parametrics make it just as easy to obtain a graph of g^{-1} as it is to get a graph of f^{-1}.

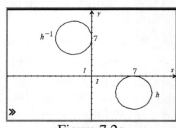

Figure 7.2g

   C. Using the results of Example 3, Cartesian relation h can be represented by the parametric function $h:\begin{cases}x=3\cos t+7\\y=3\sin t-3\end{cases}$, therefore one parameterization is $h^{-1}:\begin{cases}x=3\sin t-3\\y=3\cos t+7\end{cases}$ (Figure 7.2g).

Pros and Cons

An obvious disadvantage of parametric functions is their requirement for multiple equations to define relationships, but this is also their greatest strength. Sometimes it is actually easier to define two variables with respect to an independent variable than it is to connect them directly. Also, when x and y variables are defined independently, it is *much* easier to control, transform, and define their inverse graphical behavior. Parametric relations allow a higher order relation to be expressed as a system of simpler, two-variable relations.

Problems for 7.2:

Exercises

1. Describe the differences in how the unit circle is graphed under each of the following variations. Use $t \in [0, 2\pi]$ for each graph.

    A. $\begin{cases} x = -\cos t \\ y = \sin t \end{cases}$ *rotates CW*   *Starts at (-1, 0)*   B. $\begin{cases} x = \sin t \\ y = \cos t \end{cases}$ *starts at (0, 1)*   *rotates CW*   C. $\begin{cases} x = \sin t \\ y = -\cos t \end{cases}$ *starts at 0, -1*   *rotates ccw*

2. Identify the starting points and directions of the segment parameterizations in Example 6 of Section 7.1.

3. How long does it take (what is the width of a *t*-interval) for the parameterization in part C of Example 2 to complete one cycle of the graph of $y = x^2$ between $(-1, 1)$ and $(1, 1)$?

For each set of parametric equations in questions 4-7, draw a graph, and describe the path of the graph.

4. $a : \begin{cases} x = 2t - 1 \\ y = 3t + 5 \end{cases}$ 5. $b : \begin{cases} x = \dfrac{1}{2}t^2 + 1 \\ y = 2t^2 + 5 \end{cases}$

6. $c : \begin{cases} x = \dfrac{1}{t^2} - 1 \\ y = \dfrac{1}{t} \end{cases}$ 7. $d : \begin{cases} x = \cos^2 t \\ y = 3 - \sin^2 t \end{cases}$

Graph each set of parametric equations in questions 8-11.

8. $e : \begin{cases} x = 3\cos t \\ y = 4\sin t \end{cases}$ 9. $f : \begin{cases} x = -2\cos t + 3 \\ y = 5\sin t - 1 \end{cases}$

10. $g : \begin{cases} x = \sin t - 2 \\ y = 2\cos t + 1 \end{cases}$ 11. $h : \begin{cases} x = 3\sec t + 1 \\ y = 5\tan t - 4 \end{cases}$

For questions 12-18, define a set párametric functions for each graph described.

12. The line passing through $(5, 1)$ and $(-3, 3)$

13. The line segment passing through $(-2, 5)$ and $(4, 7)$

14. A horizontal line through $(10, 13)$

15. A vertical line though $(4, 2)$ 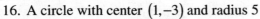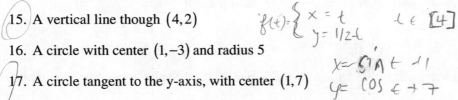

16. A circle with center $(1, -3)$ and radius 5

17. A circle tangent to the y-axis, with center $(1, 7)$ 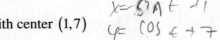

18. A hyperbola symmetric to the line $y = 2$, with asymptotes $y = \pm \dfrac{2}{5}(x + 3) + 2$

Determine parametric equations for each ellipse in questions 19-20.

19.

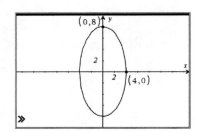

20.

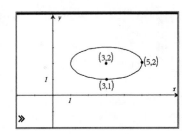

For the given functions in questions 21-24, create a graph of the function's inverse.

21. $g(x) = x^2 \cdot (x-2)$

22. $g(x) = (x^2 \cdot \sin x) + 1$

23. $g(x) = e^x - x$

24. $g(x) = \cos|2x|$

25. Determine a set of parametric equations that would graph the unit circle with a *t*-range of 1 unit.

Explorations

Identify each statement in questions 26-30 as true or false. Justify your conclusion.

26. The graphs of a function and its inverse can only intersect on the line $y = x$.

27. Every intersection point shown in Figure 7.2f occurs at an extremum of function g.

28. By changing only the algebra in a set of parametric equations, the graph of the equations can be made to appear as quickly or as slowly as desired.

29. Any relation that can be defined by a parametric function (every value of *t* determines a unique ordered pair (x,y)) has an inverse which is also a parametric *function*.

30. It is possible to obtain a complete graph of any Cartesian function by defining it parametrically and using only $t \in [0,1]$, excluding $t = 0$ and/or $t = 1$ as necessary.

31. For each situation, define a parameterization of the unit circle. When possible, parameterize without transforming the parameter.

A. Start at $(0,-1)$ and traverse clockwise.

B. Graph only the left half, initially traversing clockwise.

C. Graph all except Quadrant III, initially traversing counterclockwise.

D. Graph the entire circle, starting at point $\left(-\dfrac{1}{2}, \dfrac{\sqrt{3}}{2}\right)$ and heading down.

E. Graph Quadrant I, initially from left to right.

32. Parameterize the line segment in Example 1 to require a spread of 1000 *t* units for a complete graph.

33. Determine a parameterization of the line segment from Example 1 that begins at a *y*-value of 1.

34. Determine an equation for the curve that connects all of the intersection points of g and g^{-1} from Example 4. How many intersection points are there and how do you know?

7.3: Parametric Applications

You master mathematics if you are willing to try…. Success is a function of persistence and doggedness and the willingness to work hard for 22 minutes to make sense of something most people would give up on after 30 seconds.

– Malcolm Gladwell on Alan Schoenfeld in *Outliers*

Gravity

Remember from a physical science class that the height of a free-falling object under the influence of gravity only is described by $h(t) = \frac{1}{2}gt^2 + v_0 t + p_0$, where g is the gravitational constant (approximately $-32.2\,\text{ft}/\sec^2 \approx -9.81\,\text{m}/\sec^2$), t is time in seconds, v_0 is the initial velocity of the object, and p_0 is the initial position of the object.

Example 1:

Mark throws a baseball at 80 ft/sec at an angle of 35° from the horizontal. Assume the ball is released 5 feet above the ground and encounters no air resistance.

- Write parametric equations to model this situation,
- find the maximum height the ball reaches, and
- determine where the ball lands.

The horizontal and vertical positions of the ball at any time t constitute the x and y portions, respectively, of a parametric definition of the ball's path. Trigonometry can decompose the initial velocity into horizontal and vertical components. As suggested by Figure 7.3a, the vertical component is $y = 80\sin 35°$ and the horizontal is $x = 80\sin 35°$.

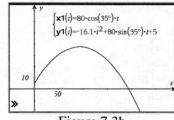

Figure 7.3a

If $h(t)$ is the vertical position of the baseball, the parametric y-component of the motion is $y(t) = -16.1t^2 + (80\sin 35°)t + 5$. The x-component is not affected by either gravity or wind resistance, so its equation is $x(t) = (80\cos 35°) \cdot t$. Together, x and y define parametric equations for the baseball's motion. It is easy to graph the function (Figure 7.3b), but there is not a maximum function tool available in parametric mode. Another approach is necessary.

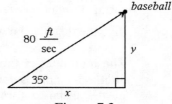

Figure 7.3b

The vertical component of the ball's motion is quadratic, so the baseball's maximum height occurs at the vertex of the y-component. The t-component of the vertex is $t = \frac{-b}{2a} = \frac{-80\sin 35°}{2(-16.1)} \approx 1.425\,\text{sec}$, so the maximum height is $y(1.425) \approx 37.695\,ft$ (top line of Figure 7.3c).

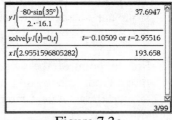

Figure 7.3c

Notice how Figure 7.3c shows the use of function names to avoid retyping the cumbersome algebra of the equations.

The ball reaches the ground when $y = 0$. Solving $y(t) = 0$ for t gives $t \approx \{-0.105\,\text{sec}, 2.955\,\text{sec}\}$. The negative answer is extraneous, so the baseball hits the ground at 2.995 seconds, and its horizontal position is $x(2.995) \approx 193.658\,\text{ft}$ (bottom two lines of Figure 7.3c).

Example 2:

Jill and Sam are playing a game of catch. Assume the following.

- Each throws a ball to the other at the exact same moment with no air resistance.
- Jill and Sam are 44 meters apart.
- Jill's release height is 1.7 meters at 21 m/sec at an angle of 0.91 radians from horizontal.
- Sam's release height is 2 meters with initial velocity of 24 m/sec at 1.1 radians.

A. Write parametric equations for the positions of the two balls.

B. What is the closest the two balls come to each other while they are in the air?

   A. Orient a number line where x is measured in meters with Jill at $x = 0$ and Sam at $x = 44$. Following the process of Example 1, the parametric equations for the balls are

$$Jill: \begin{cases} x1(t) = 21\cos(0.91) \cdot t \\ y1(t) = -4.9 \cdot t^2 + 21\sin(0.91) \cdot t + 1.7 \end{cases} \text{ and } Sam: \begin{cases} x2(t) = 44 - 24\cos(1.1) \cdot t \\ y2(t) = -4.9 \cdot t^2 + 24\sin(1.1) \cdot t + 2 \end{cases}.$$

   Notice the adjustment made in the x-equation for Sam's ball. The $24\cos(1.1) \cdot t$ term is subtracted so that the x-position of his ball decreases, making it move *toward* Jill.

   B. Points $(x1(t), y1(t))$ and $(x2(t), y2(t))$ show the positions of the balls as functions of time. The

   distance function defined on the CAS by $distance(a, b, c, d) = \sqrt{(a - c)^2 + (b - d)^2}$ can be used to

   give the distance between the balls for any t (line 1, Figure 7.3d).

   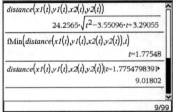

   The minimum of this result could be determined graphically, but an alternative approach uses a built-in minimum function. The minimum of the distance formula[4] (line2, Figure 7.3d) shows the balls are closest at $t \approx 1.775$ sec. To find the distance, evaluate the distance formula at $t \approx 1.776$ sec to get 9.018 m.

Figure 7.3d

Cycloids

Example 3:

A vehicle with wheels of radius 1 ft gets a stone stuck in its tread as it drives down the road. The stone is stuck in the tire in such a way that it touches the road every time the wheel completes a revolution. Write parametric equations and graph the path of the stone.

   The path any point on a circle takes as that circle travels along a line is called a **cycloid**. The path of the stone can be seen as a point traversing a circle (as in Example 3, Section 7.2) while the center of the circle is translated along a line parallel to the surface of the road.

   So, the point on the wheel experiences two motions: circular and horizontal. If the wheel is assumed to move from left to right, the stone will be picked up at the bottom of the circle and will move clockwise around the wheel. Using the lesson about the starting point and direction in parameterizations of the unit

   circle, the circular motion is given by $\begin{cases} x = -\sin t \\ y = -\cos t \end{cases}$ where t is the angle of rotation. All that remains is to

   write parametric equations for the center of the circle.

---

[4] The **fMin** command is at (menu) → (4):**Calculus** → (6):**Function Minimum,** or it could be typed directly.

Caution is required to write the non-circular component because circles tend to be parameterized in terms of angles of rotation whereas distances use length parameters. To write a single set of parametric equations for this motion, the same parameter must be used for both motions. Once the wheel completes one revolution, it will have traveled one perimeter length along the ground. Because the wheel is a unit circle and by the definition of a radian, a rotation of t radians on the wheel means it will have rolled forward t units horizontally (Figure 7.3e). So, the center of the wheel can be

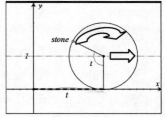

Figure 7.3e

parameterized by $\begin{cases} x = t \\ y = 1 \end{cases}$, and the path of the stone is given by

$$stone: \begin{cases} x = t - \sin t \\ y = 1 - \cos t \end{cases}.$$

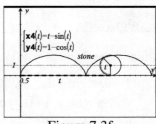

Figure 7.3f

Figure 7.3f shows the path of the stone along the cycloid with the wheel part-way through its second revolution.

Example 4:

If the wheel in Example 3 instead had radius 2 ft and the road was inclined 20° from horizontal, determine parametric equations and graph the new path of the stone.

The circular portion of the motion is a straightforward. $S_{2,2}$ dilation, resulting in: $\begin{cases} x = -2\sin t \\ y = -2\cos t \end{cases}$.

The tilted roadway makes the position of the center of the wheel more complicated. After the wheel has rotated t radians, it will have moved $2t$ units along the tilted surface of the road. If the point at which the wheel touches the road is point A, Figure 7.3g shows that the coordinates of point A are $(2t\cos 20°, 2t\sin 20°)$. From A, the center of the wheel is 2 units away, but at an angle. Using right $\triangle ABC$ in Figure 7.3g, the center of the wheel (Point C) can be defined as $2\sin 20°$ units back and $2\cos 20°$ up from point A.

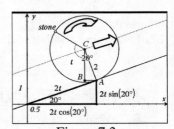

Figure 7.3g

Therefore, the position of the center of the wheel can be parameterized by $\begin{cases} x = 2t\cos 20° - 2\sin 20° \\ y = 2t\sin 20° + 2\cos 20° \end{cases}$, making the coordinates of the stone

$$stone: \begin{cases} x = 2t\cos 20° - 2\sin 20° - 2\sin t \\ y = 2t\sin 20° + 2\cos 20° - 2\cos t \end{cases}.$$ Figure 7.3h shows the tilted

roadway as a solid line, the path of the center of the wheel as a dashed line, and the path of the stone as a bold curve.

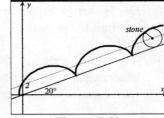

Figure 7.3h

Problems for 7.3:

Exercises

1. At what times will Mark's baseball from Example 1 be exactly 10 feet above the ground?

2. There is an 8 foot fence 183 feet away from where Mark released his baseball in Example 1. Did his throw clear the fence?

Use the information in Example 2 to answer questions 3-5.

3. Verify that Sam will not need to move to catch Jill's ball.

4. What is the closest Sam's ball comes to the point from which Jill released her ball?

5. Determine if Jill will be able to stand still to catch Sam's ball.

6. Write a Cartesian equation for the cycloid from Example 3.

Explorations

Identify each statement in questions 7-11 as true or false. Justify your conclusion.

7. Because the gravitational constant is approximately $-32.2\,\text{ft/sec}^2$ in English units, but $-9.81\,\text{m/sec}^2$ in metric, gravity is stronger in magnitude when measured in English units.

8. Even if a wind was blowing parallel to the ground in either Example 1 or Example 2, the y-components of all of the parametric equations would remain unchanged.

9. At their core, Example 3, Section 7.2; Example 3, Section 7.3; and Example 4, Section 7.3 all describe circles by adding the location of the circle's center to the motion of a point moving around the circle.

10. If you can describe in terms of the same parameter how a rotating point moves about its center and how the center of rotation moves, then you can write parametric equations for any rotating object.

11. A stone caught in the tread of a rolling tire travels a greater distance than the rolling tire itself.

12. Determine another technique to find the maximum height of the baseball in Example 1.

13. Answer these additional questions about Example 2.

    A. At what time(s) do the two balls share the same horizontal coordinates?

    B. What is the vertical distance between the balls at the time(s) identified in part A?

    C. When are the balls exactly 20 meters apart?

    D. Is there a time when the distance from Jill's ball to Sam's ball is the same as the distance from Jill's ball to the ground?

14. Two outfielders are warming up. Standing 200 ft apart, they simultaneously throw baseballs to each other, each releasing from a height of 6 ft. Player A releases at an angle of 32° from horizontal at $90\,\text{ft/sec}$, while Player B throws at $85\,\text{ft/sec}$ from an angle of 40° to the ground.

    A. What are the coordinates of the point(s) at which the paths of the balls cross?

    B. Do the balls collide? How do you know?

    C. Which player made a better throw? Justify your answer.

15. Two balls are thrown into the air simultaneously from a height of 1.8 meters. Each is thrown in the same direction at an angle of 1 radian from horizontal. The first has an initial velocity of 40 m/sec and the other has an initial velocity of 50 m/sec.

    A. What is the difference in their maximum heights?

    B. At what times are the balls at the same height?

    C. How far apart are the points at which the balls return to the ground?

16. If the wheel in Example 3 had radius r, determine possible parametric and Cartesian equations for the resulting cycloid.

17. Determine a Cartesian equation for the path of the center of the wheel in Example 4.

18. A. Transform the parametric equations from Example 4 by
- changing all of the coefficients of 2 to 1, and
- changing all of the angles from $20°$ to $0°$.

    B. Describe how the changes in the values of those two constant changes the description of the wheel and its path in Example 4.

    C. Simplify the new parametric equations you wrote in part A and explain the relationship between these equations and those from Example 3.

    D. Write parametric equations for the path of a stone on the edge of a wheel of radius r as the wheel travels up a hill of slope θ.

19. Many state fairs have a ride called a double Ferris wheel which consists of two rotating wheels attached at their centers to a bar which also rotates around the overall center of the ride (Figure 7.3i). Assume

- The height of the vertical center tower supports is 50 feet,
- The rotating arm is 64 feet long
- The radius of the wheels is 15 feet,
- The arms rotate counterclockwise every 20 seconds, and
- The wheels rotate clockwise every 16 seconds, and
- A rider boards the double Ferris wheel at $t = 0$ at the lowest possible point on the ride.

Figure 7.3i[5]

    A. What is the period of one complete cycle of a ride on this double Ferris wheel?

    B. Write parametric equations for the position of a seat positioned on the edge of one wheel.

    C. How far above the ground is a person who boards the ride at its lowest point?

    D. The ride begins immediately after the last person loads. How long will that person need to wait before reaching the highest point on the ride?

    E. Create a graph of a rider's position over time relative to the base of the center towers.

    F. Create a graph of a rider's position over time.

20. Determine parametric equations and provide a graph of the path of a point on a circle of radius r travels as that circle rolls around another circle also of radius r.[6] HINT: This situation is much like Example 4 except now the "road" is a circle.

---

[5] Photo source: http://www.flickr.com/photos/mario_rusciano/3423124190/in/pool-skywheels

[6] The path of a point on a circle that rolls around another circle without slipping is an **epicycloid**.

7.4: A Review of Vectors

The mathematician is fascinated with the marvelous beauty of the forms he constructs, and in their beauty he finds everlasting truth.

– J. B. Shaw in N. Rose *Mathematical Maxims and Minims*, 1988

Sometimes a single measure is insufficient information for a situation. For example, if you wanted to visit a friend, but were only told that the friend lived 3 blocks away, you would have some difficulty getting there. The magnitude of your travel would be three blocks, but in which direction?

Definitions and Conversions
Any quantity that can be expressed with both a **magnitude** and a **direction** is a **vector**. Real numbers are vectors, but rarely with that explicit designation. Each has a magnitude, indicating how far it is from the origin, and a direction, the $\pm$ sign indicating whether the point is to the left or right of the origin.

Ordered pairs can also be interpreted as vectors. The individual x- and y-components of an ordered pair can be thought of as the components of a vector. The Pythagorean Theorem and circular trigonometry then can be used to derive the overall magnitude and direction of the ordered pair.

New notation is advised to avoid potential confusion between vectors forms and other representations. The magnitude and direction of a vector will be noted with square brackets, $[r, \angle\theta]$, where r is the magnitude of the vector and θ is the angle between the vector and the positive x-axis as in circular trigonometry. Vectors expressed as $[r, \angle\theta]$ are in **geometric form**. Vectors defined by their horizontal and vertical coordinates, like ordered pairs, are in **component form** and are notated $\langle x, y \rangle$ with the "inequality" brackets distinguishing a vector from an ordered pair.

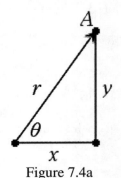

Figure 7.4a

Figure 7.4a shows a vector drawn as an arrow with its geometric and component values labeled. The rationale for drawing vectors as arrows is that vectors are dynamic "actions" whereas ordered pairs are static locations.

Like functions, vectors can be named independently of their different forms. For example, Figure 7.4a shows vector $\vec{A}$; the arrow over the name designates it as a vector.

The magnitude of a vector, $\vec{A}$, can be written several ways; different sources use $|\vec{A}|$, $\|\vec{A}\|$, or A (without the ray). For $\vec{A} = \langle x, y \rangle$, the Pythagorean Theorem shows that the magnitude of $\vec{A}$ is $|\vec{A}| = \sqrt{x^2 + y^2}$.

While both the geometric and component forms of vectors are valid representations, the component form is typically more convenient for computations. Even so, it is often easier to visualize a vector from its geometric form. For example, it is not obvious that a position defined by the vector $\langle 11, 60 \rangle$ is exactly 61 units from its starting point, a fact made obvious by its geometric form, $[61, 1.389]$. Once again, different algebraic forms tell different parts of the story.

Parametric representations of functions are exactly equivalent to vectors. The x- and y-equations of a parametric function identify the component form of a **displacement vector**—a vector used to define how far each point on the function is from the origin. A vector representation of the location of Mark's ball (Section

7.3, Example 1), $\langle(80\cos35°)t,\ -16.1t^2+(80\sin35°)t+5\rangle$, is equivalent to the previously noted parametric equations. The mathematics of vectors and parametric functions can be used interchangeably.

Section 7.3, Example 2 used expressions for Δx and Δy, the horizontal and vertical displacements between the balls, to compute the distance between the balls. Notice that this is a displacement vector, $\langle\Delta x,\Delta y\rangle=\langle(x1(t)-x2(t)),\ (y1(t)-y2(t))\rangle$, that does not originate from $(0,0)$, but from a moving point. Not all displacement vectors start at the origin.

A further application of the magnitude of a vector addresses the difference between the speed and velocity of a particular movement. In physics, speed is considered a **scalar**—a number without direction. For example, 50 mph can completely describe the speed of a car, but it is an insufficient description of the car's velocity (a vector quantity) as no direction (for instance North, East, etc.) is indicated.

Vector Addition

Let $\vec{A}=\langle2,-4\rangle$ and $\vec{B}=\langle7,3\rangle$ as shown in Figure 7.4b. Thinking of vectors as descriptions of motion, the vector defined by $\vec{C}=\vec{A}+\vec{B}$ is the result of traveling both motions, one after the other. Vectors like $\vec{C}$ defined as the sum or difference of other vectors are called **resultants.**

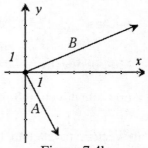

Figure 7.4b

Vectors $\vec{A}$ and $\vec{B}$ do not have to begin at the origin to describe their displacement. As long as a vector has the same magnitude and direction as $\vec{A}$, it can be placed anywhere in the coordinate plane and is considered to be equal to $\vec{A}$.

Because the motion of $\vec{C}$ is defined to be the motion along $\vec{A}$ followed by $\vec{B}$, draw $\vec{A}$ from the origin and $\vec{B}$ from the end of $\vec{A}$. Because vector addition is commutative, drawing $\vec{B}$ followed by $\vec{A}$ is exactly equivalent. The combination of the two different paths used to determine $\vec{C}$ forms a parallelogram (Figure 7.3c). Using either interpretation, the resultant, $\vec{C}$ (in Figure 7.4c), is the vector with the same *net* effect as the combined motions of $\vec{A}$ and $\vec{B}$, and is a diagonal of this parallelogram.

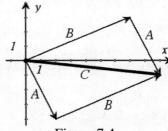

Figure 7.4c

By inspection, the component form of $\vec{C}$ is $\langle9,-1\rangle$, suggesting the algebraic approach to adding vectors: just add the corresponding components.

$$\vec{C}=\vec{A}+\vec{B}=\langle2,-4\rangle+\langle7,3\rangle=\langle2+7,-4+3\rangle=\langle9,-1\rangle.$$

Example 1:

If $\vec{C}=[8,\angle30°]$ and $\vec{D}=[4,\angle-90°]$, compute $\vec{E}=\vec{C}+\vec{D}$.

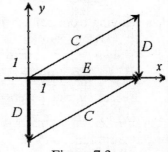

Figure 7.3c geometrically adds $\vec{C}$ and $\vec{D}$. Because $\vec{C}$, $\vec{D}$, and $\vec{E}$ form a $30°-60°-90°$ triangle, trigonometry gives $\vec{E}=[4\sqrt{3},\angle0°]=\langle4\sqrt{3},\ 0\rangle$.

Figure 7.3c

This example was unusually simple because of the $30°-60°-90°$ relationship. In more general problems, first convert to component forms, $\vec{C}=\langle4\sqrt{3},4\rangle$ and $\vec{D}=\langle0,-4\rangle$. Then, $\vec{E}=\langle4\sqrt{3},4\rangle+\langle0,-4\rangle=\langle4\sqrt{3},0\rangle$.

Multiplication of a Vector by a Scalar

Multiplication of a vector by a constant is much simpler. If $\vec{B}=\langle 7,3\rangle$, then $3\vec{B}=\vec{B}+\vec{B}+\vec{B}=\langle 7,3\rangle+\langle 7,3\rangle+\langle 7,3\rangle=\langle 21,9\rangle$. Another approach is to multiply each component by the scalar: $3\vec{B}=3\langle 7,3\rangle=\langle 3\cdot 7, 3\cdot 3\rangle=\langle 21,9\rangle$. Geometrically, adding multiples of a single vector is a dilation of the vector; it changes the magnitude, but not the direction of the vector, so $3\vec{B}=3\langle 7,3\rangle\approx 3\left[\sqrt{58},\angle 0.405\right]=\left[3\sqrt{58},\angle 0.405\right]$. Similarly, $-\vec{B}=\langle -7,-3\rangle$, a vector of the same magnitude as $\vec{B}$, pointing in the exact opposite direction.

Example 2:

Any vector with magnitude one unit is called a **unit vector,** denoted by a carat symbol such as $\hat{v}$. Find a unit vector parallel to $\vec{v}=\langle 3,4\rangle$.

Because $|\langle 3,4\rangle|=5$, dilating the given vector by 1/5 will make it a unit vector: $\frac{1}{5}\langle 3,4\rangle=\left\langle\frac{3}{5},\frac{4}{5}\right\rangle$.

Multiplying by $-\frac{1}{5}$ also would have worked because $-\frac{1}{5}\langle 3,4\rangle=\left\langle-\frac{3}{5},-\frac{4}{5}\right\rangle$ is a unit vector in the opposite direction from $\langle 3,4\rangle$, but still parallel.

Unit vectors parallel to the coordinate axes have special names. $\langle 1,0\rangle$ is parallel to the *x*-axis and is designated $\hat{i}$. Likewise, $\hat{j}=\langle 0,1\rangle$ is a unit vector parallel to the *y*-axis. Using this notation, $\langle 7,-3\rangle$ and $7\hat{i}-3\hat{j}$ are equivalent.

This concludes the discussion of multiplying vectors by scalars. There are two ways to multiply a vector by another vector—a dot product and a cross product—and each can be computed two different ways as discussed in the next two sections.

Applications
Example 3:

Find a vector representation of the line containing the points $A=(9,1)$ and $B=(3,-4)$.

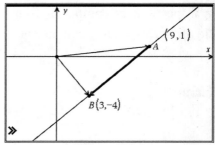

While a point-slope form of a line is easy to derive, think from a vector perspective. If O is the origin, then $\overrightarrow{OA}=\langle 9,1\rangle$ is the displacement vector from the origin to point A. Let $\overrightarrow{AB}$ be the displacement vector from point A to point B, so $\overrightarrow{AB}=\langle -6,-5\rangle$.

Moving from point A by any scalar multiple of $\overrightarrow{AB}$ is equivalent to moving parallel to the line to any other point on the line. (shown in Figure 7.3d). That means $\forall t\in\mathbb{R}$, $\overrightarrow{OA}+t\cdot\overrightarrow{AB}=\langle 9,1\rangle+t\cdot\langle -6,-5\rangle$ defines every point on the line.

Figure 7.3d

Because t can be positive or negative, this defines movement along the line in either direction, so $\overrightarrow{OA}+t\cdot\overrightarrow{AB}$ is a vector definition of $\overrightarrow{AB}$.

Algebraically, $\overrightarrow{OA}+t\cdot\overrightarrow{AB}=\langle 9,1\rangle+t\cdot\langle -6,-5\rangle=\langle 9-6t,\ 1-5t\rangle$, which can be rewritten as $\begin{cases}x=9-6t\\y=1-5t\end{cases}$, a parametric definition of $\overrightarrow{AB}$, confirming the vector-parametric connection.

Example 4:

Use the parametric equations from Example 3 to find a Cartesian equation for line $\overrightarrow{AB}$.

Given $\begin{cases} x = 9 - 6t \\ y = 1 - 5t \end{cases}$, solve for t in both the x- and y- expressions and set the t-expressions equal to each

other. This results in the point-slope form of the equation for this line.

$$t = \frac{y-1}{-5} = \frac{x-9}{-6} \Rightarrow y - 1 = \frac{-5}{-6}(x-9)$$

Note that the slope of the equation can be derived from the components of the change or displacement vector $\overrightarrow{AB} = \langle \Delta x, \Delta y \rangle = \langle -6, -5 \rangle$.

Three-Dimensional Vectors

While there are several ways to write linear equations in the xy-plane, vectors are the only option when working in three or more dimensions. This is because linear functions in higher dimensions, e.g. $x + 2y - 3z = 6$, define planes or other geometric shapes, not lines. Thankfully, the effort required to define lines in higher dimensions is barely greater than that required for 2-dimensional vector definitions. Similar to their 2-dimensional counterparts, the three-dimensional unit vector parallel to the x-axis is $\hat{i} = \langle 1, 0, 0 \rangle$, while, $\hat{j} = \langle 0, 1, 0 \rangle$ is parallel to the y-axis and $\hat{k} = \langle 0, 0, 1 \rangle$ is parallel to the z-axis.

Example 5:

Find vector and parametric representations of the line segment containing $A = (0, 9, 1)$ and $B = (3, -4, 7)$.

If $O = (0, 0, 0)$ is the 3-dimensional origin, then, $\overrightarrow{OA} = \langle 0, 9, 1 \rangle$ is the position vector to A, and $\overrightarrow{AB} = \langle 3, -13, 6 \rangle$ is a vector along the given line. Therefore, $\forall t \in \mathbb{R}$, a vector definition of this line is

$$\overrightarrow{OA} + t \cdot \overrightarrow{AB} = \langle 0, 9, 1 \rangle + t \cdot \langle 3, -13, 6 \rangle = \langle 3t, \ 9 - 13t, \ 1 + 6t \rangle.$$

This is equivalent to the parametric definition $\begin{cases} x = 3t \\ y = 9 - 13t \\ z = 1 + 6t \end{cases}$. To transform this into a definition for

segment $\overline{AB}$, recall that the starting point for this definition was point A and adding $\overrightarrow{AB}$ once resulted in moving along the line to point B. So, using the algebraic definitions above with the restriction $t \in [0, 1]$ works because t is precisely the number of additions of $\overrightarrow{AB}$ needed to move between the desired points.

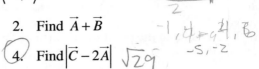

Problems for 7.4:

Exercises

Questions 1-7, use vectors $\vec{A} = \langle 2,3 \rangle$, $\vec{B} = [4,\pi]$ and $\vec{C} = \langle -1,4 \rangle$.

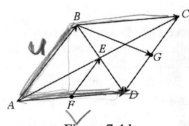

1. Find $\vec{A} + \vec{C}$ 2. Find $\vec{A} + \vec{B}$

   $-1,4 \to 4,8$
   $-5,-2$

3. Find $3\vec{B}$ $[12,\pi]$ 4. Find $|\vec{C} - 2\vec{A}|$ $\sqrt{29}$

5. Write $\vec{A}$ in geometric form $[\sqrt{13}, \tan\theta = \frac{3}{2}]$ 6. Write $\vec{B}$ in component form.

7. Find two different unit vectors parallel to $\vec{A}$.
   $\langle 6,9 \rangle$ $\langle 18,27 \rangle$

In Figure 7.4d, $ABCD$ is a parallelogram, vector $\overrightarrow{AB} = \vec{u}$ and vector $\overrightarrow{AD} = \vec{v}$. For questions 8-11, find each vector in terms of $\vec{u}$ and $\vec{v}$.

8. $\overrightarrow{AC}$ $\vec{U} + \vec{V}$

9. $\overrightarrow{BD}$

10. $\overrightarrow{FE}$, where F is the midpoint of $\overrightarrow{AD}$ $\frac{1}{2}|\vec{v}|$

11. $\overrightarrow{BG}$, where G is the midpoint of $\overrightarrow{CD}$

Figure 7.4d

Write a vector representation for each situation described in questions 12-14.

12. The line through $(6,-9)$ and $(-3,-5)$.

13. The line through $(-1,-2)$ and $(4,7)$.

14. The line segment through $(-1,-2)$ and $(9,16)$.

15. Write parametric equations for each of the lines in the previous set of problems.

16. Write a vector representation of the line through the points $\vec{A} = 3\hat{i} + 5\hat{j} - \hat{k}$ and $\vec{B} = -4\hat{i} + 4\hat{j} + 2\hat{k}$.

Explorations

Identify each statement in questions 17-21 as true or false. Justify your conclusion.

17. Any vector can be written in geometric or component form.

18. All vectors are equally easy to add in geometric or component form.

19. If $\hat{b}$ is any unit vector, then $|\hat{b}| = 1$.

20. A vector with all negative components has a negative magnitude.

21. If $|\vec{c}| = |\vec{d}|$ for any two vectors $\vec{c}$ and $\vec{d}$, then $\vec{c} = \vec{d}$.

   $-13x = 10$
   $x = \frac{10}{\sqrt{13}}$ $\frac{7}{\sqrt{13}} = \frac{7}{10}$
   $-\frac{30}{\sqrt{13}}$

22. Let $\vec{A} = \langle 3, m, -13 \rangle$ and $\vec{B} = \langle -k, 7, 10 \rangle$. What are the values of k and m if $\vec{A} \parallel \vec{B}$?

   $-91 = 10m$

   $m = -\frac{70}{13}$ $k = \frac{30}{13}$

23. A particle moves along the line $y = 7x - 13$, and is at $(-1, -20)$ at time $t = 3$.

    A. Write a displacement vector describing the particle's position at any time t.

    B. Write an expression for the distance between the particle's position and the origin at any time t.

    C. When is the particle closest to the origin?

    D. What is the closest the particle ever gets to the origin?

    E. What is the position vector of the particle when it is closest to the origin?

24. A particle moves along the parabola $y = x^2$ so that its x-coordinate is given by $x(t) = 2t - 1$.

    A. Write a displacement vector describing the particle's position at any time t.

    B. Use part A to write a parameterization of $y = x^2$.

    C. When is the particle at the origin?

    D. Write a displacement vector for the net change in the particle's position over $t \in [0, 4]$.

    E. Determine the magnitude of the vector found in part D. Is this magnitude more or less than the actual distance traveled by the particle over $t \in [0, 4]$? Explain.

    F. At what time is the particle closest to $(3, 4)$?

    G. What is the minimum distance between $y = x^2$ and $(3, 4)$, and what are the coordinates of the point on $y = x^2$ at which this minimum distance is achieved?

25. David pulls on an object with a force of 40 lbs at an angle of $40°$ from the horizontal. Hannah pulls on the same object with a force of 62 lbs at an angle of $-25°$. Find the magnitude and direction of the resultant force.

26. An airplane is flying West at a speed of 300 mph. It encounters a tailwind of 20 mph in the direction $50°$ West of North. Find the resultant speed and direction of the airplane.

27. A boat is traveling South at 10 mph on a river where the current is flowing $30°$ East of South at 2 mph. Find the actual speed and direction of the boat.

7.5: Dot Products & Work

Mathematics is the most exact science, and its conclusions are capable of absolute proof. But this is so only because mathematics does not attempt to draw absolute conclusions. All mathematical truths are relative, conditional.

– Charles P. Steinmetz in E. T. Bell *Men of Mathematics*, 1937

A **vector projection** is the shadow cast by one vector onto another from a light source arriving perpendicular to the vector receiving the shadow. It is also the portion of one vector that moves in the direction of a second vector. Figure 7.5a shows this relationship where $\overrightarrow{B_A}$ is the vector projection of $\vec{B}$ onto $\vec{A}$ and θ is the angle between the vectors. Right triangle trigonometry gives $\left|\overrightarrow{B_A}\right|$.

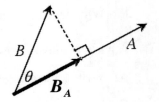

$$\cos\theta = \frac{\left|\overrightarrow{B_A}\right|}{\left|\vec{B}\right|} \quad\Rightarrow\quad \left|\overrightarrow{B_A}\right| = \left|\vec{B}\right|\cos\theta$$

Figure 7.5a

While seemingly unnecessary, multiplying the numerator and denominator by $\left|\vec{A}\right|$ gives another connection.

$$\left|\overrightarrow{B_A}\right| = \left|\vec{B}\right|\cos\theta = \frac{\left|\vec{A}\right|\left|\vec{B}\right|\cos\theta}{\left|\vec{A}\right|}$$

The numerator of this expression is defined to be the **dot product** between vectors $\vec{A}=\langle x_a, y_a\rangle$ and $\vec{B}=\langle x_b, y_b\rangle$, and it is written $\vec{A}\cdot\vec{B}$. Given its origin in vector projections. As it involves angle measures and vector lengths, this is also called the **geometric definition of the dot product**: $\vec{A}\cdot\vec{B}=\left|\vec{A}\right|\left|\vec{B}\right|\cos\theta$.

To determine an expression for the actual vector $\overrightarrow{B_A}$, remember that $\overrightarrow{B_A}$ is parallel to $\vec{A}$. Converting $\vec{A}$ to a unit vector and stretching by $\left|\overrightarrow{B_A}\right|$ establishes the desired result.

$$\overrightarrow{B_A} = \frac{\vec{A}}{\left|\vec{A}\right|}\left|\overrightarrow{B_A}\right| = \frac{\vec{A}}{\left|\vec{A}\right|}\cdot\frac{\vec{A}\cdot\vec{B}}{\left|\vec{A}\right|} = \frac{\vec{A}\cdot\vec{B}}{\left|\vec{A}\right|^2}\vec{A}$$

Example 1:
If $\vec{A}=\langle x_a, y_a\rangle$ and $\vec{B}=\langle x_b, y_b\rangle$, an **algebraic definition of the dot product** is $\vec{A}\cdot\vec{B}=x_a x_b + y_a y_b$. Prove that this definition is equivalent to the geometric definition derived from vector projections.

Consider the triangle formed by $\vec{A}$ and $\vec{B}$ (Figure 7.5b). From the Law of Cosines, $\left|\vec{A}\right|^2 + \left|\vec{B}\right|^2 - 2\left|\vec{A}\right|\left|\vec{B}\right|\cos\theta = \left|\vec{A}-\vec{B}\right|^2$. Therefore,

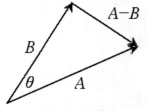

$$-2\left|\vec{A}\right|\left|\vec{B}\right|\cos\theta = \left|\langle x_a - x_b, y_a - y_b\rangle\right|^2 - \left|\langle x_a, y_a\rangle\right|^2 - \left|\langle x_b, y_b\rangle\right|^2$$

$$= \sqrt{(x_a - x_b)^2 + (y_a - y_b)^2}^{\,2} - \sqrt{x_a^2 + y_a^2}^{\,2} - \sqrt{x_b^2 + y_b^2}^{\,2}$$

$$= \left(x_a^2 - 2x_a x_b + x_b^2\right) + \left(y_a^2 - 2y_a y_b + y_b^2\right) - \left(x_a^2 + y_a^2\right) - \left(x_b^2 + y_b^2\right)$$

$$= -2x_a x_b - 2y_a y_b$$

So, $\boxed{\vec{A}\cdot\vec{B} = \left|\vec{A}\right|\left|\vec{B}\right|\cos\theta = x_a x_b + y_a y_b}$

Figure 7.5b

QED

Because its definitions produce scalar quantities, the dot product is also called the **scalar product** of vectors. This should not be confused with vector _multiplication by a scalar_.

By its format, the geometric dot product obviously applies to vectors of any dimensions. To extend the algebraic definition to higher dimensions, simply add the products of all corresponding components. Because any two vectors of any dimension define a two-dimensional plane and a two-dimensional coordinate system can be imposed on any plane, this two-dimensional proof of the equivalence of the dot product definitions suffices for vectors of any dimension.

Example 2:
Find the angle between $\vec{A} = \langle -3, 6 \rangle$ and $\vec{B} = \langle -2, -1 \rangle$.

> Using both definitions of the dot product, $\vec{A} \cdot \vec{B} = -3 \cdot -2 + 6 \cdot -1 = \sqrt{45}\sqrt{5} \cos\theta$. So, $\cos\theta = 0$ which means the angle between them is $\pi/2$.

Since $\cos\theta = 0$, the dot product of perpendicular vectors is 0. Perpendicular vectors are called **normal vectors**.

Work
A final application of the dot product is **work**. In physics, work equals the magnitude of the force in the direction of displacement of an object times the magnitude of the displacement. Work problems are elementary when the force applied is parallel to the displacement. It is only slightly more complicated when the force and displacement are not parallel.

Example 3:
Let $\vec{A}$ and $\vec{B}$ be defined as shown in Figure 7.4b. Determine the work done by force $\vec{B}$ to displace an object along $\vec{A}$.

> Because $\vec{B_A}$ is the vector projection of $\vec{B}$ along $\vec{A}$, $\vec{B_A}$ represents the portion of the force in the direction of the displacement and the work done is therefore the product of the magnitudes of the parallel vectors:

$$\text{Work} = \left|\vec{B_A}\right|\left|\vec{A}\right| = \frac{\vec{A} \cdot \vec{B}}{\left|\vec{A}\right|}\left|\vec{A}\right| = \vec{A} \cdot \vec{B}.$$

Therefore, the work done by force $\vec{F}$ on an object which moves along the vector $\vec{d}$ is $\vec{F} \cdot \vec{d}$.

Problems for 7.5:

Exercises

For questions 1-6, find the dot product of each of the given pairs of vectors.

1. $\langle 2, 3 \rangle$ and $\langle 1, 7 \rangle$ 2. $\langle -1, 2 \rangle$ and $\langle -3, -5 \rangle$

3. $\langle -3, 0 \rangle$ and $\langle 0, 9 \rangle$ 4. $\langle 5, 6 \rangle$ and $\langle -10, -12 \rangle$

5. $\langle 1, 2, 7 \rangle$ and $\langle 1, -2, 4 \rangle$ 6. $\langle 3, 4, -8 \rangle$ and $\left\langle 2, \frac{1}{2}, 1 \right\rangle$

For questions 7-10, find the angle between each of the given pairs of vectors.

7. $\langle 2,3 \rangle$ and $\langle 1,7 \rangle$

8. $\langle 3,4,-8 \rangle$ and $\langle 2,\frac{1}{2},1 \rangle$

9. $\langle -1,5,3 \rangle$ and $\langle 1,0,-3 \rangle$

10. $\langle 2,-3 \rangle$ and $\langle -4,6 \rangle$

11. Which pairs of vectors are normal?

A. $\langle 1,2 \rangle$ and $\langle -6,2 \rangle$

B. $\langle 1,-5,7,9 \rangle$ and $\langle 3,-4,-2,-1 \rangle$

12. For what value(s) of k, if any, are $\vec{u} = \langle k,2,k \rangle$ and $\vec{v} = \langle k,3,-5 \rangle$ perpendicular?

13. For what value(s) of k, if any, are $\vec{u} = \langle k,2,k \rangle$ and $\vec{v} = \langle k,3,-5 \rangle$ parallel?

14. Determine if the points $A(1,2,-2)$, $B(2,-3,4)$ and $C(0,6,1)$ define the vertices of a right triangle in three-dimensional space.

15. Find the vector projection of a force $\vec{F} = \langle 4,-3 \rangle$ in the direction of $\vec{d} = \langle 5,1 \rangle$.

16. A block is being pulled along the floor by a force of 50 lbs acting at an angle of $38°$ to the floor. What is the component of the force acting in the horizontal direction?

17. How much work is done in pushing a 60 lb suitcase up a 10 ft ramp at an angle of $25°$ to the floor?

18. How much work is done by force $\langle 2,3 \rangle$ when an object moves along $\langle 1,7 \rangle$?

19. How much work is done by force $\langle 2,3 \rangle$ when an object moves along $\langle 2,3 \rangle$? Explain.

20. How much work is done by force $\langle 2,3 \rangle$ when an object moves along $\langle -3,2 \rangle$? Explain.

Explorations

Identify each statement in questions 21-25 as true or false. Justify your conclusion.

21. For any vectors $\vec{A}$ and $\vec{B}$, $\vec{A} \cdot \vec{B} = \vec{B} \cdot \vec{A}$.

22. For any constant k and any vectors $\vec{A}$ and $\vec{B}$, $(k\vec{A}) \cdot \vec{B} = \vec{A} \cdot (k\vec{B}) = k(\vec{A} \cdot \vec{B})$.

23. For any vectors $\vec{A}$ and $\vec{B}$, $\vec{A} \cdot \vec{A} = |\vec{A}|^2$.

24. For any vectors $\vec{A}$, $\vec{B}$, and $\vec{C}$, $\vec{A} \cdot (\vec{B} + \vec{C}) = \vec{A} \cdot \vec{B} + \vec{A} \cdot \vec{C}$.

25. Negative work happens when an object's displacement vector points in the opposite direction of the vector projection of the force vector onto the displacement vector.

26. $\forall \theta \in \mathbb{R}$, determine the angle between vectors $\vec{A} = \langle \sin(2\theta), 4\sin^2 \theta \rangle$ and $\vec{B} = \langle \sin(2\theta), -\cos^2 \theta \rangle$.

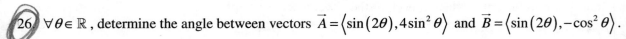

27. A right triangle's hypotenuse has endpoints $A(1,2,-3)$ and $B(-5,0,7)$. As precisely as possible, describe the set of all points, (x,y), for which $C(x,y,0)$ is the third vertex of the triangle.

7.6: Cross Products

Mathematics, as much as music or any other art, is one of the means by which we rise to a complete self-consciousness. The significance of mathematics resides precisely in the fact that it is an art; by informing us of the nature of our own minds it informs us of much that depends on our minds.
— John William Navin Sullivan in *Aspects of Science*, 1925.

Its meaning and application will be explained momentarily, but this section begins with the definition and notation of the other product of two vectors. The **cross product** of vectors $\vec{A} = \langle x_a, y_a, z_a \rangle$ and $\vec{B} = \langle x_b, y_b, z_b \rangle$ is written $\vec{A} \times \vec{B}$, and its algebraic definition is:

$$\vec{A} \times \vec{B} = \langle y_a z_b - y_b z_a, \ z_a x_b - z_b x_a, \ x_a y_b - x_b y_a \rangle$$

While the dot product produces a scalar, the cross product is a vector. For this reason, the cross product is also called the **vector product**.

Geometrically, $\vec{A} \times \vec{B}$ is a vector of magnitude $\left| \vec{A} \times \vec{B} \right|$ normal to the plane defined by $\vec{A}$ and $\vec{B}$. The direction of $\vec{A} \times \vec{B}$ is determined by the **right-hand rule**. If the right hand is pointed in the direction of $\vec{A}$ and the fingers are curled in the direction of $\vec{B}$, the thumb will point in the direction of $\vec{A} \times \vec{B}$ (Figure 7.6a).

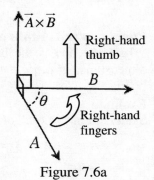

The magnitude of the cross product is defined geometrically by
$$\left| \vec{A} \times \vec{B} \right| = \left| \vec{A} \right| \left| \vec{B} \right| \sin \theta$$
where θ is the angle between the two vectors.

Figure 7.6a

If $\vec{A} \| \vec{B}$, then $\theta = 0$ or $\theta = \pi$, so $\left| \vec{A} \times \vec{B} \right| = 0$. Because its magnitude is zero, the cross product of any two parallel vectors is $\vec{A} \times \vec{B} = \vec{0}$, a zero vector.

Example 1:
If θ is the angle between vectors $\vec{A}$ and $\vec{B}$, then find an expression for the area of the parallelogram formed using $\vec{A}$ and $\vec{B}$ as edges (Figure 7.6b).

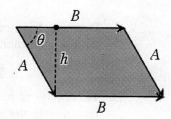

The area of a parallelogram is base times height. In Figure 7.6b, if $\left| \vec{B} \right|$ is the magnitude of the base and h is the height, $\sin \theta = \dfrac{h}{\left| \vec{A} \right|} \Rightarrow h = \left| \vec{A} \right| \sin \theta$, so

Figure 7.6b

base $\cdot$ height $= \left| \vec{B} \right| \left| \vec{A} \right| \sin \theta = \left| \vec{A} \times \vec{B} \right|$. The area of the parallelogram formed using $\vec{A}$ and $\vec{B}$ is the magnitude of $\vec{A} \times \vec{B}$.

Example 2:

Find the volume of the parallelepiped formed using vectors $\vec{A}$, $\vec{B}$, and $\vec{C}$ as edges (Figure 7.6c).

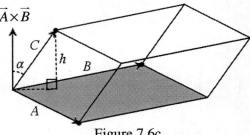

Let h be the height of the parallelepiped. Because both are normal to the base $h \| (\vec{A} \times \vec{B})$. If α is the angle between h and $\vec{C}$, then α is also the angle between $\vec{C}$ and $\vec{A} \times \vec{B}$ because parallel lines create congruent alternate interior angles. The volume of a parallelepiped is base area times height. From Example 1, the base area is $\left| \vec{A} \times \vec{B} \right|$, so the volume is $\left| \vec{A} \times \vec{B} \right| \cdot h = \left| \vec{A} \times \vec{B} \right| \left| \vec{C} \right| \cos \alpha$. Because α is the angle between $\vec{C}$ and $\vec{A} \times \vec{B}$, this volume is also equivalent to a dot product:

$$\text{volume} = \left| \vec{A} \times \vec{B} \right| \left| \vec{C} \right| \cos \alpha = \left(\vec{A} \times \vec{B} \right) \cdot \vec{C}.$$

Figure 7.6c

The result $\left(\vec{A} \times \vec{B} \right) \cdot \vec{C}$ sometimes is called a **triple product** and is denoted $\left[\vec{A},\ \vec{B},\ \vec{C} \right]$.

Equations and graphs of planes

Where the line is the basic 2-dimensional linear function, the plane is the basic linear function in three dimensions. $\forall\, a,b,c,d \in \mathbb{R}$, the graph of $d = ax + by + cz$ is a plane.

Example 3:

Sketch graphs of each of the following planes.

      A. $x + 2y + 3z = 6$ B. $z = 2x - 5y + 10$

      C. $y = x$ D. $z = -1$

    Lines are easy to graph using their intercepts; so are planes. The intercepts of $x + 2y + 3z = 6$ are $(6,0,0)$, $(0,3,0)$, and $(0,0,2)$. A standard practice for graphing planes is to draw and shade in the triangle formed by the three intercepts. Using this, Figure 7.6d shows a graph of $x + 2y + 3z = 6$. The intercepts of graph B are $(-5,0,0)$, $(0,2,0)$, and $(0,0,10)$ (Figure 7.6e).

    C.   Because z is not mentioned in this equation, ordered pairs on this plane only need satisfy the equality of their x- and y-coordinates. The result is a plane perpendicular to the xy-plane containing the line $y = x$ in the xy-plane (Figure 7.6f).

    D.   This final example only requires a z-component of -1. The result is a horizontal plane intersecting the z-axis at -1 (Figure 7.6g). Therefore, $z = -1$ is parallel to and one unit below the xy-plane.

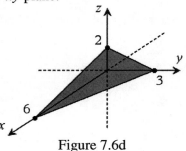

Figure 7.6d

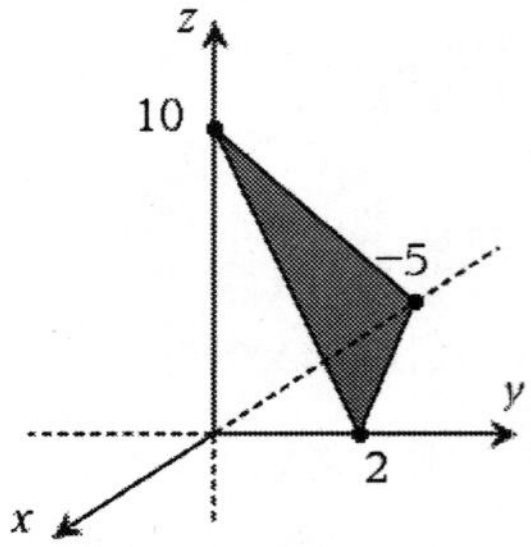

Figure 7.6e

Harrow & Merchant © 2010

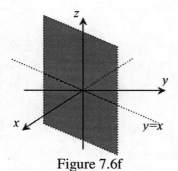

Figure 7.6f

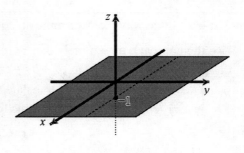

Figure 7.6g

Example 4:

Find two vectors in the plane $x + 2y + 3z = 6$. Then take the cross product of those vectors and determine its relationship to the given equation.

There are an infinite number of vectors in any given plane. Because vectors can be used to show movement from one point to another, an easy way to find them is to use ordered triples from the function. Three such points for $x + 2y + 3z = 6$ are $(6,0,0)$, $(0,3,0)$, and $(1,1,1)$. The vector from $(6,0,0)$ to $(0,3,0)$ is $\langle -6,3,0 \rangle$. The vector from $(0,3,0)$ to $(1,1,1)$ is $\langle 1,-2,1 \rangle$.

The cross product, $\langle -6,3,0 \rangle \times \langle 1,-2,1 \rangle$, can be computed via the algebraic definition or with a CAS (Figure 7.6h).[7] Either way, the result is $\langle -6,3,0 \rangle \times \langle 1,-2,1 \rangle = \langle 3,6,9 \rangle$. Because a cross product is perpendicular to the plane containing the two given vectors, $\langle 3,6,9 \rangle$ must be perpendicular to $x + 2y + 3z = 6$. Such a vector is called a **normal vector to the plane**.

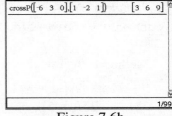

Figure 7.6h

Normal vector $\langle 3,6,9 \rangle$ is equivalent to $3 \cdot \langle 1,2,3 \rangle$, so $\langle 1,2,3 \rangle$ is another normal vector to the plane. Notice that the components of $\langle 1,2,3 \rangle$ are the respective coefficients of x, y, and z from $x + 2y + 3z = 6$.

The result of Example 4 is not a coincidence. The cross product of non-parallel vectors from a plane always produces a normal vector to the plane, and that normal vector is always some scalar multiple of the coefficients of x, y, and z in an equation of the plane.

PROOF: Let $\langle a,b,c \rangle$ be a normal vector to some plane, (x_0, y_0, z_0) be any specific point in the plane, and let (x,y,z) be a random point in the plane. Because (x,y,z) and (x_0, y_0, z_0) are points in the plane, then $\langle x - x_0, y - y_0, z - z_0 \rangle$ is a vector in the plane. Therefore, the angle between $\langle a,b,c \rangle$ and $\langle x - x_0, y - y_0, z - z_0 \rangle$ is $\pi/2$, making $\langle a,b,c \rangle \cdot \langle x - x_0, y - y_0, z - z_0 \rangle = 0$. This is equivalent to $a(x - x_0) + b(y - y_0) + c(z - z_0) = 0$. Collecting the constant terms to the right side of the equation completes the connection above:

$$ax + by + cz = ax_0 + by_0 + cz_0$$

The right side is a constant and is equivalent to the value of d in the equation: $ax + by + cz = d$. Also notice that $d = ax_0 + by_0 + cz_0 = \langle a,b,c \rangle \cdot \langle x_0, y_0, z_0 \rangle$.

---

[7] The **CrossP** command is at (menu) → ⑦:**Matrix & Vector** → ©:**Vector** → ②:**Cross Product**. Note that vectors are entered in square brackets with their components separated by commas.

There are at least three ways to determine an equation for a plane:

I. Use the approach of Example 4: A cross product between two vectors in the plane gives a normal to the plane, $\langle a,b,c \rangle$. Then $ax+by+cz=0$ passes through the origin and is *parallel* to the desired plane. To get a final answer, translate this plane from origin to any of the known points in the plane, (x_0, y_0, z_0), ending with a point slope form of a plane,

$$a(x-x_0)+b(y-y_0)+c(z-z_0)=0.$$

II. Use a cross product between two vectors in the plane to find normal vector $\langle a,b,c \rangle$. If $\langle x_0, y_0, z_0 \rangle$ is the position vector for some specific point in the plane, then $d=\langle a,b,c \rangle \cdot \langle x_0, y_0, z_0 \rangle$ gives the final constant for an equation of the plane in the form $ax+by+cz=d$.

III. Another form of an equation for a plane is $ax+by+cz=1$. Because there are only three parameters, a tedious approach substitutes the ordered triples and solves the resulting system.

Example 5:

Find an equation for the plane containing the points $A(5,-8,6)$, $B(-7,3,0)$, and $C(2,-1,-1)$.

The vector from A to B is $\langle -12,11,-6 \rangle$. Likewise $\overrightarrow{AC}=\langle -3,7,-7 \rangle$. Therefore, $\langle -12,11,-6 \rangle \times \langle -3,7,-7 \rangle = \langle -35,-66,-51 \rangle$ defines coefficients of a potential equation for the desired plane. Using method II, an initial equation is $-35x+-66y+-51z=d$, and d can be found using any of the given ordered triples: $d=\langle -35,-66,-51 \rangle \cdot \langle -7,3,0 \rangle = 47$. Therefore, two equations for the plane are
1) $-35x+-66y+-51z=47$ and 2) $35x+66y+51z=-47$.

Method I suggests that no additional work is required after the computation of $\langle -35,-66,-51 \rangle$. Just slide the parallel plane, $-35x-66y-51z=0$, to one of the given ordered triples to get
3) $-35(x-5)-66(y+8)-51(z-6)=0$, or 4) $-35(x+7)-66(y-3)-51z=0$, or
5) $-35(x-2)-66(y+1)-51(z+1)=0$. While not as "clean-looking" as Method II, this approach dodges the additional dot product.

Method III produces a system of three equations: $\begin{cases} 1=5a-8b+6c \\ 1=-7a+3b \\ 1=2a-b-c \end{cases}$.

Solving gives $a=\dfrac{-35}{47}$, $b=\dfrac{-66}{47}$, and $c=\dfrac{-51}{47}$ (Figure 4.6i).

Therefore a final equation of the plane is 6) $\dfrac{-35}{47}x-\dfrac{66}{47}y-\dfrac{51}{47}z=1$.

All six equations represent the same plane, so they must be equivalent.

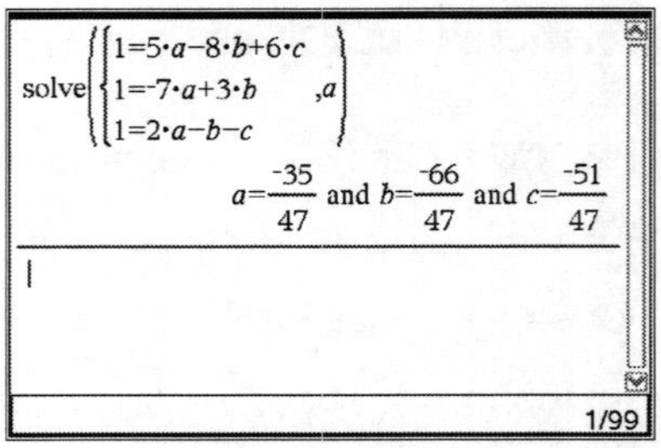

Figure 7.6i

Problems for 7.6:

Exercises

For questions 1-4, sketch a graph of each plane.

1. $4x - y + 2z = 8$

2. $z = -3x + 2y - 7$

3. $z = 2y - 1$

4. $x = \pi$

For questions 5-6, find the cross product of each pair of vectors.

5. $\langle 1, -1, 5 \rangle$ and $\langle 2, 4, 3 \rangle$

6. $\langle 0, 4, 7 \rangle$ and $\langle 5, -2, 8 \rangle$

7. Let $\vec{u} = \langle 1, a, 5 \rangle$ and $\vec{v} = \langle b, -5, a^2 \rangle$. Find values for a and b if $\vec{u} \times \vec{v} = \langle 52, 1, -11 \rangle$.

8. Find a vector perpendicular to the plane containing the vectors $\langle 8, 3, -2 \rangle$ and $\langle -2, 1, 1 \rangle$.

9. Find a vector perpendicular to the plane containing the points $A(1, 2, -2)$, $B(2, -3, 4)$ and $C(0, 6, 1)$.

10. Prove that all six equations for the plane in Example 5 are equivalent.

Find an equation for each plane described in questions 11-14.

11. Contains $(1, -5, 7)$, $(9, 6, 7)$, and $(0, 0, 3)$.

12. The x, y, and z-intercepts are 3, 4, -9, respectively.

13. Contains the point $(-2, -8, 6)$ and is parallel to $5x - 7y + 1 = 5$.

14. Contains the point $(1, -1, 7)$ and is parallel to $5x - y + 4z = 6$.

Explorations

Identify each statement in questions 15-19 as true or false. Justify your conclusion.

15. For any vectors $\vec{A}$ and $\vec{B}$, $\vec{A} \times \vec{B} = -\left(\vec{B} \times \vec{A} \right)$

16. For any vectors $\vec{A}$ and $\vec{B}$, $\left(k\vec{A} \right) \times \vec{B} = \vec{A} \times \left(k\vec{B} \right) = k \left(\vec{A} \times \vec{B} \right)$ for any constant k

17. For any vectors $\vec{A}$ and $\vec{B}$, $\vec{A} \times \vec{A} = \vec{0}$

18. For any vectors $\vec{A}$, $\vec{B}$, and $\vec{C}$, $\vec{A} \times \left(\vec{B} + \vec{C} \right) = \vec{A} \times \vec{B} + \vec{A} \times \vec{C}$

19. The equation of a plane can be found using any two points in that plane.

20. Use the right hand rule to explain why $\vec{A} \times \vec{B} = -\left(\vec{B} \times \vec{A} \right)$ is true.

For questions 20-21, find the angle between each pair of planes.

21. $x + 2y + 3z = 6$ and $z = 2x - 5y + 10$

22. $2x - 9y + 3z = -4$ and $3x + y + z = 12$

23. From Example 4, verify that using the first and third given ordered triples in the dot product computation of d would have produced the same results.

24. Given the points $A(0,1,-1)$, $B(2,-4,1)$ and $C(3,-1,1)$.

    A. Find a vector perpendicular to the plane containing A, B and C.

    B. Find the area of triangle ABC.

    C. Find all possible coordinates for a point D such that $ABCD$ would form a parallelogram.

25. Find an equation for the plane normal to $5x - y + 4z = 6$ that contains $(1,-1,7)$ and $(0,5,4)$.

26. Find k if $A = \langle 4,-k,-1 \rangle$ and $B = \langle k-4,1,1 \rangle$ form two sides of a triangle with area $\dfrac{\sqrt{269}}{2}$.

27. Prove $\left(\vec{A} \times \vec{B}\right) \cdot \vec{C} = \vec{A} \cdot \left(\vec{B} \times \vec{C}\right)$.

28. Prove $\left[\vec{A},\ \vec{B},\ \vec{C}\right] = \left[\vec{C},\ \vec{A},\ \vec{B}\right] = \left[\vec{B},\ \vec{C},\ \vec{A}\right]$.

29. Explain how a plane can be defined using 2 vectors similar to the vector definition of a line.

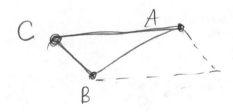

UNIT 8: EXTENSIONS OF TRIGONOMETRY

As a language, mathematics lies somewhere between the worlds that can be translated to language and those that are described by art.

– Priya Hemenway in *Divine Proportion*

Enduring Understandings:

- Defining relations and transformations in different coordinate systems offers a deeper understanding of graphs while revealing new mathematical behaviors and patterns.

- Polar coordinates sometimes offer simpler ways to analyze functions and complex numbers relative to their Cartesian equivalents, and vice versa.

- Complex numbers enable unexpected connections between the trigonometries of circles and hyperbolas.

This section assumes you have:

- Comfort with trigonometric functions, graphs and analysis.

- A basic understanding of complex numbers, especially including four-function operations.

8.1: A New Way to Define Functions

It can be of no practical use to know that Pi is irrational, but if we can know, it surely would be intolerable not to know.

        – E. C. Titchmarsh in N. Rose *Mathematical Maxims and Minims*, 1988

The common theme underlying all of the functions studied thus far is that they all graph the same way—using (x, y) coordinates in the Cartesian Coordinate System.

Another method of graphing is suggested by the alternative method of defining vectors—using magnitude and direction, $[r, \angle\theta]$, rather than by components, $\langle x, y \rangle$. Defining points using magnitude and direction is the foundation of the **Polar Coordinate System**.

Polar Coordinates

The common reference point in the polar coordinate system is called the **pole**. Its location is identical to the origin in Cartesian coordinates.

The variables used to define vectors as $\langle x, y \rangle$ and $[r, \angle\theta]$ are the same in the Cartesian and Polar coordinate systems. Like the magnitude-direction form of vectors, polar coordinates are enclosed in square brackets. Geometrically, the Cartesian coordinates form the legs of a right triangle, and the polar coordinates denote the hypotenuse and angle measure from the origin/pole (Figure 8.1a). The Pythagorean Theorem and right triangle trigonometry allow simple conversions between the coordinate systems (Figure 8.1b).

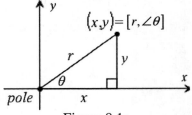

Figure 8.1a

Cartesian to Polar	Polar to Cartesian
$x^2 + y^2 = r^2$	$r\cos\theta = x$
$\dfrac{y}{x} = \tan\theta$	$r\sin\theta = y$

Figure 8.1b

Example 1:

Graph the points with polar coordinates $A = [3, \angle 75°]$, $B = [-2, \angle 45°]$, and

$$C = \left[-5, \angle\left(-\frac{\pi}{6}\right)\right].$$

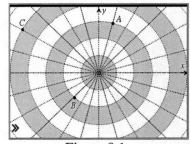

Figure 8.1c

  All three points are graphed in Figure 8.1c.
- Point A is three units away from the origin, rotated 75° from the positive *x*-axis.
- Point B has a radius of -2. Start from -2 on the *x*-axis and rotate 45° to arrive at the proper location.
- Point C has a negative radius and angle. Start at −5 on the *x*-axis and rotate $\dfrac{\pi}{6}$ radians clockwise.

Example 2:
Convert $(2,-5)$ to polar coordinates.

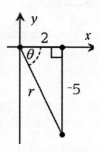

Figure 8.1d

From Figure 8.1d, $r = \sqrt{(2)^2 + (5)^2} = \sqrt{29}$ and $\tan\theta = \dfrac{-5}{2}$. Because $\tan\theta = -2.5$ for an infinite number of values of θ, there are countless polar coordinate pairs coincident with $(2,-5)$. The smallest positive value of θ is 5.093 radians and the largest negative value is -1.190, giving two potential pairs of polar coordinates: $\left[\sqrt{29}, \angle -1.190\right]$ and $\left[\sqrt{29}, \angle 5.093\right]$. The complete set of polar coordinate equivalents is $\left[\sqrt{29}, \angle(5.093 + 2\pi k)\right] \ni k \in \mathbb{Z}$.

Converting Equations between Coordinate Systems

When defining functions in polar coordinates, it is standard practice to consider θ as the independent variable and r as the dependent variable. Therefore, polar functions are typically written $r = f(\theta)$.

Example 3:
Convert each equation to polar form.

     A. $x^2 + y^2 = 1$ B. $y = 2x + 3$

       A.   The Pythagorean Theorem (Figures 8.1a and 8.1b) gives
$$x^2 + y^2 = 1 \quad \Rightarrow \quad r^2 = 1 \quad \Rightarrow \quad r = 1 \text{ or } r = -1.$$
In both cases, θ does not feature in the polar form and can assume all values. A radius of 1 and a radius of -1 both produce the unit circle when $\theta \in \mathbb{R}$ with $r = 1$ starting at $\theta = 0$ from $(1,0)$ and $r = -1$ starting from $(-1,0)$. The algebraic definition for this curve is much cleaner in polar coordinates. It is also a function in its polar form while it is not a Cartesian function.

       B.   Replacing x and y with their polar equivalents and solving for r gives
$$y = 2x + 3 \quad \Rightarrow \quad r\sin\theta = 2r\cos\theta + 3 \quad \Rightarrow \quad r(\sin\theta - 2\cos\theta) = 3 \quad \Rightarrow \quad r = \frac{3}{\sin\theta - 2\cos\theta}$$
In this case, the Cartesian form of the line is *much* cleaner than its polar counterpart.

Example 4:
Convert $r = 2\cos\theta - 2\sin\theta$ from polar to Cartesian form.

       A useful strategy when converting from polar coordinates is to make sure all trigonometric functions have a coefficient of r. Multiplying both sides by r and substituting gives
$$r = 2\cos\theta - 2\sin\theta \quad \Rightarrow \quad r^2 = 2r\cos\theta - 2r\sin\theta \quad \Rightarrow \quad x^2 + y^2 = 2x - 2y.$$
Moving the variables to one side and completing the square finishes the problem.
$$x^2 + y^2 = 2x - 2y \quad \Rightarrow \quad (x^2 - 2x) + (y^2 + 2y) = 0$$
$$\Rightarrow \quad (x^2 - 2x + 1) + (y^2 + 2y + 1) = 2$$
$$\Rightarrow \quad (x-1)^2 + (y+1)^2 = 2$$
The Cartesian form shows that this is a graph of a circle of radius $\sqrt{2}$ centered at $(1,-1)$, facts not as obvious from the polar form. Again, each coordinate system has advantages and types of functions for which it has clearer expressions.

Problems for Section 8.1:

Exercises

1. Give three different polar coordinates equivalent to the Cartesian point $(-3,3)$.

2. Four polar points are given below, but one is different. Which one is it, and why?

$$\left[3, \angle -\frac{8\pi}{5}\right] \quad \left[-3, \angle \frac{7\pi}{5}\right] \quad \left[3, \angle \frac{12\pi}{5}\right] \quad \left[-3, \angle -\frac{2\pi}{5}\right]$$

For questions 3-6, convert each Cartesian point into polar coordinates.

3. $\left(1, \sqrt{3}\right)$

4. $(5, -3)$

5. $(-1, -2)$　　　$\left[\sqrt{5}, 1.1\right]$

6. $(-5, 5)$

For questions 7-10, convert each polar point to Cartesian coordinates.

7. $[2, \angle 100°]$

8. $\left[-3, \angle -\frac{\pi}{4}\right]$　　$-4 \cdot \cos 450° = x$
$-4 \cdot \sin 450° = y$

9. $[8, \angle 5\pi]$

10. $[-4, \angle 450°]$　　$\langle 0, -4 \rangle$

For questions 11-16, convert the given Cartesian equation into polar coordinates, and if possible, into the form $r = f(\theta)$.

11. $x + y = 3$

12. $y = 4x - 1$

13. $x^2 + y^2 = 9$

14. $(x-1)^2 + (y+2)^2 = 25$

15. $\frac{x^2}{4} + \frac{y^2}{9} = 1$

16. $y = x^2$

17. Find the length of the line segment whose endpoints are $[6, \angle 50°]$ and $[-3, \angle -100°]$

For questions 18-25, convert the given polar equation to Cartesian form and classify the type of curve.

18. $r = 4$

19. $r = 8\sin\theta$

20. $r = 2\cos\theta + \sin\theta$

21. $r = 2\cos\theta - 4\sin\theta + \frac{1}{r}$

22. $r = \dfrac{1}{\sin\theta - 4\cos\theta}$

23. $r = 5\sec\theta$

24. $r = -3\csc\theta$

25. $r^2 = \dfrac{1}{\cos(2\theta)}$

Explorations

Identify each statement in questions 26-30 as true or false. Justify your conclusion.

26. $r = 5\sec\theta$ is a polar function. *yes, many θ two one r*

27. $[5, \angle 70°]$ and $[-5, \angle -70°]$ represent the same Cartesian point. *False, the first is in the I quadrant and the second is in the II*

28. Every polar function is equivalent to some Cartesian function. *false*

29. While there is exactly one Cartesian representation of every point in the coordinate plane, there is an infinite number of polar representations of every point. *true, there is an infinite # of equal angles*

30. Every Cartesian *function* is equivalent to some polar *function*. *false*

31. $\forall a \in \mathbb{R}$, a polar function of the form $r = a \cdot \cos\theta$ always graphs as a circle.

    A. Prove the above claim by converting the equation to a recognizable Cartesian equivalent.

    B. What is the area contained within any polar graph defined by $r = a\cos\theta$?

32. Example 3, part A claimed that $r = 1$ was a polar function, but its Cartesian equivalent, $x^2 + y^2 = 1$, was not a Cartesian function. Justify this claim.

33. Find an equation of a relation that is neither a polar nor a Cartesian function.

8.2: Polar Graphing I

I believe we can find answers to these questions. They may not be full and complete and perfect answers, but there at least will be initial answers, tentative answers, working answers. And each step along the way will teach us more. What is essential is that we search.

— His Highness The Aga Khan IV

Basic graphs[8]

In the polar coordinate system, one of the simplest equations is $r = k$ for $k \in \mathbb{R}$. Because θ is not explicitly defined, the equation implies that the radius remains constant regardless of the value of θ. The angle can take on all possible values while the radius is fixed at k. This describes a circle of radius k centered at the pole (Figure 8.2a).

Another easy polar equation is $r = \theta$, the Spiral of Archimedes. The variable r is a distance, so it makes sense to use the distance-related units for θ, radians, for the graph. The radius and the angle grow at the same rate, starting from the pole, so $\forall \theta \geq 0$, $r = \theta$ is a constantly widening spiral (Figure 8.2b).

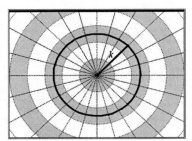

Figure 8.2a: $r = k$

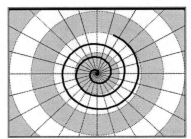

Figure 8.2b: $r = \theta$, $\theta \in [0, 20]$

These graphs clearly indicate that $r = \theta$ and $r = k$ are not Cartesian functions. They are, however, polar functions, because each polar input value (θ) corresponds to a single, predictable output value (r). The basic idea of a function appears dependent on the coordinate system in which its domain and range are defined.

Example 1:
$\forall \theta \in [0, 2\pi)$, graph $r(\theta) = \cos \theta$.

Figure 8.2c shows a table of r-values for selected θ-values.

θ	0	$\dfrac{\pi}{4}$	$\dfrac{\pi}{2}$	$\dfrac{3\pi}{4}$	π	$\dfrac{5\pi}{4}$
r	1	$\dfrac{1}{\sqrt{2}}$	0	$-\dfrac{1}{\sqrt{2}}$	-1	$-\dfrac{1}{\sqrt{2}}$

Figure 8.2c

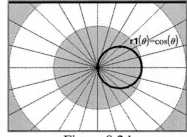

Figure 8.2d

Figure 8.2d then shows these polar points joined to form what appears to be a circle. Notice that the entire graph is completed over $\theta \in [0, \pi)$ and re-graphs over itself for $\theta \in [\pi, 2\pi)$. More about such graphical overlaps is discussed in the next section.

---

[8]To change to polar graphing mode on a TI-Nspire, start in a (⌂) → ②:**Graphs & Geometry** window. Then change to polar mode with (menu) → ③:**Graph Type** → ③:**Polar**. Equations and θ-domains then can be entered as usual.

Example 2:
Prove that the graph of the polar function from Example 1 is actually a circle.

Converting $r(\theta) = \cos\theta$ to Cartesian form accomplishes this. First,

$$r = \cos\theta \;\Rightarrow\; r^2 = r\cos\theta \;\Rightarrow\; x^2 + y^2 = x\,.$$

Then, completing the square gives

$$x^2 - x + \left(\frac{1}{2}\right)^2 + y^2 = \frac{1}{4} \;\Rightarrow\; \left(x - \frac{1}{2}\right)^2 + y^2 = \frac{1}{4}\,.$$

This is the standard form of an equation for a circle with center $\left(\frac{1}{2}, 0\right)$ and radius $\frac{1}{2}$.   *QED*

Example 3:

$\forall \theta \in [0, 2\pi)$, graph $r = \dfrac{2}{\cos\theta - \sin\theta}$ and verify the shape of the graph.

As with Example 1, a table of sample θ-values for $\theta \in \left(\dfrac{\pi}{4}, \dfrac{5\pi}{4}\right)$ suggests the function's pattern (Figure 8.2e).

θ	$\frac{\pi}{4}^{+}$	$\frac{\pi}{2}$	$\frac{3\pi}{4}$	π	$\frac{5\pi}{4}^{-}$
r	$-\infty$	-2	$-\sqrt{2}$	-2	$+\infty$

Figure 8.2e

A graph of the non-infinite points (Figure 8.2f) appears linear at an angle of $\theta = \dfrac{\pi}{4}$ to the Cartesian axes. To confirm, convert the given equation to Cartesian form.

$$r = \frac{2}{\cos\theta - \sin\theta} \;\Rightarrow\; r\cos\theta - r\sin\theta = 2 \;\Rightarrow\; x - y = 2$$

Thus, $r = \dfrac{2}{\cos\theta - \sin\theta}$ and $x - y = 2$ are equivalent forms (Figure 8.2g), confirming both the linearity of the function and, via its slope, its angle of incidence with the Cartesian axes.

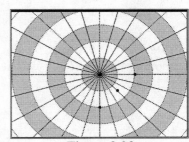

Figure 8.2f

Also notice that $\displaystyle\lim_{\theta \to \left(\frac{\pi}{4}\right)^{+}} \frac{2}{\cos\theta - \sin\theta} = -\infty$ and

$\displaystyle\lim_{\theta \to \left(\frac{5\pi}{4}\right)^{-}} \frac{2}{\cos\theta - \sin\theta} = +\infty$, allowing the polar version of the equation to reach the infinite "ends" of the line. As the θ-values continue to grow, the polar equation completely re-graphs the line every π units.

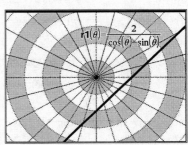

Figure 8.2g

Polar Curve Symmetry
Any symmetries of a polar function's graph often depend on the trigonometric function(s) used in its definition. Commonly discussed polar symmetries are those with respect to the x-axis, the y-axis, and the pole.

x-axis symmetry: One way to define symmetry with respect to the *x*-axis is to say that the point (x, y) lies on the graph iff $(x, -y)$ also lies on the graph (Figure 8.2h). Polar coordinates suggest that a graph with *x*-axis symmetry contains $[r, \angle \theta]$ iff it also contains $[r, \angle -\theta]$. Cosine is an even function and $\forall \theta \in \mathbb{R}$, $\cos(-\theta) = \cos \theta$ ensures that any polar function defined using only the cosine function and its simple transformations has *x*-axis symmetry.

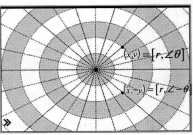

Figure 8.2h

y-axis symmetry: A graph is *y*-axis symmetric when the point (x, y) lies on the graph iff $(-x, y)$ does, too. Figure 8.2i shows these *y*-axis symmetric points along with their equivalent polar coordinates. The polar coordinates show that a graph is symmetric to the *y*-axis iff both $[r, \angle \theta]$ and $[r, \angle(\pi - \theta)]$ lie on the graph. In other words, a polar function *r* has *y*-axis symmetry iff $r(\pi - \theta) = r(\theta)$. An identity quickly demonstrates that any polar function defined using only the sine function and its simple transformations has *y*-axis symmetry.

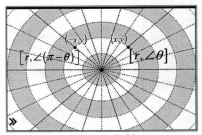

Figure 8.2i

$$\sin(\pi - \theta) = \sin \pi \cos \theta - \cos \pi \sin \theta = \sin \theta$$

pole symmetry: Graphs are symmetric to the pole iff (x, y) and $(-x, -y)$ are both on the graph. In polar coordinates, a graph is pole-symmetric iff $[r, \angle \theta]$ and $[r, \angle(\theta + \pi)]$ are both on the graph.

Example 4:
Determine if $r = 4\cos \theta - 3\sin \theta$ is symmetric to either Cartesian axis.

Test for x-axis symmetry:

$$r(-\theta) = 4\cos(-\theta) - 3\sin(-\theta)$$
$$= 4\cos \theta + 3\sin \theta$$
$$\neq r(\theta)$$

$r(-\theta) \neq r(\theta)$, so the graph is not symmetric to the *x*-axis.

Test for y-axis symmetry:

$$r(\pi - \theta) = 4\cos(\pi - \theta) - 3\sin(\pi - \theta)$$
$$= 4(\cos \pi \cos \theta + \sin \pi \sin \theta) - 3\sin \theta$$
$$= -4\cos \theta - 3\sin \theta$$
$$\neq r(\theta)$$

$r(-\theta) \neq r(\theta)$, so the graph is not symmetric to the *y*-axis.

Remember that in the polar coordinate system, every point has an infinite number of coordinate representations. The implication of this fact is that there are many more ways to identify types of symmetry using polar coordinates other than the ones given above.

Example 5:
Confirm that the polar function $r = 4\sin\theta$ is symmetric to the y-axis by using a pair of coordinates other than the ones used in Example 4.

In Figure 8.2g, the points also could have been labeled as $\left[r, \angle\left(\dfrac{\pi}{2} - \theta\right)\right]$ and $\left[r, \angle\left(\dfrac{\pi}{2} + \theta\right)\right]$. For a function with y-axis symmetry, $r\left(\dfrac{\pi}{2} - \theta\right) = r\left(\dfrac{\pi}{2} + \theta\right)$ should also be true.

$$r\left(\frac{\pi}{2} - \theta\right) = 4\sin\left(\frac{\pi}{2} - \theta\right)$$
$$= 4\cos\theta \qquad \text{(cofunctions)}$$
$$= 4\left(1 \cdot \cos\theta + 0 \cdot \sin\theta\right)$$
$$= 4\left(\sin\frac{\pi}{2}\cos\theta + \cos\frac{\pi}{2}\sin\theta\right)$$
$$= 4\sin\left(\frac{\pi}{2} + \theta\right)$$

This confirms the y-axis symmetry of $r = 4\sin\theta$ using the relationship $r\left(\dfrac{\pi}{2} - \theta\right) = r\left(\dfrac{\pi}{2} + \theta\right)$.

Basic Transformations on Polar Graphs

Rotations
With Cartesian functions, when a quantity is added to the input, the result is a horizontal translation. By contrast, for polar functions of the form $r = \sin(\theta + h)$, the value of h actually creates a *rotation* about the pole of the entire graph, since it is the angle that is being increased or decreased.

Example 6:
Graph $r = \sin\left(\theta + \dfrac{\pi}{2}\right)$, and explain the graph's location.

Plotting and joining the polar points (Figures 8.2j and 8.2k) gives a graph identical to that of $r = \cos\theta$. , since $\sin\left(\theta + \dfrac{\pi}{2}\right) = \cos\theta$. Graphically, $r = \sin\theta$ was rotated clockwise $\dfrac{\pi}{2}$ units.

θ	0	$\dfrac{\pi}{4}$	$\dfrac{\pi}{2}$	$\dfrac{3\pi}{4}$	π	$\dfrac{5\pi}{4}$
$\theta + \dfrac{\pi}{2}$	$\dfrac{\pi}{2}$	$\dfrac{3\pi}{4}$	π	$\dfrac{5\pi}{4}$	$\dfrac{3\pi}{2}$	$\dfrac{7\pi}{4}$
r	1	$\dfrac{1}{\sqrt{2}}$	0	$-\dfrac{1}{\sqrt{2}}$	-1	$-\dfrac{1}{\sqrt{2}}$

Figure 8.2j

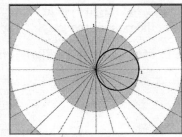

Figure 8.2k

Stretches
An output scale change on a polar graph changes the *r*-value, or the distance from the pole, resulting in an overall magnitude change on the entire graph. While similar in algebraic form to vertical scale changes for Cartesian functions, *r*-scale changes dilate the *r*-values radially outward from the pole (that is, in every direction), not just vertically.

Example 7:
Compare the polar function $r = 3\sin\theta$ to the Cartesian function $y = 3\sin x$ and use the Cartesian function pattern to graph the polar function.

A big difference between the Cartesian and polar graphs is that Cartesian outputs (*y*-values) are measured vertically from the *x*-axis, while polar outputs (*r*-values) measure outward from the pole. A polar graph is akin to wrapping the corresponding Cartesian graph around the pole.

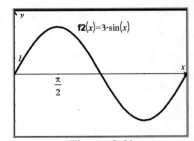

Figure 8.2l

When $x = 0$, the Cartesian graph (Figure 8.2l) is zero units from its reference point. The polar graph behaves the same way at $\theta = 0$. At $\theta = \dfrac{\pi}{2}$ (or $x = \dfrac{\pi}{2}$), both graphs are their maximum distances from their respective reference points. The polar graph returns the pole, at $\theta = \pi$. After $\theta = \pi$, the outputs are all negative, causing the Cartesian graph to dip below the *y*-axis but, the polar graph loops back and superimposes another circle for $\theta \in [\pi, 2\pi]$ (Figure 8.2j).

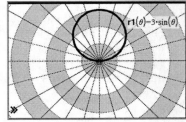

Figure 8.2m

The final result is a new circle with a radius scaled up to 3 units.

Example 8:
Graph $r = -3\sin\theta$ and explain its position using transformations.

One approach is to take $r = 3\sin\theta$ (Figure 8.3m) and makes all the *r*-values negative, resulting in the function graphed in the opposite direction on the opposite side of the pole (Figure 8.2n).

Another approach is to think of the transformation as a rotation. When shifted $\pm\pi$ units, the untransformed sine curve becomes the opposite of sine. Rotating Figure 8.2m $\pm\pi$ units creates Figure 8.2n.

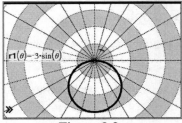

Figure 8.2n

In general, what might appear as a reflection over the *x*- or *y*- axis of a polar graph is, in reality, a rotation of $\pm\pi$ about the pole of the graph. The appearance of reflection over the *x*- or *y*- axis is a result of the original symmetry of the graph. Similarly, sine graphs are related to cosine graphs by rotations of $\pm\dfrac{\pi}{2}$ about the pole.

Problems for Section 8.2:

Exercises

For questions 1-10, graph each polar function. Then convert each to Cartesian form to verify the graphs.

1. $r = 5\sin\theta$

2. $r = -\pi\cos\theta$

3. $r = \dfrac{7}{2\cos\theta - e\sin\theta}$

4. $r = 4$

5. $r = -5$

6. $r = 4\cos\theta + 6\sin\theta$

7. $r = 2\cos\theta + 8\sin(-\theta)$

8. $r = -\theta$

9. $r = 3\theta$

10. $r = 2^\theta$

Describe the transformations on $r = \sin\theta$ to produce each function in questions 11-14, and provide its graph.

11. $r = \sin\left(\theta + \dfrac{\pi}{3}\right)$

12. $r = \sin\left(\dfrac{\pi}{4} - \theta\right)$

13. $r = -\cos\theta$

14. $r = 3\cos\left(\theta + \dfrac{\pi}{6}\right)$

For questions 15-18, give a possible polar equation for each graph.

15.

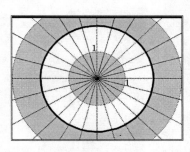

16.

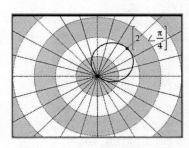

17.

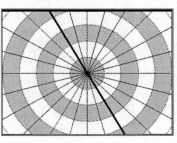

18.

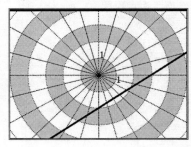

Without graphing, determine if the graph of each polar function in questions 19-22 is symmetric to the *x*-axis, the *y*-axis and/or the pole.

19. $r = 5\sin 5\theta$ ˅ axis

20. $r = \cos 4\theta$ x axis y axis pole

21. $r = \sin\theta - \cos\theta$ none

22. $r = \theta - \sin\theta$

Explorations

Identify each statement in questions 23-27 as true or false. Justify your conclusion.

23. The graphs of $r = \theta$ for $\theta \geq 0$ and $r = \theta$ for $\theta \leq 0$ are identical.

24. If a line with slope m intersects the x-axis at angle θ, then $m = \tan \theta$.

25. A polar graph is symmetric with respect to the pole iff $r(\theta + \pi) = r(\theta - \pi)$.

False that is symmetric about the x axis

26. A polar graph is symmetric with respect to the line $y = \sqrt{3} \cdot x$ iff $r(\theta) = r\left(\theta - \dfrac{2\pi}{3}\right)$.

27. The graph of $r = \cos\left(\theta + \dfrac{\pi}{6}\right)$ is identical to the graph of $r = -\cos\left(\theta - \dfrac{5\pi}{6}\right)$.

28. What happens to the graph of $r = \theta$ for $\theta \in [-20, 0]$? How does this compare to the graph of $r = \theta$ for $\theta \in [0, 20]$? Explain.

29. If $a \in \mathbb{R}$, how far is the line $r = \dfrac{a}{\cos \theta - \sin \theta}$ from the pole?

30. What is the period of the polar graphing of the line in Example 3?

31. Graph $r = |5 \cos \theta|$.

8.3: Polar Graphing II

Believe nothing, no matter where you read it, or who said it, no matter if I have said it, unless it agrees with your own reason and your own common sense.

– Buddha

Section 8.2 explored transformations of the forms $r = \sin(\theta + h)$ and $r = B\sin\theta$. This section covers transformations of the form $r = B\sin\theta + A$ and $r = C\sin(D\theta)$ in detail.

Limaçons
Limaçons are polar functions defined $\forall A, B \in \mathbb{R}$ by $r(\theta) = A + B\cos\theta$ and $r(\theta) = A + B\sin\theta$. Limaçon graphs fall into three sub-categories: **cardioids, limaçons with loops,** and **limaçons without loops.**

Example 1:
Graph $r = 4 + 4\cos\theta$.

Figure 8.3a provides a table of values for the function. Beginning at its maximum distance of 8 for $\theta = 0$, the curve transitions through positive r-values to $\left[4, \angle\dfrac{\pi}{2}\right]$ (Figure 8.3b). The radius decreases to 0 at $\theta = \pi$, and returns to 8 at $\theta = 2\pi$ (Figure 8.3c).

There are no restrictions on the angle, so the function's domain is $\theta \in (-\infty, \infty)$. The range of the cosine is $[-1, 1]$, making the range of the function $r \in [0, 8]$. Note how the range is entirely non-negative. The values of A and B are the same in absolute value, resulting in $r = 0$ for some θ and a corresponding "corner" at the pole. Because of its heart-like shape and the characteristic corner at the pole, limaçons of this type are called **cardioids.**

θ	0	$\dfrac{\pi}{2}$	π	$\dfrac{3\pi}{2}$	2π
r	8	4	0	4	8

Figure 8.3a

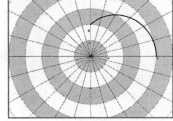

Figure 8.3b

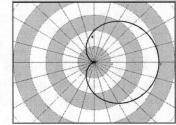

Figure 8.3c

Example 2:
Graph $r = 2 - 3\sin\theta$.

Sample values for θ and r are provided in Figure 8.3d. Notice that r changes sign from positive to negative and back during $\theta \in [0, \pi]$. Because sine is a continuous function, the Intermediate Value Theorem guarantees that there are two values of θ within the interval where $r = 0$. In other words, the function passes through the pole in each of $\theta \in \left(0, \dfrac{\pi}{2}\right)$ and $\theta \in \left(\dfrac{\pi}{2}, \pi\right)$.

Using this idea, Figure 8.3e shows the first part of the graph for $\theta \in \left[0, \dfrac{\pi}{2}\right]$, starting at $[2, \angle 0]$, rotating through the pole at some point during the interval, and ending at $\left[-1, \angle\dfrac{\pi}{2}\right]$. Connecting the remainder of

the points creates the final graph (Figure 8.3f).

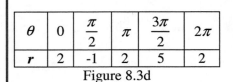

θ	0	$\dfrac{\pi}{2}$	π	$\dfrac{3\pi}{2}$	2π
r	2	-1	2	5	2

Figure 8.3d

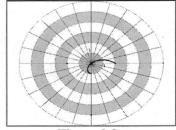

Figure 8.3e

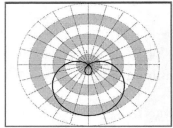

Figure 8.3f

The range of the function is $r \in [-1,5]$. The changing sign in the r-values causes the inner loop in Figure 8.3f. A limaçon with this graphical characteristic is called a **limaçon with a loop**. Because the range of the sine and cosine functions is $[-1,1]$, the r-value sign change is guaranteed whenever $|A| < |B|$, creating polar graphs that pass through the pole. Also notice the y-axis symmetry for non-translated polar functions containing the sine function, as predicted in Section 8.2.

The polar equation from Example 2 could be converted to Cartesian form, but the result is not pleasant.

$$r = 2 - 3\sin\theta \;\Rightarrow\; r^2 = 2r - 3r\sin\theta$$
$$\Rightarrow\; x^2 + y^2 = 2\sqrt{x^2 + y^2} - 3y$$
$$\Rightarrow\; \left(x^2 + y^2 + 3y\right)^2 = 4\left(x^2 + y^2\right)$$

This could be further expanded to a fourth-degree polynomial in terms of x and y. Given specific values of y, the Fundamental Theorem of Algebra guarantees this equation will have exactly four solutions for given values of x or y. For this reason, any horizontal or vertical line through this graph will intersect the curve *at most* four times. For example, if the line $y = -0.5$ were drawn in Figure 8.3f, it would intersect $r = 2 - 3\sin\theta$ four times, corresponding exactly to the four solutions guaranteed by the Fundamental Theorem of Algebra.

The function for Example 2 has a cleaner polar expression, but its Cartesian equivalent clearly showed that there are up to four real points on the graph given any fixed x or y-value—a characteristic not obvious from the polar form. Again, each algebraic form reveals different characteristics about the underlying function.

Example 3:
Graph $r = 5 + 3\cos\theta$.

Figure 8.3g lists some ordered pairs for $[r, \theta]$. The graph initially transitions from a maximum radius at $[8, \angle 0]$ to a minimum radius at $[2, \angle\pi]$ (Figure 8.3h). For $\theta \in [\pi, 2\pi]$, the radius again increases, reaching its maximum at $\theta = 2\pi$ (Figure 8.3i).

θ	0	$\dfrac{\pi}{2}$	π	$\dfrac{3\pi}{2}$	2π
r	8	5	2	5	8

Figure 8.3g

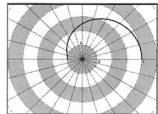

Figure 8.3h

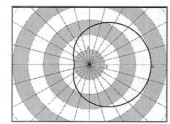

Figure 8.3i

The range of $y = \cos\theta$ is $[-1,1]$, so the polar function's range is $r \in [2,8]$. In this case, the range is entirely positive with no zeros or sign changes in r-values, so the curve never touches the pole. A limaçon will always have a positive range whenever $|A| > |B|$. This type of curve is a **limaçon without a loop**.

The range of a limaçon is especially helpful when classifying a limaçon. It can also be used to find the Cartesian axes intercepts of the graph. Specifically, the endpoints of the range are the intercepts of the axis of symmetry, while the midpoint of the range gives the intercepts on the other axis.

Rose Curves

It was noted in Section 8.2 that an output stretch on a polar equation magnified the graph in all directions. So what happens when a dilation is applied to the input variable of a polar function? The answer is an angle stretch. To explore this, consider the family of polar curves with equations of the form $r = C\cos(D\theta)$ or $r = C\sin(D\theta)$, where $C \in \mathbb{R}$ and $D \in \mathbb{Z}$. As already explored, the value of C stretches polar graphs relative to the pole, so the following discussion analyzes the effects of D by letting $C = 1$.

Example 1, Section 8.2 showed that $r = \cos\theta$ completes a full circle for $\theta \in [0,\pi]$. Over $\theta \in [\pi, 2\pi]$, the function re-graphs, passing through the pole at $\theta = \dfrac{3\pi}{2}$. One complete cycle of the graph, as is also the case in the Cartesian system, goes from maximum → zero → minimum → zero → maximum. Any angular compression would make the graph complete its cycle faster, going through more cycles in $[0, 2\pi]$. Integer compressions of the angle create multiple loops that look very similar to the petals of a flower, earning such graphs the name **rose curves**.

Example 4:
Graph $r = 2\cos(3\theta)$.

The Cartesian function $y = 2\cos(3x)$ is horizontally compressed by a factor of 3 from its parent graph, creating three cycles over $x \in [0, 2\pi]$. This still happens when the function is expressed in polar coordinates.

The output of $r = 2\cos(3\theta)$ is $r \in [-2,2]$. Combined with the θ-compression, this means the polar graph should oscillate through $r \in [-2,2]$ three times for $\theta \in [0, 2\pi]$. Each oscillation occuring over an equal θ-interval. The behavior of $r = 2\cos(3\theta)$ for $\theta \in \left[0, \dfrac{\pi}{3}\right]$ explains how the function behaves for $\theta \in [0, 2\pi]$.

Figure 8.3j is a table of values for the interval $\theta \in \left[0, \dfrac{\pi}{3}\right]$. Figure 8.3k plots and connects these to show the transition of $r = 2\cos 3\theta$ from a maximum radius of 2 at $\theta = 0$ to its minimum radius of -2 at $\theta = \dfrac{\pi}{3}$.

Over the interval $\theta \in \left[\dfrac{\pi}{3}, \dfrac{2\pi}{3}\right]$, r grows from -2 back to 2 in another smooth transition, ending at

$\left[2,\dfrac{2\pi}{3}\right]$. This corresponds to one complete Cartesian cycle. Both the Cartesian and polar versions of the

function begin at their maximum values, pass through zero at $\theta=\dfrac{\pi}{6}$ and reach their minima at $\theta=\dfrac{\pi}{3}$.

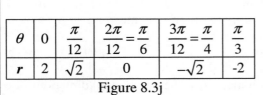

θ	0	$\dfrac{\pi}{12}$	$\dfrac{2\pi}{12}=\dfrac{\pi}{6}$	$\dfrac{3\pi}{12}=\dfrac{\pi}{4}$	$\dfrac{\pi}{3}$
r	2	$\sqrt{2}$	0	$-\sqrt{2}$	-2

Figure 8.3j

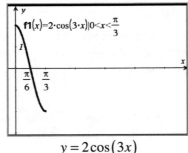

$y=2\cos(3x)$

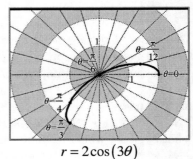

$r=2\cos(3\theta)$

Figure 8.3k

Repeating the process for $\theta\in\left[\dfrac{2\pi}{3},\pi\right]$ creates the final leg of the polar graph (Figure 8.3l). Just like the circles in Section 8.2, the function re-graphs for $\theta\in[\pi,2\pi]$.

This is a rose curve with three distinct petals.

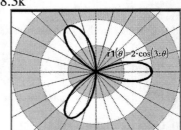

Figure 8.3l

Interestingly, the angle compression factor happens to be the same as the number of petals on the rose curve. Is this *always* the case?

Example 5:
Analyze and graph the polar function $r=\sin(2\theta)$.

$r=\sin(2\theta)$ completes two full cycles in $\theta\in[0,2\pi]$, going from zero to maximum to zero to minimum to zero over $\theta\in[0,\pi]$. Figure 8.3m shows this portion of the graph with the five key θ-values marked. The cycle repeats for $\theta\in[\pi,2\pi]$, but given the positive r-values for these θ-values, there is no overlap in the graph, and two more distinct petals are formed, resulting in a four-petaled rose curve. Figure 8.3n shows the curve along with its direction of graphing.

With a mental picture of the Cartesian version of the graph and all its key points, it is easier to translate the key points into their polar counterparts and get a picture of the corresponding polar graph.

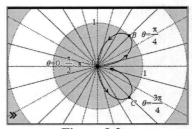

Figure 8.3m

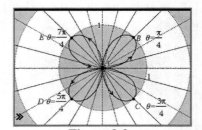

Figure 8.3n

In this case, the angle compression factor is not the same as the number of petals appearing on the rose curve, so the conjecture prior to Example 5 is disproven.

Example 6:

Graph $r = \sin\left(3\left(\theta - \dfrac{\pi}{6}\right)\right)$.

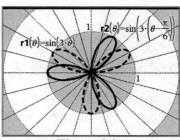

This is the graph of $r = \sin(3\theta)$ rotated $\dfrac{\pi}{6}$ counterclockwise. Figure 8.3o
shows both the parent (dashed) and rotated (solid) graphs.

Figure 8.3o

Intersections of polar curves.

The intersection of graphs is equivalent to the solution set of a corresponding algebraic system of equations. Unfortunately, intersections of curves are not uniquely defined in polar coordinates, so direct computation of the solution cannot happen from a calculator graph. Even so, a graph can still provide valuable information.

Example 7:

Determine the point(s) of intersection $r_1 = \dfrac{1}{\cos\theta - \sin\theta}$ and $r_2 = \cos\theta + \sin\theta$.

The graph of r_1 looks suspiciously like a tangent line to the circle formed by r_2 (Figure 8.3p).

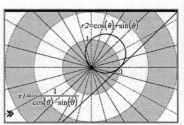

Figure 8.3p

Method 1: Solve the system defined by the functions' Cartesian forms.
Here, r_1 converts to $x - y = 1$, and r_2 becomes $x^2 + y^2 = x + y$.
Substitution gives

$$x = y + 1 \;\Rightarrow\; (y+1)^2 + y^2 = (y+1) + y$$
$$\Rightarrow\; 2y^2 + 2y + 1 = 2y + 1$$
$$\Rightarrow\; 2y^2 = 0$$
$$\Rightarrow\; y = 0 \text{ and } x = 1.$$

Method 2: Solve the system directly in polar coordinates.

$$\cos\theta + \sin\theta = \dfrac{1}{\cos\theta - \sin\theta}$$
$$\cos^2\theta - \sin^2\theta = 1$$
$$\cos 2\theta = 1$$

Therefore $\forall k \in \mathbb{Z}$, $2\theta = 0 + 2\pi k \;\Rightarrow\; \theta = \pi k \;\Rightarrow\; \theta = \{0, \pi\} \;\Rightarrow\; r = \{1, -1\}$, giving two solutions with coordinates $[1, \angle 0]$ and $[-1, \angle \pi]$. These are equivalent to the Cartesian $(1, 0)$, confirming the Method 1 solution.

Finally, the Cartesian equivalent of $r_2 = \cos\theta + \sin\theta$ is $\left(x - \dfrac{1}{2}\right)^2 + \left(y - \dfrac{1}{2}\right)^2 = \dfrac{1}{2}$, a circle centered at

$\left(\dfrac{1}{2}, \dfrac{1}{2}\right)$. The slope from this center to $(1, 0)$ is -1, and from its Cartesian equivalent, the graph of

$r_1 = \dfrac{1}{\cos\theta - \sin\theta}$ has a slope of 1. Since both curves contain $(1, 0)$ and the radius segment is normal to

r_2, $r_1 = \dfrac{1}{\cos\theta - \sin\theta}$ is a tangent line to $r_2 = \cos\theta + \sin\theta$.

In Example 7, all solutions were found algebraically and directly. However, there are occasions when the *graphs* of two polar functions may coincide at a certain point, but algebraically, that point does not satisfy the system of polar equations. Example 8 explores what happens in such a situation.

Example 8:

$\forall \theta \in [0, 2\pi)$, determine the polar coordinates of all points where the graphs of $r_1 = \cos(2\theta)$ and $r_2 = \sin(2\theta)$ intersect simultaneously.

Figure 8.3q shows nine intersection points: the pole and eight others. Then,

$$r_1 = r_2 \quad \Rightarrow \quad \cos 2\theta = \sin 2\theta \quad \Rightarrow \quad \tan 2\theta = 1$$

Adjusting for the doubled input values leads to

$$\theta \in [0, 2\pi) \quad \Rightarrow \quad 2\theta \in [0, 4\pi)$$

$$2\theta = \frac{\pi}{4}, \frac{5\pi}{4}, \frac{9\pi}{4}, \frac{13\pi}{4}$$

$$\theta = \frac{\pi}{8}, \frac{5\pi}{8}, \frac{9\pi}{8}, \frac{13\pi}{8}.$$

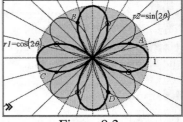

Figure 8.3q

So, the intersection points are $A\left[\dfrac{\sqrt{2}}{2}, \angle\dfrac{\pi}{8}\right]$, $B\left[-\dfrac{\sqrt{2}}{2}, \angle\dfrac{5\pi}{8}\right]$, $C\left[\dfrac{\sqrt{2}}{2}, \angle\dfrac{9\pi}{8}\right]$, and $D\left[-\dfrac{\sqrt{2}}{2}, \angle\dfrac{13\pi}{8}\right]$

(labeled in figure 8.3q). Interestingly, the pole and four other points were not identified. To understand this, it is important to grasp what is meant for a point to be an intersection point of two graphs. For polar graphs, both graphs must share the same distance from the pole *at precisely the same angle*.

For example, at point A, at $\theta = \pi/8$, $r = \sqrt{2}/2$ for both functions. This is a **simultaneous intersection point**. While the remaining five points certainly have Cartesian coordinates where the *paths* cross, the intersections do not happen at the same angle for both graphs; these are **non-simultaneous intersection points**. While the graphs have equivalent Cartesian coordinates at those points, they do not share equivalent *polar* coordinates, and hence those particular points do not satisfy the system of polar equations. For this reason calculators do not have intersection operations for polar graphs.

Non-Simultaneous Intersection Points

Consider what is happening when the graphs of two polar functions intersect. When functions share the same r-output at the same θ-input, those points can be determined as in Example 8. Recall, however, that the coordinates of points are not uniquely defined in the polar graphing system.

For any two polar functions $r = f(\theta)$ and $r = g(\theta)$, if there is an angle $\theta \ni g(\theta) = -f(\theta + \pi)$ where θ is the input to function g, the graphs of the functions will cross. If $f(\theta) = -f(\theta + \pi)$, this is a simultaneous point of intersection, otherwise it is a point of non-simultaneous intersection. Equivalently, all $\theta \ni f(\theta) = -g(\theta + \pi)$ where θ is now the input to function f, also define points of non-simultaneous intersection.

While $g(\theta) = -f(\theta + \pi)$ seems a bit complicated, it is relatively straightforward when the nature of graphing polar points is considered. The equation suggests that if a point on g can be defined by $[r, \angle\theta]$ and if there is a point $[-r, \angle(\theta + \pi)]$ on f, then the curves overlap. This makes sense because for polar coordinates and $\forall \theta \in \mathbb{R}$, the angles θ and $\theta + \pi$ are always on opposite sides of the pole and the corresponding opposite output r-values force the curves to overlap.

The pole is the exception to the $g(\theta) = -f(\theta + \pi)$ rule. If $r = 0$ is in the range of any two polar functions, then the graphs of those two functions will intersect at the pole. The value of θ that creates $r = 0$ in each function is irrelevant, and the intersection point could be simultaneous or non-simultaneous depending on the functions.

Example 9:
Determine polar coordinates of all points of intersection of $r_1 = \cos(2\theta)$ and $r_2 = \sin(2\theta)$.

- The simultaneous points of intersection were found in Example 8.
- Because both curves are continuous with range $r \in [-1, 1]$, the Intermediate Value Theorem guarantees that each curve will have value $r = 0$ for some values of θ. The fact that this point was not identified in Example 8 makes the pole a point of non-simultaneous intersection.
- Relative to θ-values on r_1, the other non-simultaneous intersection points can be located using

$$r_1(\theta) = -r_2(\theta + \pi)$$
$$\cos(2\theta) = -\sin(2(\theta + \pi))$$
$$\cos(2\theta) = -\sin(2\theta)$$

Assuming $\theta \in [0, 2\pi) \Rightarrow 2\theta \in [0, 4\pi)$, then $2\theta = \dfrac{3\pi}{4}, \dfrac{7\pi}{4}, \dfrac{11\pi}{4}, \dfrac{15\pi}{4} \Rightarrow \theta = \dfrac{3\pi}{8}, \dfrac{7\pi}{8}, \dfrac{11\pi}{8}, \dfrac{15\pi}{8}$.

Relative to r_1, the intersection points are $\left[-\dfrac{\sqrt{2}}{2}, \angle\dfrac{3\pi}{8} \right]$, $\left[\dfrac{\sqrt{2}}{2}, \angle\dfrac{7\pi}{8} \right]$, $\left[-\dfrac{\sqrt{2}}{2}, \angle\dfrac{11\pi}{8} \right]$, and

$\left[\dfrac{\sqrt{2}}{2}, \angle\dfrac{15\pi}{8} \right]$. The coordinates of the points can also be expressed relative to $r_2 = \sin(2\theta)$.

Another *purely* algebraic way to determine all points of intersection is to convert the equations to their Cartesian equivalents and solves the resulting system. Such systems can be quite complicated to solve, but a CAS simplifies the exploration of this option, and allows you to focus on the mathematics behind what is happening. Example 10 repeats the results of Example 9 using a Cartesian CAS solution.

Example 10:
Solve the system of equations from Example 8 by first converting the equations to Cartesian form.

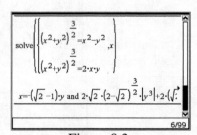

Figure 8.3r

Converting the polar equations gives
$$\begin{cases} r = \cos 2\theta \\ r = \sin 2\theta \end{cases} \Rightarrow \begin{cases} r = \cos^2\theta - \sin^2\theta \\ r = 2\sin\theta\cos\theta \end{cases}$$

$$\Rightarrow \begin{cases} r^3 = r^2\cos^2\theta - r^2\sin^2\theta \\ r^3 = 2(r\cos\theta)(r\sin\theta) \end{cases} \Rightarrow \begin{cases} \left(\sqrt{x^2 + y^2}\right)^3 = x^2 - y^2 \\ \left(\sqrt{x^2 + y^2}\right)^3 = 2xy \end{cases}.$$

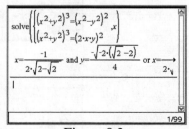

Figure 8.3s

An attempt to solve the Cartesian system in this form (Figure 8.3r) suggests that the CAS is unable to completely solve the system in its current form. Figure 8.3r does not show the entire solution line, but does show that answers from this version of the system are not constant.

While squaring both sides of an equation sometimes creates extraneous solutions, doing so here enables the CAS solutions (Figure 8.3s): $\left(\pm\dfrac{1}{2\sqrt{2-\sqrt{2}}},\pm\dfrac{\sqrt{4-2\sqrt{2}}}{4}\right)$, $\left(\pm\dfrac{1}{2\sqrt{2+\sqrt{2}}},\pm\dfrac{\sqrt{4+2\sqrt{2}}}{4}\right)$, and $(0,0)$.

All nine Cartesian solutions were determined by solving the system in the second form. Remember that, despite the computational power of a CAS, sometimes it is necessary to manipulate the form of the input.

Problems for Section 8.3:

Exercises

For each polar function in questions 1-14, provide the polar range, axis intercepts, and a graph.

1. $r = 3 + 5\sin\theta$

2. $r = 5 - 2\sin\theta$

3. $r = 4 + \cos\theta$

4. $r = 3\cos\theta + 1$

5. $r = 3\sin\theta - 1$

6. $r = 1 - 8\cos\theta$

7. $r = 2 - 2\cos\theta$

8. $r = 4 + 4\sin\theta$

9. $r = -3\cos 4\theta$

10. $r = \sin 4\theta$

11. $r = 4\cos 5\theta$

12. $r = -\pi\sin 9\theta$

13. $r = 1 + \sin\left(\theta + \dfrac{\pi}{2}\right)$

14. $r = 2\sin\left(\theta - \dfrac{\pi}{4}\right)\cos\left(\theta - \dfrac{\pi}{4}\right)$

For questions 15-20, give a possible equation for each polar graph.

15.

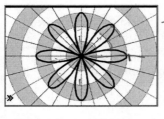

16.

$3\cos 4\theta$

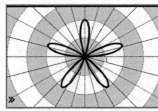

$2\sin 5\theta$

$r = 2\sin 5\theta$

17.

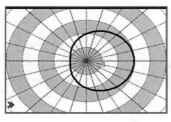

18.

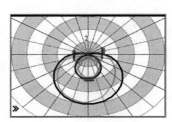

$2 + \cos\theta$

19.

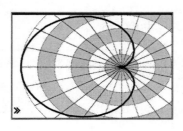

20.

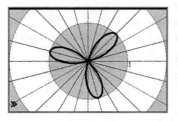

21. Find a possible equation for the graph in Figure 8.3p. Determine a Cartesian equation for its axis of symmetry.

22. Find polar coordinates for the points of intersection of the function in Figure 8.3p with its axis of symmetry.

23. Find an equation for a non-rotated polar function that would match the graph in Example 6.

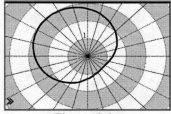

Figure 8.3p

24. The graph of $r = 2 - 3\sin\theta$ in Example 2 passed through the pole twice. Determine polar coordinates for these points.

Find all simultaneous polar points of intersection for each set of equations in questions 25 and 26.

25. $\begin{cases} r = 3 \\ r = 6\cos\theta \end{cases}$
26. $\begin{cases} r = -2\csc\theta \\ r = 5\sin(3\theta) \end{cases}$

27. Determine the Cartesian coordinates of the four solutions to the system $\begin{cases} y = -\dfrac{1}{2} \\ r = 2 - 3\sin\theta \end{cases}$ predicted in the

paragraph following Example 2.

Explorations

Identify each statement in questions 28-32 as true or false. Justify your conclusion.

28. A cardioid is a limaçon. *True — both are $r = a + b\cos\theta$*

29. The graph of the rose curve defined by $r = 4\cos 8\theta$ has eight distinct petals.

30. For any limaçon of the form $r = A + B\sin\theta$, the y-intercepts are always at $(0, B \pm A)$.

31. The graphs of $r = A + B\cos\theta$ and $r = A - B\cos\theta$ are $\pm\pi$ rotations of each other about the pole. *true*

32. The graphs of $r = 2\sin\theta\cos\theta$ and $r = \cos 2\left(\theta + \dfrac{\pi}{4}\right)$ are identical.

33. Under what conditions for A and B will a limaçon never touch the pole?

34. Explain how to use the range of a limaçon to determine what category of limaçon it is.

35. In terms of A and B, how long is the inner loop in a limaçon with a loop?

36. In terms of A and B, what is the greatest distance a limaçon ever gets from the pole?

37. Prove that the x-intercepts of $r = A + B\sin\theta$ are at $(\pm A, 0)$. Show that A is also the midpoint of the range of the function.

38. Under what conditions for A and B will an equation of the form $r = A + B\sin\theta$ be symmetric to both Cartesian coordinate axes?

39. Given $r = C\cos(D\theta)$ or $r = C\sin(D\theta)$, where $D \in \mathbb{Z}$, specifically describe how D affects the number of distinct petals on the corresponding rose curve.

40. Under what conditions for C or D will a rose curve be symmetric to both coordinate axes?

41. [NC] Graph $r = 1 - \cos(2\theta)$.

42. Determine without using a calculator what the graphs of the following polar equations look like. (You may check your answers on a calculator.) These graphs are called **lemniscates**.
    A. $r^2 = \cos(2\theta)$
    B. $r^2 = \sin(2\theta)$

43. Give a second possible equation for the graph in question 15. Explain using transformations and algebraically how the two equations you found are equivalent.

44. How many real solutions could the system $\begin{cases} y = k \\ r = 2 - 3\sin\theta \end{cases}$ have if k can assume any real value? Explain.

8.4: The Polar Form of Complex Numbers

One merit of mathematics few will deny: it says more in fewer words than any other science. The formula, $e^{i\pi} = -1$ expressed a world of thought, of truth, of poetry, and of the religious spirit "God eternally geometrizes."

– David Eugene Smith in N. Rose *Mathematical Maxims and Minims*, 1988

Graphs of complex numbers (Section 1.2) look exactly like scatter plots of real ordered pairs, (a,b) without axis labels. This suggests an alternative representation of complex numbers.

The first sections of this unit, in part, discussed the conversion of points and functions between Cartesian and polar forms. The same idea allows complex numbers to be rewritten in a polar form. That is, every complex number of the form $z = a + bi$ is a horizontal units and b vertical units from the point 0, and can also be thought of as being r units in a straight line from 0 at an angle θ from the positive x-axis. Therefore, z can be re-expressed as

$$z = a + bi = r\cos\theta + (r\sin\theta)i = r(\cos\theta + i\sin\theta).$$

The last expression is sometimes written

$$r \cdot \operatorname{cis}\theta = \underline{r} \cdot (\underline{c}\cos\theta + \underline{is}\sin\underline{\theta}).$$

Example 1:
  A. Convert $3 + 3i$ to polar form.
  B. Convert $4\operatorname{cis}(-2)$ to Cartesian form.

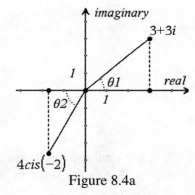
Figure 8.4a

  A. $3^2 + 3^2 = r^2 \implies r = 3\sqrt{2}$, and $\tan\theta = 3/3 \implies \theta = \pi/4$.
      Therefore, $3 + 3i = 3\sqrt{2}\operatorname{cis}\left(\dfrac{\pi}{4}\right)$.

  B. $a = 4\cos(-2) \approx -1.665$ and $b = 4\sin(-2) \approx -3.637$. Therefore,
      $4\operatorname{cis}(-2) \approx -1.665 - 3.637i$

Basic operations on Complex numbers are easy using Cartesian representations, but what happens when the numbers are in polar form? One option is to change back to Cartesian form and perform the operations there. Even so, some operations are much more difficult in Cartesian form. While $(2-i)^{10}$ is cumbersome, how do you approach $\sqrt{2-i}$? Figure 8.4b suggests that $a + bi$ forms are possible for both of these and perhaps for even a wider range of complex number operations.

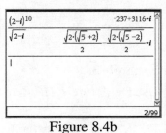
Figure 8.4b

Example 2:

  A. Determine a product rule for the polar form of complex numbers $z_1 = r_1\operatorname{cis}\theta_1$ and $z_2 = r_2\operatorname{cis}\theta_2$.

  B. Verify the rule from part A by computing $(8-2i)(3+i)$ two different ways.

  A. Expand the product $z_1 \cdot z_2$ and apply trigonometric identities.

$$z_1 \cdot z_2 = (r_1 \operatorname{cis}\theta_1)(r_2 \operatorname{cis}\theta_2)$$
$$= r_1 \cdot r_2 \cdot (\cos\theta_1 + i\sin\theta_1)(\cos\theta_2 + i\sin\theta_2)$$
$$= r_1 \cdot r_2 \cdot (\cos\theta_1\cos\theta_2 + i\sin\theta_1\cos\theta_2 + i\cos\theta_1\sin\theta_2 + i^2\sin\theta_1\sin\theta_2)$$
$$= r_1 \cdot r_2 \cdot ((\cos\theta_1\cos\theta_2 - \sin\theta_1\sin\theta_2) + i(\sin\theta_1\cos\theta_2 + \cos\theta_1\sin\theta_2))$$
$$= r_1 \cdot r_2 \cdot (\cos(\theta_1 + \theta_2) + i\sin(\theta_1 + \theta_2))$$
$$= r_1 \cdot r_2 \cdot \operatorname{cis}(\theta_1 + \theta_2).$$

B. In Cartesian form, $(8-2i)(3+i) = 26+2i$. Then convert all three complex numbers to polar form.

$$8 - 2i \approx \sqrt{68}\operatorname{cis}(-0.245) \qquad 3+i \approx \sqrt{10}\operatorname{cis}(0.322) \qquad 26+2i \approx \sqrt{680}\operatorname{cis}(0.077)$$

Because $\sqrt{68} \cdot \sqrt{10} = \sqrt{680}$ and $-0.245 + 0.322 = 0.077$, the polar multiplication rule is verified.

Geometrically, $z_1 \cdot z_2 = r_1 \cdot r_2 \cdot \operatorname{cis}(\theta_1 + \theta_2)$ means that the product of two complex numbers has a radius equal to the product of the radii of its factors, and is positioned at an angle equal to the sum of its composite angles. Correspondingly, two complex numbers $z_1 = r_1\operatorname{cis}\theta_1$ and $z_2 = r_2\operatorname{cis}\theta_2$, can be divided using:

$$\frac{z_1}{z_2} = \frac{r_1}{r_2}\operatorname{cis}(\theta_1 - \theta_2).$$

A significant use of the product rule for multiplication of complex numbers in polar form is raising complex numbers to integer powers. For example,

$$z^2 = (r\operatorname{cis}\theta)^2 = (r\operatorname{cis}\theta)(r\operatorname{cis}\theta) = r^2\operatorname{cis}(2\theta)$$
$$z^3 = (r\operatorname{cis}\theta)^3 = (r\operatorname{cis}\theta)(r\operatorname{cis}\theta)(r\operatorname{cis}\theta) = r^3\operatorname{cis}(3\theta)$$
$$\vdots$$
$$z^n = (r\operatorname{cis}\theta)^n = \underbrace{(r\operatorname{cis}\theta)(r\operatorname{cis}\theta)\cdots(r\operatorname{cis}\theta)}_{n \text{ times}} = r^n\operatorname{cis}(n\theta)$$

While this establishes a generality of the multiplication rule for polar complex numbers only for $\forall n \in \mathbb{N}$, this rule actually applies $\forall n \in \mathbb{R}$. This is **DeMoivre's Theorem**.

$$\boxed{\text{If } z = r\operatorname{cis}\theta \text{, then } z^n = r^n \cdot \operatorname{cis}(n \cdot \theta) \ \forall n \in \mathbb{R}}$$

Example 3:

Use DeMoivre's Theorem to verify that $(2-i)^{10} = -237 + 3116i$.

A CAS verifies the Cartesian result. To use DeMoivre's Theorem, first convert the complex number to polar form and expand.

$$(2-i) \approx \sqrt{5}\operatorname{cis}(-0.464)$$
$$(2-i)^{10} \approx \left(\sqrt{5}\operatorname{cis}(-0.464)\right)^{10} = 3125\operatorname{cis}(-4.636)$$

Then convert back to Cartesian form.

$$3125\operatorname{cis}(-4.636) = 3125(\cos(-4.636) + i\sin(-4.636))$$
$$= 3125(-0.076 + 0.997i)$$
$$= -237 + 3116i$$

Example 4:

Rewrite $\sqrt{i}$ in $a+bi$ form.

This is equivalent to finding some number $z \ni z^2 = i = 1\mathrm{cis}(\pi/2)$. Because the polar expressions of complex numbers are not unique, $\forall k \in \mathbb{Z}$ this initial equation also can be written as

$$z^2 = i = 1\mathrm{cis}\left(\frac{\pi}{2} + 2\pi k\right)$$

Re-applying the square root and using DeMoivre's Theorem gives

$$z = \left[1\mathrm{cis}\left(\frac{\pi}{2} + 2\pi k\right)\right]^{1/2} = 1^{1/2}\mathrm{cis}\left[\frac{1}{2}\left(\frac{\pi}{2} + 2\pi k\right)\right] = 1\mathrm{cis}\left(\frac{\pi}{4} + \pi k\right).$$

$k = \{0,1\}$ produces all of the unique results: $z = \left\{1\mathrm{cis}\left(\frac{\pi}{4}\right),\ 1\mathrm{cis}\left(\frac{5\pi}{4}\right)\right\} = \left\{\frac{1}{\sqrt{2}} + \frac{1}{\sqrt{2}}i,\ \frac{-1}{\sqrt{2}} - \frac{1}{\sqrt{2}}i\right\}.$

Notice that this approach produces two solutions to the initial equation, $z^2 = i$, satisfying the Fundamental Theorem of Algebra. But, the square root operation is a *function*, and therefore only one solution is permitted. The **principal solution** of an equation with multiple, complex solutions is typically considered to be the solution with the smallest positive angle when in polar form. Therefore, $\sqrt{i} = \frac{1}{\sqrt{2}} + \frac{1}{\sqrt{2}}i$ is the solution.

Notice two things. With DeMoivre's Theorem, any complex number can be raised to any real power with the result being a complex number in the form $a+bi$. Second, roots of complex numbers have multiple solutions. In fact, the Fundamental Theorem of Algebra guarantees that there will be n solutions to the nth root of any complex number.

Example 5:

Compute the five fifth roots of $1-i$.

This asks for all values of z where $z^5 = 1 - i = \sqrt{2}\,\mathrm{cis}\left(-\frac{\pi}{4}\right)$. $\forall k \in \mathbb{Z}$, this is equivalent to $z^5 = \sqrt{2}\,\mathrm{cis}\left(-\frac{\pi}{4} + 2\pi k\right)$. Re-applying the root and applying DeMoivre's Theorem gives

$$z = \left[\sqrt{2}\,\mathrm{cis}\left(-\frac{\pi}{4} + 2\pi k\right)\right]^{1/5} = 2^{1/10}\mathrm{cis}\left[\frac{1}{5}\left(-\frac{\pi}{4} + 2\pi k\right)\right] = 2^{1/10}\mathrm{cis}\left(-\frac{\pi}{20} + \frac{2\pi}{5}k\right)$$

after which $k \in \{0,1,2,3,4\}$ gives the five unique solutions.

$$1)\ z_1 = 2^{1/10}\mathrm{cis}\left(-\frac{\pi}{20}\right) \qquad 2)\ z_2 = 2^{1/10}\mathrm{cis}\left(\frac{7\pi}{20}\right) \qquad 3)\ z_3 = 2^{1/10}\mathrm{cis}\left(\frac{15\pi}{20}\right)$$

$$4)\ z_4 = 2^{1/10}\mathrm{cis}\left(\frac{23\pi}{20}\right) \qquad 5)\ z_5 = 2^{1/10}\mathrm{cis}\left(\frac{31\pi}{20}\right)$$

Geometrically, notice that all five roots are spaced from each other by equal angles around a full circle of radius $r = \sqrt[10]{2}$. These roots could be converted to Cartesian form, but there is not much point to this unless further manipulations will be performed on the results.

Problems for Section 8.4:

Exercises

For questions 1-6, convert the given expression to a simplified polar form and plot it on the complex plane.

1. $4+4i$

2. $1-\sqrt{3}i$

3. $i-\sqrt{3}$

4. $-2-5i$

5. 2

6. $-7i$

For questions 7-10, convert the expression to Cartesian form.

7. $5\text{cis}\left(\dfrac{2\pi}{3}\right)$

8. $-2\text{cis}\left(\dfrac{\pi}{5}\right)$

9. $\pi\text{cis}(-4)$

10. $6\text{cis}(42°)$

For questions 11-18, simplify the expression and where possible, evaluate without a calculator.

11. $\left|4cis\theta\right| \ni \theta \in \mathbb{R}$

12. $\left|(1+i)(4-3i)\right|$

13. $(1-i)(1+i)(2-2\sqrt{3}\cdot i)$

14. $\left(\sqrt{3}-i\right)^{7}$ $2^{7}\ cis\text{-}\pi/6$
$128\left(-\sqrt{3}/2+1i/2\right)$
$-64\sqrt{3}+64i$

15. $(1+i)^{-8}$

16. $\dfrac{8\text{cis}(158°)}{-4\text{cis}(22°)}$ $-2cis136°$

17. $\left(2\text{cis}\left(\dfrac{\pi}{3}\right)\right)\left(4\text{cis}\left(\dfrac{\pi}{6}\right)\right)$

18. $\left(-\text{cis}\left(\dfrac{\pi}{6}\right)\right)^{10}$

For questions 19-21, determine the stated roots of the given complex number and graph the solutions on a complex plane.

19. Cube roots of $125\text{cis}(\pi)$

20. Fourth roots of $4\text{cis}\left(\dfrac{\pi}{2}\right)$

21. Sixth roots of $\sqrt{3}-i$

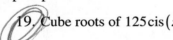 $2\ cis\left(\dfrac{\pi}{24}\right)$

Explorations

Identify each statement in questions 22-26 as true or false. Justify your conclusion.

22. Every complex number has a Cartesian and a polar form

23. $5\text{cis}\left(\dfrac{\pi}{4}\right)=\dfrac{5}{\sqrt{2}}(1+i)$

24. $\left|(a+ai)^{100}\right|=2^{100}\cdot a^{100}$

25. $r_{1}\text{cis}(\theta_{1})+r_{2}\text{cis}(\theta_{2})=(r_{1}+r_{2})\text{cis}(\theta_{1}+\theta_{2})$

26. There are n distinct solutions to $z^{n}=k\ \ \forall k\in\mathbb{C}$.

27. Let $z_1 = r_1 \operatorname{cis} \theta_1$ and $z_2 = r_2 \operatorname{cis} \theta_2$ be two complex numbers. Without using DeMoivre's Theorem, prove $\dfrac{z_1}{z_2} = \dfrac{r_1}{r_2} \operatorname{cis}(\theta_1 - \theta_2)$.

28. Let z_1, z_2, ..., z_n be the n n^{th} roots of a complex number, z. What is unique about $\displaystyle\sum_{k=1}^{n} z_k$ $\forall n \in \mathbb{N}$, $z \in \mathbb{C}$?

29. What is the relationship between the roots of a complex number and the tips of the graph of a rose curve? Explain.

8.5: A New Trigonometry

If you disregard the very simplest cases, there is in all of mathematics not a single infinite series whose sum has been rigorously determined. In other words, the most important parts of mathematics stand without a foundation.

– Niels H. Abel in G. F. Simmons, *Calculus Gems* (1992)

There are several ways to define functions: Cartesian, parametric, vector, and polar, to name a few. A variation on the Cartesian approach presented thus far is with Maclaurin Series, a method of representing functions using infinite polynomial series. This topic is covered in much greater detail in calculus.

Example 1:

Let $f(x) = 1 - \dfrac{x^2}{2!} + \dfrac{x^4}{4!} - \dfrac{x^6}{6!} + \dfrac{x^8}{8!} - \ldots$ where the terms in the denominator are factorials. Graph several terms of this function, and hypothesize a more familiar equation for f.

All of the terms cannot be entered on a calculator, but the graphs in Figure 8.5a show the result of graphing the 8th, 14th, and 20th degree approximations, respectively, of f.

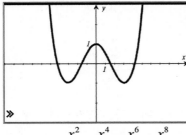

$$y_1 = 1 - \frac{x^2}{2!} + \frac{x^4}{4!} - \frac{x^6}{6!} + \frac{x^8}{8!}$$

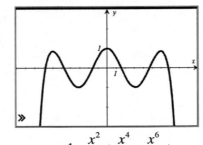

$$y_2 = 1 - \frac{x^2}{2!} + \frac{x^4}{4!} - \frac{x^6}{6!} +$$
$$+ \frac{x^8}{8!} - \frac{x^{10}}{10!} + \frac{x^{12}}{12!} - \frac{x^{14}}{14!}$$

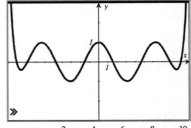

$$y_3 = 1 - \frac{x^2}{2!} + \frac{x^4}{4!} - \frac{x^6}{6!} + \frac{x^8}{8!} - \frac{x^{10}}{10!}$$
$$+ \frac{x^{12}}{12!} - \frac{x^{14}}{14!} + \frac{x^{16}}{16!} - \frac{x^{18}}{18!} + \frac{x^{20}}{20!}$$

Figure 8.5a

By adding terms to the approximation, several patterns become obvious.

1) The end behavior of the approximation flips up and down as the center portion expands.

2) The center of the graph seems to become increasingly stable with additional terms.

3) As the center stabilizes, the range of the middle part of the graph becomes $y \in [-1, 1]$.

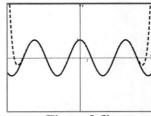

As the number of terms increases, the middle portion increasingly looks like $y = \cos x$. This could be tested by overlaying a graph of $y = \cos x$ (Figure 8.5b). Approximations for f appear to reach their minima closest to the y-axis just outside $x = \pm \pi$, precisely where they should for $y = \cos x$. While a proof requires calculus, these facts strongly suggest that $f(x) = \cos x$.

Figure 8.5b

The equation from Example 1, $\cos(x) = 1 - \dfrac{x^2}{2!} + \dfrac{x^4}{4!} - \dfrac{x^6}{6!} + \dfrac{x^8}{8!} - \ldots$, is called the Maclaurin series for cosine. The Maclaurin Series for cosine is deeply related to a new family of functions defined parametrically by $x = \dfrac{e^t + e^{-t}}{2}$ and $y = \dfrac{e^t - e^{-t}}{2}$.

Example 2:

For x and y defined parametrically as above, prove $x^2 - y^2 = 1$.

$$x^2 - y^2 = \left(\frac{e^t + e^{-t}}{2}\right)^2 - \left(\frac{e^t - e^{-t}}{2}\right)^2 = \frac{\left(e^{2t} + 2 + e^{-2t}\right)}{4} - \frac{\left(e^{2t} - 2 + e^{-2t}\right)}{4} = \frac{4}{4} = 1 \quad QED$$

This means ordered pairs of the form $(x, y) = \left(\dfrac{e^t + e^{-t}}{2}, \dfrac{e^t - e^{-t}}{2}\right)$ represent parametric coordinates of points on the graph of $x^2 - y^2 = 1$, but does this parameterization represent all points on $x^2 - y^2 = 1$?

Example 3:

Determine the t-domain and the x- and y-ranges of $(x, y) = \left(\dfrac{e^t + e^{-t}}{2}, \dfrac{e^t - e^{-t}}{2}\right)$.

Exponential functions are defined for all real values of their exponents, so the t-domain is $t \in \mathbb{R}$.

$x = \dfrac{e^t + e^{-t}}{2}$ is an even function of t with $\lim\limits_{t \to -\infty} \dfrac{e^t + e^{-t}}{2} = \infty$, so its minimum occurs at $t = 0$ and $x_{min} = \dfrac{e^0 + e^{-0}}{2} = 1$. The x-range is $x \in [1, \infty)$.

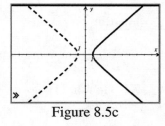

y is the sum of continuous functions, so y is also a continuous function. Because $\lim\limits_{t \to -\infty} \dfrac{e^t - e^{-t}}{2} = -\infty$ and $\lim\limits_{t \to \infty} \dfrac{e^t - e^{-t}}{2} = \infty$, the y-range is $y \in \mathbb{R}$.

The graph of $(x, y) = \left(\dfrac{e^t + e^{-t}}{2}, \dfrac{e^t - e^{-t}}{2}\right)$ is the solid curve on $x^2 - y^2 = 1$ (dashed curve) in Figure 8.5c.

Figure 8.5c

As these two functions define coordinates on the right side of the unit hyperbola ($x^2 - y^2 = 1$), they are used to define the basis of a variation on trigonometry. To parallel circular trigonometry, in **hyperbolic trigonometry,** the basic two functions are labeled as versions of sine and cosine.

$$\cosh x = \frac{e^x + e^{-x}}{2} \qquad \sinh x = \frac{e^x - e^{-x}}{2}$$

Here, "cosh" (hyperbolic cosine) rhymes with "gosh", and "sinh" (hyperbolic sine) is pronounced "sinch." Where cosine and sine from circular trigonometry define the respective x- and y-coordinates of all points on the unit circle, Examples 2 and 3 established that the hyperbolic cosine and hyperbolic sine define the x- and y-coordinates on the right side of the unit hyperbola, $x^2 - y^2 = 1$.

There are four other hyperbolic functions: $\tanh x$, $\coth x$, $\text{sech}\, x$, and $\text{csch}\, x$. In the same way as their circular trigonometry counterparts, these four functions are defined in terms of $y = \cosh x$ and $y = \sinh x$.

$$\tanh(x) = \frac{\sinh(x)}{\cosh(x)} \qquad\qquad \coth(x) = \frac{\cosh(x)}{\sinh(x)}$$

$$\text{sech}(x) = \frac{1}{\cosh(x)} \qquad\qquad \text{csch}(x) = \frac{1}{\sinh(x)}$$

Just as the hyperbolic trigonometric definitions parallel circular trigonometric definitions, identities in hyperbolic trigonometry also behave much like their circular counterparts. The basic Pythagorean identity from circular trigonometry is $\cos^2\theta + \sin^2\theta = 1$. Combining the results of Example 2 with the definition of the hyperbolic functions gives the equivalent basic Pythagorean identity from hyperbolic trigonometry: $\cosh^2 t - \sinh^2 t = 1$.

Problems for Section 8.5:

Exercises

Evaluate each expression in questions 1-3.

    1. $\sinh 0$ 2. $\tanh 0$ 3. $\cosh(\ln 3)$

4. Determine if each of the six hyperbolic trig functions is even, odd, or neither.

5. [NC] State a simplified expression for $\tanh(\ln x)$.

6. Use the hyperbolic Pythagorean identity, $\cosh^2 t - \sinh^2 t = 1$, to derive identities involving $\tanh t$, $\coth t$, $\text{sech}\, t$, and $\text{csch}\, t$. How are these related to the circular Pythagorean identities?

Explorations

Identify each statement in questions 7-11 as true or false. Justify your conclusion.

7. $\tanh(\ln 7) = \dfrac{24}{25}$

8. Every base hyperbolic trigonometry function is either even or odd.

9. The range of $y = \tanh x$ is $y \in \mathbb{R}$.

10. In the parametric definition of the right side of the unit hyperbola, $\begin{cases} x = \cosh t \\ y = \sinh t \end{cases}$, $t \in \mathbb{R}$ represents the angle at which any point (x, y) has been rotated from the positive x-axis.

11. Other than the point $(1,0)$, there are no other points anywhere on the graph of $\begin{cases} x = \cosh t \\ y = \sinh t \end{cases}$ for which both coordinates are integers.[9]

---

[9] HINT: Consider the Cartesian equivalent of the curve and what happens if one coordinate is an integer.

12. Another function defined by a Maclaurin series is $g(x) = \sum_{n=1}^{\infty} (-1)^{n+1} \dfrac{x^{2n-1}}{(2n-1)!}$. Write the first five terms of this series. Then graph several terms of g and predict its more common name.

13. The final Maclaurin series for this section is $h(x) = 1 + x + \dfrac{x^2}{2!} + \dfrac{x^3}{3!} + \dfrac{x^4}{4!} + \ldots$. Determine the more common name for h.

14. Use the Maclaurin Series from Example 1 and the previous two questions to prove the following Euler formulas for sine and cosine.

    A. $\sin x = \dfrac{1}{2i}\left(e^{ix} - e^{-ix}\right)$ B. $\cos x = \dfrac{1}{2}\left(e^{ix} + e^{-ix}\right)$

15. Prove.

    A. $e^{ix} = \cos x + i\sin x$ B. $e^{-ix} = \cos x - i\sin x$

16. A.   Prove $e^{i\cdot(x+2\pi)} = e^{ix}$.

    B.   Why are exponential functions with imaginary exponents periodic?

    C.   What is the period of exponential functions with imaginary exponents? Explain.

17. One of the most beautiful equations in all of mathematics includes five of the most important mathematical constants: $1 + e^{i\pi} = 0$. Prove this equation is true.

18. [NC] Determine values for $\cosh(ix)$ and $\sinh(ix)$ in terms of $\cos x$ and $\sin x$.

19. [NC] Determine values for $\cos(ix)$ and $\sin(ix)$ in terms of $\cosh x$ and $\sinh x$.

20. Determine hyperbolic identities for $\cosh(\alpha+\beta)$, $\sinh(\alpha+\beta)$, and $\tanh(\alpha+\beta)$. The results of the previous three questions may be helpful. Compare and contrast each with its circular counterpart.

21. Find an identity analogous to $\sin(2x) = 2\sin x\cos x$ for the hyperbolic functions.

22. Find an identity analogous to $\cos(2x) = \cos^2 x - \sin^2 x$ for the hyperbolic functions.

23. Determine a hyperbolic tangent double angle identity.

24. Give a simplified expression for $\dfrac{1 + \tanh x}{1 - \tanh x}$.

8.6: Graphs & Inverses of Hyperbolic Functions

We must rise above the antiquated approaches of earlier days and instead infuse our students with what I would call three "A's" of modern learning - the spirit of anticipation, the spirit of adaptation and the spirit of adventure.

— His Highness The Aga Khan, IV

All progress is precarious, and the solution of one problem brings us face to face with another problem.

— Martin Luther King, Jr.

One way to graph $y = \cosh x$ is to start with $y = \dfrac{e^x}{2}$ and bend its end behavior asymptote into the shape of $y = \dfrac{e^{-x}}{2}$. Then, the image of $y = \dfrac{e^x}{2}$ is the graph of $y = \cosh x$ (Figure 8.6a). But there is nothing in the equation $y = \dfrac{e^x}{2} + \dfrac{e^{-x}}{2}$ to say that $y = \dfrac{e^{-x}}{2}$ was the end behavior asymptote. Bending $y = \dfrac{e^{-x}}{2}$ into the shape of $y = \dfrac{e^x}{2}$ as the end behavior asymptote produces the same final result (Figure 8.6b). Both approaches are valid views of the graphical behavior of $y = \cosh x$. Therefore, $y = \cosh x$ has two end behavior asymptotes, $y = \dfrac{e^x}{2}$ and $y = \dfrac{e^{-x}}{2}$ (Figure 8.6c).

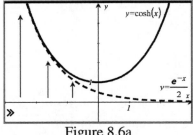

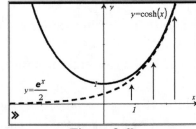

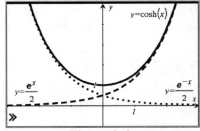

| Figure 8.6a | Figure 8.6b | Figure 8.6c |

The graph of $y = \cosh x$ is called a **catenary.** Physically, it is the shape a chain takes when hanging under its own weight while suspended between two end supports. Architect Eero Saarinen designed the Gateway to the West monument in St. Louis, MO as an inverted catenary. Figure 8.6d shows the Gateway and its reflection in the Mississippi River, an image of the graph of $y = \cosh x$.

Figure 8.6d[10]

Example 1:
Use algebraic analysis to determine the graph, domain, and range of $y = \tanh x$.

First, $y = \tanh x = \dfrac{e^x - e^{-x}}{e^x + e^{-x}} = \dfrac{e^{2x} - 1}{e^{2x} + 1}$. This function is odd, so analyzing $x \geq 0$ is sufficient.

- $y(0) = \tanh(0) = 0$

---

[10] Image source: http://en.wikipedia.org/wiki/Jefferson_National_Expansion_Memorial

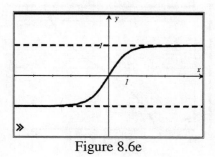

- $\lim\limits_{x \to \infty} \tanh(x) = \lim\limits_{x \to \infty} \dfrac{e^{2x} - 1}{e^{2x} + 1} = 1^-$ because the ± 1 in the numerator and denominator become irrelevant as e^{2x} approaches infinity. The function approaches $y = 1$ from below because the numerator is less than the denominator $\forall x \ge 0$.

- The graph is monotonically increasing $\forall x \ge 0$ because the numerator and denominator change at exactly the same rate.

Figure 8.6e

The left side of $y = \tanh x$ is the $180°$ rotation image about the origin of its right side (Figure 8.6e).

From the second bullet point, $\lim\limits_{x \to \infty} \tanh x = 1^-$, so the rotation gives $\lim\limits_{x \to -\infty} \tanh x = -1^+$ making the range $y \in (-1, 1)$. Finally, $y = e^{2x}$ is continuous $\forall x \in \mathbb{R}$ and the denominator of $\tanh x = \dfrac{e^{2x} - 1}{e^{2x} + 1}$ can never be zero, so the domain of $y = \tanh x$ is $x \in \mathbb{R}$.

Example 2:

Just as the circular trigonometry functions have inverse functions, so do the hyperbolic functions. Find an expression for $y = \text{arcsinh}\, x$ that does not involve any trigonometric functions.

Use the definition of $y = \sinh x$, switch x and y, and move all the terms to one side.

$$y = \sinh x = \frac{e^x - e^{-x}}{2} \implies x = \frac{e^y - e^{-y}}{2} \implies 2x = e^y - \frac{1}{e^y} \implies 0 = \left(e^y\right)^2 - 2xe^y - 1$$

The last line is a quadratic function in e^y, so

$$e^y = \frac{(2x) \pm \sqrt{(2x)^2 - 4 \cdot 1 \cdot (-1)}}{2 \cdot 1} = \frac{2x \pm 2\sqrt{x^2 + 1}}{2}.$$

This means $y = \ln\left(x \pm \sqrt{x^2 + 1}\right)$. Because $\sqrt{x^2 + 1} > x \quad \forall x \in \mathbb{R}$, $x - \sqrt{x^2 + 1} < 0$ and is outside the domain of the logarithm. Therefore, the only solution is $y = \text{arcsinh}\, x = \ln\left(x + \sqrt{x^2 + 1}\right)$.

Example 3:

Determine the domain, range, and a graph of $y = \sinh^{-1}(x)$.

$y = \sinh x$ (Figure 8.6f) is a one-to-one function with domain and range $\mathbb{R}$; so is $y = \text{arcsinh}\, x$. The graph of the inverse of a function is the reflection image of the original function over $y = x$.

The image of the asymptote of a function under a transformation is the asymptote of the image of the function under the same transformation, so reflecting $y = \frac{1}{2}e^x$ and $y = -\frac{1}{2}e^{-x}$ (the end behavior asymptotes of $y = \sinh x$) over $y = x$ creates the end behavior asymptotes of $y = \sinh^{-1}(x)$: $y = \ln(2x)$ and $y = -\ln(-2x)$, (Figure 8.6g). Reflecting $y = \sinh x$ over $y = x$ completes the graph (Figure 8.6h).

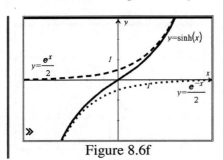
Figure 8.6f

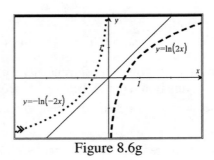
Figure 8.6g

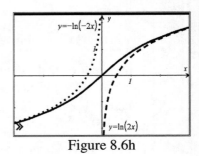
Figure 8.6h

Problems for Section 8.6:

Exercises

1. Use the reciprocal transformation to create a graph of $y = \operatorname{sech} x$.

2. Graph $y = \coth x$ using the reciprocal transformation.

3. Find an expression for $y = \tanh^{-1}(x)$ that does not involve any trigonometric functions. The answer involves a natural logarithm.

4. State the domain and range of $y = \sinh x$, $y = \operatorname{sech} x$, $y = \operatorname{csch} x$, and $y = \coth x$. Be sure to discuss x- or y-intercepts, and any asymptotes.

Explorations

Identify each statement in questions 5-9 as true or false. Justify your conclusion.

5. $y = \tanh^{-1}(x)$ is an odd function.

6. The graph of every inverse hyperbolic trigonometric function contains at least one point whose coordinates are both integers.

7. [NC] The equation $\cosh x = \sinh x$ has no real solutions.

8. All circular trigonometric functions are periodic, but no hyperbolic trigonometric functions are.

9. $f(x) = 1 + \tanh x$ is a logistic function.

10. From circular trigonometry that there is a difference in definition between $y = \cos^{-1}(x)$ and the inverse relation defined by $x = \cos y$.

   A. Explain why there is a similar distinction between $y = \cosh^{-1}(x)$ and the inverse relation defined by $x = \cosh y$.

   B. State an expression for $y = \cosh^{-1}(x)$ that does not involve any trigonometric functions.

   C. Graph $y = \cosh^{-1}(x)$.

   D. Determine the domain and range for $y = \cosh^{-1}(x)$.

11. Solve $\tanh x = e^x - 1$ for x. Confirm that you have found all of the solutions.

 Harrow & Merchant © 2010

12. Consider the family of curves defined by $y = a \cosh\left(\dfrac{x}{a}\right)$ for $a > 0$. Sketch graphs for $a = \{1, 2, 3\}$. Describe the effect on the family of curves caused by varying a.

8.7: Further Explorations and Projects

1. $\forall A, B \in \mathbb{R}$, explore the polar equation: $r(\theta) = \dfrac{A \cdot B}{1 + B\cos\theta}$.

   A. What types of graphs does r generate?

   B. What is the effect of varying B on the graphs of r?

   C. What appears to be significant about the pole for the graphs of r?

   D. What happens when you change cosine to sine in the equation of r?

   E. What is the effect of varying A on the graphs of r?

   F. What sort of graph is r? Prove your claim by changing the equation into its standard Cartesian form.

APPENDIX

Appendix A: From Minimizing Residuals to Linear Regressions

This Appendix explores the effects of fitting a line to a data set (Figure App.A1) and what can be learned by analyzing the residuals of a model for this data set.

xlist	1	2	2	3	4	5	6
ylist	5	5	3	4	1	1	2

Figure App.A1

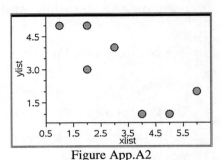

Figure App.A2 provides a scatter plot of the data showing a general downward trend from left to right. While no line could perfectly fit this data, lines are among the simplest functions and are easy to manipulate. As such, they are often among the first choices when initially exploring data sets.

Figure App.A2

An obvious goal for any model to data is for all of its residuals to have a sum of zero. This would signify that the model runs more-or-less through the center of the data set, but is it enough?

Example 1:
Based on the general pattern of the data, David thought the slope of the linear model should be -1. What y-intercept would make the sum of his residuals as close to zero as possible?

Rather than guessing-and-checking numerically, David decided to use a slider for the value of his y-intercept. Figure App.A3 shows the result of his construction with individual residuals and their overall sum displayed. The sum of residuals was 3.75 at his initial guess of a y-intercept of 5.75. By experimentally adjusting the intercept, David discovered that the sum of residuals for his line was essentially zero by using approximately $b = 6.2857$ as his y-intercept (Figure App.A4).

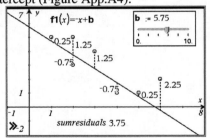

Figure App.A3

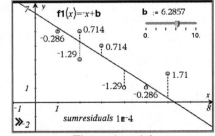

Figure App.A4

Example 2:
Lilly wondered what would happen if she used a slope of 1.5 for the linear model. What was the y-intercept of her line when her sum of residuals was zero?

While a slope of 1.5 certainly does not fit the given data pattern, Figure App.A5 shows that the sum of residuals was zero if the y-intercept was $b = -1.9286$.

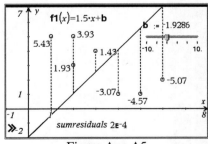

Figure App.A5

If a slope running nearly normal to the data pattern as in Example 2 could have a zero sum of residuals, one might wonder if a zero residual sum is possible with any slope. In fact, the Intermediate Value Theorem guarantees exactly this.

For any data set, if the graph of a linear model lies entirely below the scatter plot with y-intercept at some real-valued $y = b_1$, all residuals will be positive and therefore the sum of residuals is positive. Likewise, the sum of residuals must be negative for any linear model (with y-intercept at some real-valued $y = b_2$) which lies entirely above the same scatter plot. Without proof, it is reasonable to assume that the sum of residuals for a linear model with a fixed slope as in (Examples 1 and 2) is a continuous function. The Intermediate Value Theorem therefore guarantees that for any real-valued slope of a linear model to data that $\exists \hat{b} \in (b_1, b_2) \ni$ the sum of residuals is exactly zero. Starting with the results of Examples 1 and 2, Figure App.A6 gives several such values of $\hat{b}$.

Slope of Line	-1	1.5	3.5	0	-2	π
y-intercept $= \hat{b}$	6.2857	-1.9286	-8.5	3	9.5714	-7.3224

Figure App.A6

While a zero sum of residuals obviously remains an important goal for function models of data, the infinitely many lines with zero sum of residuals suggested by the preceding paragraphs makes it obviously clear that this condition may be necessary, but certainly is not sufficient for determining a "best fit model." So what does a zero sum of residuals guarantee?

Figure App.A7 shows a scatter plot of the current data set with all of the lines described in Figure App.A6 plotted. The relationship between all of the lines with a zero sum of residuals visually seems obvious. They coincide at a single point whose coordinates for this data set are about $(3.29, 3)$. Its location near the visual "center" of the data set suggests an algebraic derivation of the point.

A centroid of any data set is a point whose coordinates are

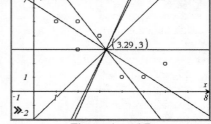

Figure App.A7

determined by the mean of its corresponding component variables.[11] For the data set in this Appendix, the centroid is determined by $\left(\overline{x}, \overline{y} \right) = \left(\dfrac{1+2+2+3+4+5+6}{7}, \dfrac{5+5+3+4+1+1+2}{7} \right) = \left(\dfrac{23}{7}, 3 \right) \approx (3.2857, 3)$, exactly the same coordinates as the point of intersection of the zero sum of residuals linear models in Figure App.A7.

Apparently, any line through the centroid will automatically have a zero sum of residuals, but obviously many of these lines are not appropriate models for the data. So which is "best"? To answer this, understand that "best" is entirely a qualitative judgment and what is best in one situation may not be best in another.

If there are infinitely many lines with a zero sum of residuals for any data set, *one way* to determine a "best fit line" from the infinite possibilities is to choose the line through the centroid whose sum of *squared* residuals is as small as possible. This way, the sum of residuals still can still be guaranteed to be zero, while for any non-collinear data set the squaring eliminates the possibility of the adjusted residuals numerically canceling. *Minimizing* the sum of squared residuals is clearly better than other alternatives.[12]

Figure App.A8 shows the results of varying the slope of all possible lines through the centroid of the data set to determine which had the smallest sum of squared residuals. For the given data, this happens with a slope of approximately -0.7721. Most graphing calculators can compute linear regressions on data. Figure App.A9 shows the graph and equation of the linear regression on the data set for this Appendix.

Notice that the directly computed slope of the linear regression in Figure App.A9 is exactly the same as the experimentally derived slope in Figure App.A8. Figure App.A10 expands the experimentally determined equation

---

[11] In physics, the centroid is the center of gravity of the set of points, if each has equal mass.
[12] There are certainly other ways the canceling could be avoided, but for reasons related to calculus, the sum of squared residuals offers one way to achieve a "best-fit line" with minimal computations.

showing that the two equations appear to be identical. In fact, a linear regression for a set of data is exactly the line through the median of the data that has the smallest sum of squared residuals.

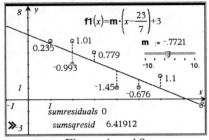

Figure App.A8

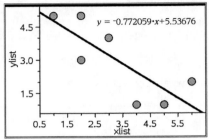

Figure App.A9

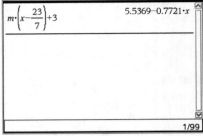

Figure App.A10

Unit 0 – Algebra & CAS Fundamentals
0.1 Mathematical Grammar & A Toolbox

...

1.

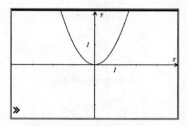

3.

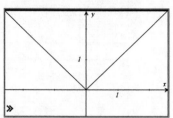

5.

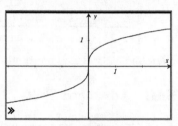

7.

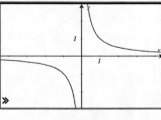

9.

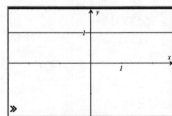

11. A. x is a non-negative integer
    B. x is a negative integer
    C. x is any rational number in simplified form whose denominator is not 1.
    D. x is the set of the squares of all integers.
    E. x is the set of all positive rational numbers which do not reduce to integers.
    F. x is the set of real-valued x-intercepts of the graph of $y = ax^2 + bx + c$.

13. False
15. True
17. $y = x$

19. $y = -x \cdot (x+1)(x-1)$
21. $y = \tan(x)$. NOTE: $y = x^3$ is not acceptable due to the way the graph crosses the origin.
23. A. $1 + 2 - 3 - 456789$
    B. $1 - 2 - 3 + 456789$
    C. $123 - 45 - 67 + 89$
25. $bc > 0$
27. $f(2) = 7/4$

0.2 CAS I: Basic Skills

1. A. $3 \cdot 42096529$ is not prime
    B. $2^2 \cdot 881 \cdot 35837$ is not prime
    C. Prime
    D. The numerator is prime, but the denominator is not.

3. The CAS returns $(\sqrt{x} - 4)(\sqrt{x} - 1) = 0$ which is easily solved with the zero product property. Application of the distributive property on the left returns the original equation, confirming the CAS results.

5. The CAS returns all 198 digits and the last five are ...32437.

7. $7a^4 - 56a^3 + 168a^2 - 223a + 105$

9. $-2729024 - 1597632i$

11. There are two such points where $y = -1 \pm 2\sqrt{6}$ as CAS-computed below.

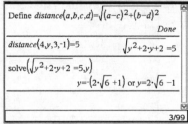

13. False
15. True
17. $7625597484987 = 3^{27}$, so there are 28 distinct factors (including the number 1), of which 3 is the only prime.

19. The CAS says "true," indicating that the two expressions are equivalent. Both are obviously positive, so one way to prove the equivalence is to square both expressions giving $2 + \sqrt{3}$ on both sides.

21. 599

23. Among the CAS factorization of 200! are 2^{197} and 5^{49}, so there are 49 zeros at the end.

25. $\log(300!) \approx 614.5$, so there are 615 digits in the expansion of 300!.

27. $x^2 + y^2 = 1$. This could be derived using a CAS distance function.

0.3 CAS II: Graphing Skills

1. The three points of intersection can be found simultaneously using the intersection tool:

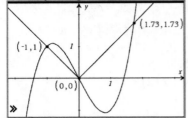

3. $y = x^4 - x^2 + 173$ is even:

$$y(-x) = (-x)^4 - (-x)^2 + 173$$
$$= x^4 - x^2 + 173$$
$$= y(x)$$

5. $y = \sqrt[3]{x}$ is odd:

$$y(-x) = \sqrt[3]{(-x)} = -\sqrt[3]{x} = -y(x)$$

7.

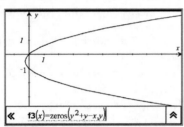

9. There are two possible values: $a = \pm 2\sqrt{6}$.

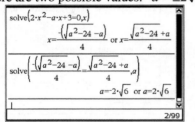

11. True

13. True

15. $a = -\dfrac{10}{7}$ and $b = \dfrac{31}{10}$. The other point of intersection is about $(-2.034, -14.375)$.

17. Assume $(0, a) \ni a \neq 0$ is a y-intercept of a continuous odd function. By point-symmetry, $(0, -a)$ must also be on the graph. But, this gives two outputs for one input, contradicting the definition ofr a function. Therefore $a = 0$, and every continuous odd function contains the origin.

19.

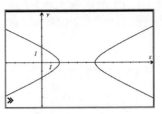

21.

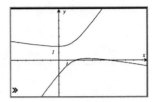

23.

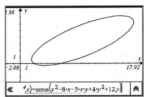

25. If $h \in (-2, 2)$, there are 2 solutions. There is 1 solution if $|h| = 2$ and no solutions otherwise.

0.4 CAS III: Advanced Skills

1. $(x, y, z) = \left(\dfrac{5\pi - 1}{4}, -\dfrac{43\pi + 1}{44}, \dfrac{117\pi + 15}{44} \right)$

3. A. quotient $= x - 2$ with remainder 3
   B. quotient $= x^3 + x^2 + x - 7$ with remainder -14
   C. quotient $= 4x - \dfrac{9}{2}$ with remainder
   $$-\dfrac{7}{2}x + \dfrac{29}{2}$$

5. A. $\dfrac{55835135}{15519504}$
   B. 1
   C. e
   D. $\dfrac{\pi^2}{6}$
   E. e^x
   F. $\sqrt{e}$
   G. $\dfrac{1073741823}{1073741824}$
   H. $\dfrac{1}{32} - i$

7. True

9. True

11. There are two such points: $\left(0, -7 \pm \dfrac{52\sqrt{5}}{5} \right)$.

13. Pi notation for products ((menu) → (4):Calculus → (5):Product) has virtually the same syntax as sigma notation for sums.

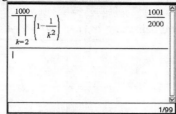

15. A. $\dfrac{67}{70} = \dfrac{6}{7} + \dfrac{1}{10}$

   B. $\dfrac{35}{26} = \dfrac{11}{13} + \dfrac{1}{2}$

17. A. $\dfrac{-10x-18}{x^2-9x} = \dfrac{2}{x} + \dfrac{-12}{x-9}$

   B. $\dfrac{-11x^2+x+64}{x^3+x^2-9x-9} = \dfrac{-19/6}{x+3} + \dfrac{-4/3}{x-3} + \dfrac{-13/2}{x+1}$

Unit 1 – Powers, Polynomials, & Functions
1.1 Power Functions

1. Power function. Single term with real-valued coefficient (3) and exponent (2).

3. Power function. Single term with real-valued coefficient (1/5) and exponent (1).

5. Not a power function; x is in the exponent.

7. Power function. This reduces to $y = 4$ which fits the power function constraints.

9. Power function. This is equivalent to $y = x^4$ which fits the power function constraints.

11. Domain: $x \in \mathbb{R}$, Range: $y \in \mathbb{R}$, odd function symmetry, continuous $\forall x \in \mathbb{R}$, end behavior: $x \to -\infty$, $y \to -\infty$ and as $x \to \infty$, $y \to \infty$.

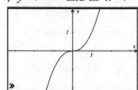

13. Domain: $x \in \mathbb{R} \ni x \neq 0$, Range: $y \in (0, \infty)$, even function symmetry, continuous $\forall x \in \mathbb{R} \ni x \neq 0$, end behavior: $x \to -\infty$, $y \to 0$ and as $x \to \infty$, $y \to 0$.

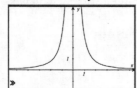

15. Domain: $x \in \mathbb{R}$, Range: $y \in [0, \infty)$, even function symmetry, continuous $\forall x \in \mathbb{R}$, end behavior: $x \to \pm\infty$, $y \to \infty$.

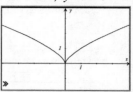

17. Domain: $x \in \mathbb{R}$, Range: $y \in \{4\}$, even function symmetry, continuous $\forall x \in \mathbb{R}$, end behavior: $x \to \pm\infty$, $y = 4$.

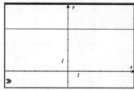

19. Power functions $\left(y = a \cdot x^b \right)$ are concave down $\forall b \in (0,1)$ and concave up $\forall b \notin [0,1]$.

21. $d = \dfrac{3200}{23} ft$

23. $y = 3x^{2/3}$

25. $y = -2x^{1/4}$

27. True

29. True

31. False

33. A. a and b are both odd.
   B. a is even and b is odd.
   C. a has no restrictions, while b is non-zero and even.
   D. $a < 0$ and b is non-zero.

35. Let $f(x) = a \cdot x^b$. Then,
$$f(-x) = \begin{cases} a \cdot (-x)^b = a \cdot x^b \text{ if } b \text{ is even} \\ a \cdot (-x)^b = -a \cdot x^b \text{ if } b \text{ is odd} \end{cases}.$$
So, f is even if b is even and odd if b is odd.

1.2 Linear & Quadratic Functions

1. parallel: $2x + 5y = 13$,
   perpendicular: $5x - 2y = -11$

3. A. I or III
   B. II only
   C. none
   D. I only
   E. IV only
   F. I only

5. A. There must be two real solutions. The graph is a parabola, facing upward, and with a negative y-intercept, so it must cross the x-axis twice.
   B. $x \in \{-2, 4\}$

7. A. Vertex form because the vertex is given.

    B. $y = \frac{3}{4}(x-1)^2 - 1$

9. A. No special points given, so standard form.

    B. $y = x^2 + 7x - 2$

11. zeros: $x = \pm 5/4$, Vertex $= (0, -25)$

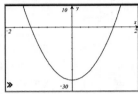

13. zeros: $x \in \left\{ -\frac{1}{4}, \frac{2}{3} \right\}$, Vertex $= \left(\frac{5}{24}, -\frac{121}{48} \right)$

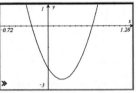

15. $4 + 5i$

17. $-25 - 5i$

19. $-i$

21. $\left(4 + 5\sqrt{26} \right) + 7i$

23. $\sqrt{17}$

25. 8

27. True

29. True

31. True

33. $S_{\frac{1}{m}, 1}$ followed by $T_{-\frac{b}{m}, 0}$

35. A is the x-intercept and B is the y-intercept. Only oblique lines can use this linear form.

37. A. $T_{-\frac{B}{2A}, C - \frac{B^2}{4A}}$, sliding the vertex from the origin to its new position.

    B. $S_{\frac{1}{\sqrt{A}}, 1}$

39. $x = \frac{1 + \sqrt{5}}{2} = \phi$

41. A. Sometimes

    B. Sometimes

    C. Sometimes

    D. Sometimes

    E. Sometimes

1.3 Generic Polynomial Behavior

1. Remainder $= 0$, so $x = 2$ is an x-intercept

3. $y = \frac{29}{672}x^3 + \frac{9}{112}x^2 - \frac{1319}{672}x + 0$

5. $y = Ax^2(x-4)^2 \ni A > 0$

7. $y = B(x+3)(x+1)^2(x-1)(x-3)^3 \ni B > 0$

9. $y = C(x+3)(x+1)(x-1) \ni C < 0$

11. zeros: $x \in \left\{ -\frac{1}{2}, 0, \frac{1}{2} \right\}$, y-intercept $= (0,0)$,

    end behavior: as $x \to \pm\infty$, $y \to \infty$.

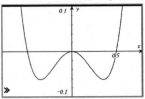

13. zero at $x = \frac{2}{3}$, y-intercept $= (0,8)$, end behavior: as $x \to -\infty$, $y \to \infty$, and as $x \to \infty$, $y \to -\infty$.

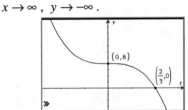

15. Choice A only

17. A. $x \to -\infty$, $y \to -\infty$, and as $x \to \infty$, $y \to \infty$.

    B. All real roots $\in (2,3)$

    C. Extrema are near $(0,-3)$ and $(1,-4)$.

    D. The inflection point is near $(0.5, -3.5)$

19. False

21. False

23. False (but could be)

25. True

27. One root must be 0 with even multiplicity. The other two roots must be numerical opposites with identical multiplicity.

29. A. $(2,10)$ is on every such curve.

    B. There are infinitely many answers, or $y = 7(x-1)^3 + 3$ and
$y = -\frac{7}{3}\left((x-1)^3 - 4(x-1) \right) + 3$ are two.

31. The extrema are all on $y = -2x^3$.

33. The common difference property, as stated, applies only to polynomials.

1.4 Limit Notation

1. 3

3. 1

5. ∞

7. 0

9. 0^+

11. 4

13. 1/8

15. ∞

17. -6
19. DNE
21. ∞
23. 1
25. 1
27. DNE
29. False
31. False
33. $-|m|$
35. Power functions with ∞ end behavior as
   $x \to \infty$, have exponents ≥ 1. So, $a \leq \dfrac{4}{7}$.
37. e
39. A. $h(x) = 1$
   B. 1

1.5 Piecewise Functions

1. $f(x) = \begin{cases} 3x - 2, & x \geq 2/3 \\ 2 - 3x, & x < 2/3 \end{cases}$

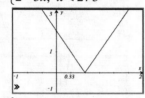

3. $g(x) = \begin{cases} x^2 - 5x + 6, & x \leq 2 \\ -x^2 + 5x - 6, & 2 < x \leq 3 \\ x^2 - 5x + 6, & 3 < x \end{cases}$

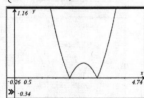

5. $h(x) = \begin{cases} -1, & x < -4 \\ 1, & x > -4 \end{cases}$

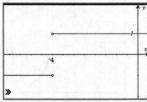

7. $f(x) = \begin{cases} 2(x+4) - 2, & x < -3 \\ -3 - x, & x \geq -3 \end{cases}$

9. $g(x) = \begin{cases} -(x-2)^2 + 3, & x \leq -2 \\ -1.5x, & -2 < x < 2 \\ (x-2)^2 - 3, & 2 \leq x \end{cases}$

11. $x = 2$ or $x = 8$
13. $x = 0$ or $x = -2$ or $x = -1 \pm \sqrt{17}$

15. $x \in \left(-\infty, \dfrac{1}{2}\right) \cup (1, \infty)$
17. A. 1
   B. 0
   C. DNE
   D. No solutions
19. True
21. True
23. $x \in [0, 1]$
25. $x \in [-1, 1] \cup [2, 4]$
27. $x = -\dfrac{2}{5}$ or $x = \dfrac{5 \pm \sqrt{41}}{4}$
29. $x = 2.5$
31. $x < -3$
33. 0
35. $a^{5/2}$

1.6 Continuity & the IVT

1. Continuous $\forall x \in \mathbb{R}$
3. Vertical asymptote discontinuity at $x = 0$
5. Vertical asymptote at $x = 0$ and a hole at $x = 8$.
7. Jump discontinuity at $x = 1$.
9. Continuous $\forall x \in \mathbb{R}$
11. Continuous $\forall x \in (-\infty, 4) \cup (4, \infty)$
13. No. For this example, the IVT guarantees outputs only in the interval $y \in [1, 16]$.
15. True
17. False
19. False
21. 0
23. $a = -17$

<u>Unit 2 – Transformations</u>
2.1 Basic & Variable Transformations

1. $f(-1) = 5$, $g(3) = -5$, $g(8) = 13$, $g(9) = 7$, and $h(1) = 8$
3. A. $T_{0,2}$ and $T_{-2,0}$
   B. $T_{0,x}$
5. $f(x) = T_{0, -\frac{13}{3}} \circ S_{1, \frac{7}{3}} \circ g(x)$
7. $S_{1,8}$ and $S_{\frac{1}{2}, 1}$
9. False
11. True
13. False

15.

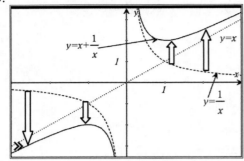

17.

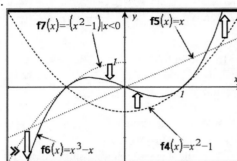

19.

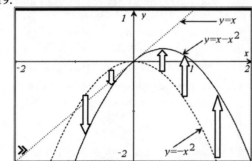

21. $g_{MAX} = T_{2,5}(-0.63, 0.38) = (1.37, 5.38)$, and
$g_{MIN} = T_{2,5}(0.63, -0.38) = (2.63, 4.62)$.

2.2 Absolute Value Transformations

1. A.

x	-4	-3	-1	0	4
$h(x)$	3	9	0	9	5

B.

x	-4	0	4
$i(x)$	-5	9	-5

C.

x	-2	2	6
$j(x)$	-5	9	-5

D.

x	-4	-3	-1	0	4
$k(x)$	9	3	6	15	1

3. $y = \sqrt[3]{|x|}$ and $y = |\sqrt[3]{x}|$ are identical.

5. A.

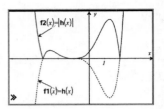

B.

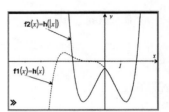

C.

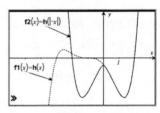

D.

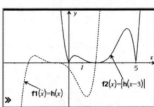

7.

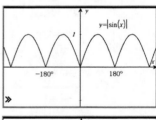

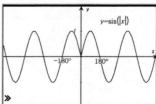

9. False

11. True

13. $2n$

15. In both cases, the solution is the y-intercept of $y = m(x)$.

17. $y = |f(x)|$ could have as few as 0 corners and as many as the number of x-intercepts of f.

2.3 The Squaring Transformation

1.

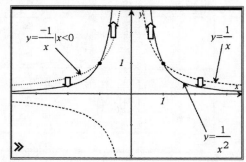

3. A.

x	-4	-3	-1	0	4
$h(x)$	9	81	0	81	25

B.

x	-2	0
$i(x)$	0	-5

C.

x	-1	0	2	3	7
$j(x)$	9	81	0	81	25

D.

x	-4	0	4
$k(x)$	25	81	25

5.

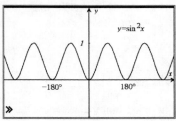

7. False
9. True
11. True
13. One possibility is $i(x) = x^2 - 1$.

15. $x = ABS(x) \Rightarrow x = \begin{cases} x, & x \geq 0 \\ -x, & x < 0 \end{cases}$, and this is

true for $x \geq 0$ only.

17. One possibility is $y = \sqrt[3]{x}$.

2.4 The Square Root Transformation

1. $\forall x \geq 0$, $x = SR(x) \Rightarrow x^2 = \left(SR(x)\right)^2 = x$, so
at 1 and 0.

3. A.

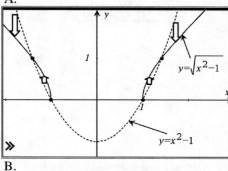

B.

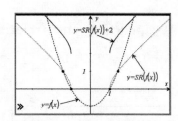

C.

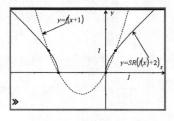

D.

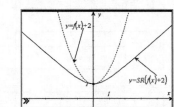

5. Domain = $[-3, \infty) \cap \left((-\infty, -2] \cup [2, \infty)\right)$, so

Domain = $x \in [-3, -2] \cup [2, \infty)$, and

Range = $y \in [0, \infty)$.

7. The graphs for parts A and B are identical.
Depending on the assumed degree of
$y = g(x)$, the x-intercepts of the image may
be corners or bounces.

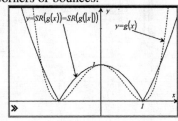

9. $y = \sqrt{\sin^2 x} = |\sin x|$

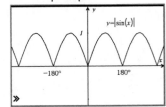

11. False
13. True

15. $\dfrac{\sqrt{10}+\sqrt{2}}{2} = \sqrt{\left(\dfrac{\sqrt{10}+\sqrt{2}}{2}\right)^2} = \sqrt{\dfrac{12+2\sqrt{20}}{4}} = \sqrt{3+\sqrt{5}}$

17. A possibility: $f(x) = (x+1)^4 (x-1)^2 (x-2)^2$

19. For any function f whose range is a subset of

$y \in (0,\infty)$, then let $h(x) = \dfrac{|f(x)|}{f(x)} = \dfrac{f(x)}{|f(x)|}$.

$h(x) = 1$ is an easy example.

2.5 The Reciprocal Transformation

1. $\forall x \neq 0$, $x = REC(x) = \dfrac{1}{x} \Rightarrow x^2 = 1$, so ± 1.

3. A.

x	-4	-3	0	4
$h(x)$	1/3	-1/9	1/9	-1/5

B.

x	-1/4	-1/3	-1	1/4
$i(x)$	3	-9	0	-5

C.

x	-7	-6	-3	1
$j(x)$	1/3	-1/9	1/9	-1/5

D.

x	-1/4	-1/3	-1	1/4
$k(x)$	1/4	-1/8	1	-1/4

5. A.

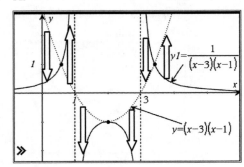

B.

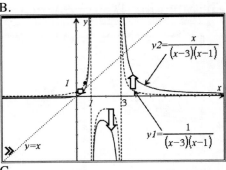

C.

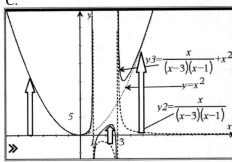

7. True
9. True
11. True
13. All have $(x-2)$ and $(x+2)$ as factors of the denominator, and all have a higher degree expression in the denominator.
15. A. *SR* to eliminate the portion of the domain $\forall x < -4$ and *REC* to create the vertical asymptote at $x = -4$.

B. $y = \dfrac{1}{2}(x+4)$

17. $\dfrac{y-3}{2} = \dfrac{1}{x-4}$

19. There are many answers. One is

$y = -g(x) + 2$ where $g(x) = 2 + \dfrac{2}{x^2}$.

2.6 Inverses of Functions

1. A. $f^{-1}(3) = 2$ B. $f(4) = 5$
    C. Cannot be determined

3. A. $f^{-1}(x) = \dfrac{x+7}{3}$ B. $j^{-1}(x) = (x^2 - 1)^2$

5. A.

X	3	-9	0	9	-5
$h(x)$	-4	-3	-1	0	4

B.

x	3	-9	0	9	-5
$i(x)$	16	9	1	0	16

C.

x	3	-9	0	9	-5
$j(x)$	4	3	1	0	4

D.

x	3	-9	0	9	-5
$k(x)$	-1/4	-1/3	-1	0	1/4

7. False
9. True
11.

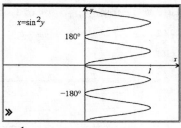

13. $x = \dfrac{1}{(y-4)^2} + 2$

15. $z^{-1}(x) = \begin{cases} 2x+3, & x \in (-\infty,-1) \\ \frac{1}{3}(x+4), & x \in [-1,2) \\ x, & x \in [2,\infty) \end{cases}$

Unit 3 – Exponentials, Logs, & Logistics
3.1 Exponential Functions
1. A. All three are geometric with common ratios 2, 3, and -2, respectively.
　B. With its negative ratio, Data Set III cannot form an exponential function.
3.

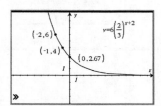

5.

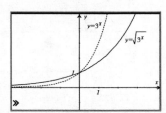

7.

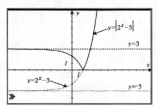

9. Graph I could be $y = -2 + 3^x$, and Graph V could be $y = -\pi^x$.

11. $y = 3 - \left(\dfrac{1}{4}\right)^x$

13. $y = -3 \cdot \left(\dfrac{1}{2}\right)^x + 2$

15. $y = 120 \cdot \left(\dfrac{1}{6}\right)^{\frac{x+1}{3}}$

17. $y = \dfrac{2}{3} \cdot \left(\dfrac{\sqrt{2}}{4}\right)^{\frac{x+2}{3}}$

19. 190147590034234410224505480864
21. False
23. False
25. True
27. Method 3: $y = 3 \cdot 2^{x+2} = 3 \cdot 2^x \cdot 2^2 = 12 \cdot 2^x$

Method 4: $y = 3 \cdot 8^{\left(\frac{x+2}{3}\right)} = 3 \cdot 2^{x+2} = 12 \cdot 2^x$

29. b, 1, a, c
31. $y = \dfrac{1}{10^x} = 10^{-x}$ which is the reflection of $y = 10^x$ over the y-axis.

3.2. Polynomials versus Exponentials
1. $\left(\dfrac{1}{2}\right)^x, \dfrac{1}{x}, \dfrac{1}{\sqrt[3]{x}}, 1, 10^{10} \cdot x^2, x^3, 3^x, 200 \cdot 5^x, 10^x$
3. $(3, 2048)$
5. There are three solutions.
7. A. $y = \{5, 25, 125, 625, 3125\}$
　B. $\Delta y = \{20, 100, 500, 2500\}$
　C. $\Delta y = 4 \cdot 5^x$
9. True
11. True
13. True
15. $d = 3$ and $a \approx 1.25606$
17. $b = 2$ and $a \in \mathbb{R}$

3.3 The Number e
1. 7610 or 7611
3. A. \$577.18
　B. \$584.88
　C. \$586.45
5. −0.0088 or 0.88% decay per day
7. True
9. True
11. True
13.

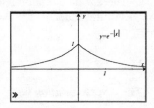

15.

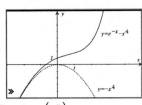

17. $y = e^{-2x} = S_{-\frac{1}{2},1}(e^x)$

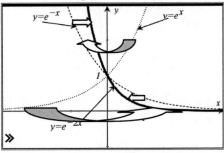

19. $S_{2,1}(e^x) = e^{x/2} = (e^x)^{1/2} = SR(e^x)$

21. e

3.4 Inverses of Exponential Functions

1. 5

3. -4

5. $\log_6 216 = 3$

7. $\left(\dfrac{3}{4}\right)^{-2} = \dfrac{16}{9}$

9. $\log\left(\dfrac{25}{0.5^2}\right) = \log(100) = 2$

11. $6^{\log_6 3^2} = 6^{\log_6 9} = 9$

13. $\log_{2^2} 5^2 + \log_4 5^3 = \log_4 (5^2 \cdot 5^3) = \log_4 (5^5)$

15. A. $x + 2y + z$

    B. $\dfrac{1}{2}x + \dfrac{3}{2}y$

    C. $x - 2y + \dfrac{z}{2}$

17. $\dfrac{8e^2 - 3}{1 + e^2}$

19. 10000

21. 1/64

23. 1/2

25. $\log_3 4$

27. False. There is no common real values in the domains of the addends.

29. True

31. True

33. $h(x) = 3 \cdot 2^{\log_7 x}$

35. A. The inverse *relation* is $y = x \pm \sqrt{x^2 - 1}$.

    B. $4 \pm \sqrt{15}$

37. A. $f(x) = \log_{1.05}\left(\dfrac{x}{200}\right) - 1$

    B. $\log_{1.05}\left(\dfrac{5}{4}\right) - 1$

    C. 210

3.5 Graphs & Transformations of Logarithms

1. $y(-x) = \ln(-x)^2 = \ln x^2 = y(x)$ *QED*

3. $5^{-1/12}$

5.

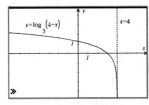

7.

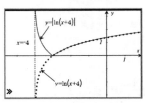

9.

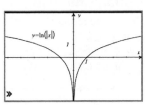

11. This is the right side of the graph of #10.

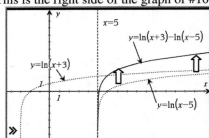

13.

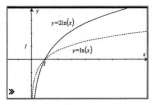

15. This is the right side of question 14.

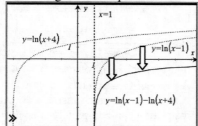

17.

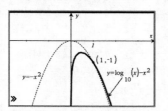

19.

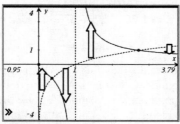

21. False. Not true if n is odd.

23. False. There are three points of intersection.

25. $\ln(x) = \dfrac{\log(x)}{\log(e)} = S_{1,\frac{1}{\log(e)}}\left(\log(x)\right)$

27. A.

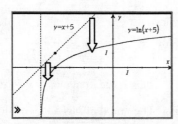

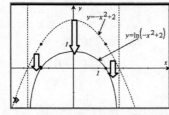

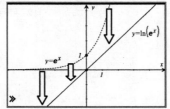

B. Intercepts: $(-1.5,0)$, $(0.5,0)$, $(0,\ln 2)$

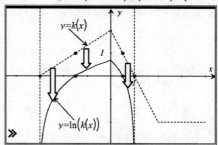

29. There are *many* possibilities. One pair is
$f(x) = x - 2$ and $g(x) = -x - 2$.

3.6 Logistic Functions

1. $y1 = 3$, $y2 = 10$, $y3 = 0.1$, $y4 = 0.5$

3. $(14,36)$

5. $(-2,12)$

7.

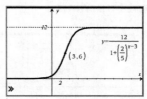

9.

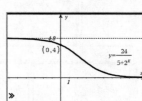

11. $y = \dfrac{15}{1 + \dfrac{3}{2} \cdot \left(\dfrac{4}{3}\right)^x}$

13. $y = \dfrac{30}{1 + 0.25 \cdot 2^x}$

15. True

17. True

19. True

21. $\left(\log_2 5, \dfrac{12}{5}\right)$

23. There are many answers, all of the form

$y = \dfrac{c}{1 + a \cdot b^x} \ni a = \dfrac{c-2}{2}$ and $b = \sqrt[3]{\dfrac{c-4}{2c-4}}$

25. $y = \dfrac{a}{1 + b^{x-c}} = \dfrac{a}{1 + b^{-c} \cdot b^x}$, so the forms are

equal for $d = b^{-c}$.

27. The two forms are equivalent, so it suffices to show for one. $\forall b \in (0,1)$, $\lim\limits_{x \to -\infty} (1 + b^{x-c}) = \infty$

and $\lim\limits_{x \to \infty} (1 + b^{x-c}) = 1$. Therefore,

$$\lim\limits_{x \to -\infty} \left(\frac{a}{1 + b^{x-c}} \right) \to \frac{a}{\infty} \to 0 \text{ and}$$

$$\lim\limits_{x \to \infty} \left(\frac{a}{1 + b^{x-c}} \right) = \frac{a}{1} = a. \text{ The limits are}$$

reversed $\forall b \in (1, \infty)$.

Unit 4 – Data & Rational Functions
4.1 A Review of Regressions

1. $MODEL(x) = 10 - x$
3. True
5. False
7. True

4.2 Rational Functions I

1.

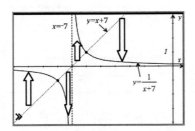

3.

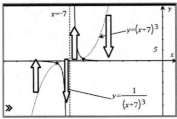

5. $y = \dfrac{-3x^2 + 6x}{(x-1)^2} = -3 + \dfrac{3}{(x-1)^2}$

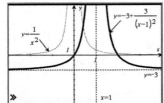

7. $y = \dfrac{x+2}{x^3 + 6x^2 + 12x + 8} = \dfrac{x+2}{(x+2)^3} = \dfrac{1}{(x+2)^2}$

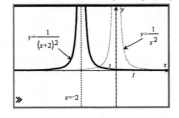

9. True
11. False
13. False

15. $f(x) = \dfrac{3x-2}{x^2 + x - 12} = \dfrac{2}{x+4} + \dfrac{1}{x-3}$. So, f has odd vertical asymptotes at $x = -4$ and $x = 3$.

17. A rational function has a hole when a denominator factor causing a domain restriction can completely cancel out. Otherwise, the domain restriction is a vertical asymptote, even if it partially cancels.

19. One of *many*: $y = \dfrac{(x+1)^2}{(x+1)(x-3)^2}$

21. (high+low)/2 = 79
23. 6 miles

4.3 Rational Functions II

1. $x = -1.753$, $x = 0.5756$, and $x = 5.386$

3. $y = \dfrac{1}{2}x - \dfrac{5}{4}$ and $x = \dfrac{3}{2}$

5. If the degree of n is less than that of d, then the quotient must be zero by the rules of polynomial division. Because the quotient is always the EBA of a rational function, that means the x-axis ($y = 0$) is always the EBA of such rational functions.

7. True
9. False

11. $\left(-1, -\dfrac{1}{12} \right)$

13. Sample: $y = x^2 + 3 + \dfrac{1}{(x-3)^2}$

15. A. This is the image of $y = \dfrac{1}{x}$ under $T_{-\pi, 0}$.

    B. The graph of b is a line of slope 1 with a hole at $(-\pi, 0)$.

    C. c has no domain restrictions, and its graph is the parabola defined by $y = x^2 + \pi^2$.

17. Sample: $y = \dfrac{3(x-2)(x+3)(x+5)}{(x-4)^2 (x+5)}$

19. Sample: $y = (x-5)^2 - \dfrac{(x-5)^2 (x-1)}{x^4 (x-2)}$

21.

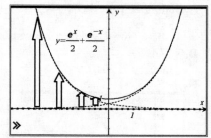

23. EBA: $y = \log x$, odd VA at $x = 10$, even VA
at $x = 100$, and linear EBA crossing at
$x = 10^{1.5} \approx 31.6$.

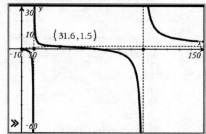

<u>**Unit 5 – Trigonometry I**</u>

5.1 Angle Measure, Arc Length, & …

1. $5\pi/3$

3. $7\pi/2$

5. $420°$

7. $257.831°$

9. $510°$, $-210°$, $150° + 360°n \ni n \in \mathbb{Z}$

11. $\dfrac{6\pi}{5}$, $-\dfrac{4\pi}{5}$, $-\dfrac{14\pi}{5} + 2\pi n \ni n \in \mathbb{Z}$

13. Quadrant I

15. $10\pi/9$

17. $49\pi/16$

19. $\theta = 16/5$, $r = 5$

21. True

23. True

25. 5.162

27. A. $390°$

   B. $7/12$ sec

29. 0.1005

5.2 The Trigonometric Functions

1. $3/\sqrt{7}$

3. $\sin\theta = 2/\sqrt{13}$, $\cos\theta = -3/\sqrt{13}$,
   $\tan\theta = -2/3$, $\csc\theta = \sqrt{13}/2$, $\cot\theta = -3/2$

5. $\pi/4$

7. 1

9. $\sqrt{2}$

11. $1/2$

13. $\sqrt{3}/2$

15. undefined

17. $-\sqrt{2}$

19. 0

21. undefined

23. $\pi/4$

25. $\pm\dfrac{\pi}{4} + 2\pi n \ni n \in \mathbb{Z}$

27. $-a$

29. $1/a$

31. $\sqrt{1-a^2}$

33. $\dfrac{\sqrt{1-a^2}}{a}$

35. 0.848

37. $\csc 6$, $\cos 2$, $\sin 3$, $\cot 4$, $\tan 1$, $\sec 5$

39. $35.754°$

41. True

43. False

45. False

47. $\overline{OA} = \cos\theta$, $\overline{AB} = \sin\theta$, $\overline{OD} = \sec\theta$
   $\overline{CD} = \tan\theta$, $\overline{OF} = \csc\theta$, $\overline{EF} = \cot\theta$

5.3 Graphs of Trigonometric Functions

1.

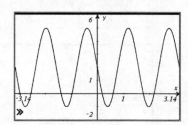

3.

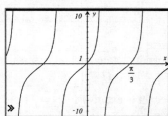

5. $y = 2\cos\left(3\left(x + \dfrac{\pi}{12}\right)\right) + 2$

7. $y = \sec\left(\dfrac{x}{3}\right) + 5$

9.

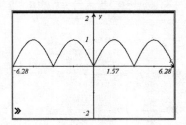

11.

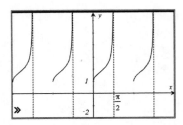

13.

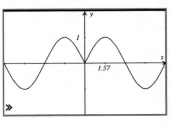

15.

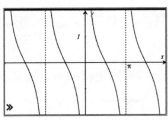

17.

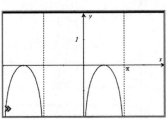

19. True

21. False

23. Tangent, cotangent and cosecant are odd;

$$f(-x) = -f(x).$$

Secant is even; $f(-x) = f(x)$.

25. Example: $y = |\tan x|$

5.4 Solving equations and sinusoidal applications

1. $x = \pm\dfrac{\pi}{3} + 2\pi n$ or $x = \pi + 2\pi n \ni n \in \mathbb{Z}$.

3. $x = \dfrac{\pi}{4} + \dfrac{\pi}{2}n \ni n \in \mathbb{Z}$.

5. $\forall n \in \mathbb{Z}$, $x = \dfrac{7\pi}{6} + 2\pi n$, $x = \dfrac{11\pi}{6} + 2\pi n$, or

$x = \dfrac{5\pi}{4} + \pi n$

7. $x = \dfrac{\pi}{12}, \dfrac{5\pi}{12}, \dfrac{13\pi}{12}, \dfrac{17\pi}{12}$

9. A. 52.5 m

   B. 8.461 sec., descending.

11. A. Sample: $I(t) = \dfrac{1}{2}\cos\left(\dfrac{\pi}{15}(t-7)\right) + \dfrac{1}{2}$

   where t the days since $t = 1$ at June 1.

   B. 1 / 4

   C. January 7, 2010.

13. Nome:

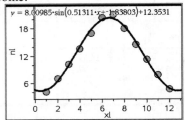

Stuttgart:

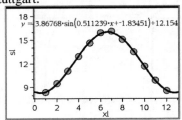

Dar-Es-Salaam:

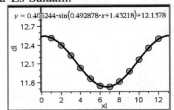

15. South; longest days are in December/January.

17. At about 22 minutes after midnight.

19. True

21. True

23. True

5.5 Advanced transformations

1.

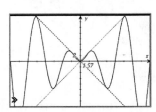

3.

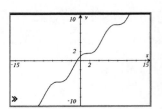

5.

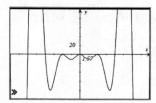

7.

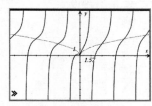

9.

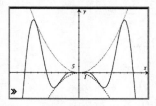

11.

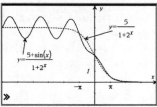

13. $y = |x| + \cos x$

15. True

17. False

19. $\tan \theta$

21. A.

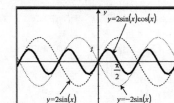

B.

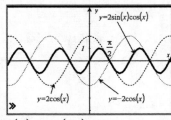

C. $f(x) = \sin(2x)$

23. A.

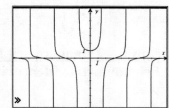

B. Even; odd divided by odd

C. $\displaystyle\lim_{x \to \infty} \frac{\tan x}{x}$ does not exist.

5.6 Inverse Trigonometric functions

1. $x = -\pi / 3$

3. A. $x = \dfrac{\pi}{3}, \dfrac{5\pi}{3}$

   B. $x = \dfrac{7\pi}{18}, \dfrac{11\pi}{18}, \dfrac{19\pi}{18}, \dfrac{23\pi}{18}, \dfrac{31\pi}{18}, \dfrac{35\pi}{18}$

5. $\pi / 3$

7. $-\pi / 6$

9. $25 / 24$

11. $\pi / 2$

13. $25 / 144$

15. $-\pi / 6$

17. $1 / 2$

19. Domain: $x \in [-1, 1]$; Range: $y \in [0, 3\pi]$

21. Domain: $x \in [-2, 2]$; Range: $y \in \mathbb{R}$

23. Domain: $x \in [-1/3, 1/3]$; Range: $y \in \mathbb{R}$

25. $S_{2,3}$; Domain: $x \in [-2, 2]$; Range: $y \in [0, 3\pi]$

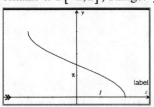

27. $S_{3,5}$; Domain: $x \in \mathbb{R}$; Range: $y \in \left[-\dfrac{5\pi}{2}, \dfrac{5\pi}{2}\right]$

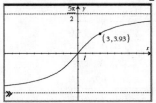

29. S_{1,x^2}; Domain: $x \in \mathbb{R}$; Range: $y \in \mathbb{R}$

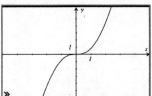

31. True

33. False

35. Domain: $x \in \left[-1,1\right]$; Range: $y \in \mathbb{R}$

Unit 6 – Trigonometry II
6.1 Basic Trigonometric Identities

1. $1/a$

3. $\dfrac{a}{\sqrt{1-a^2}}$

5. $-\sqrt{1-a^2}$

7. The graphs are the same.

9. Identity

11. Not an identity

13. Not an identity

15. Identity

17. $\cos\left(x-\dfrac{\pi}{2}\right)=\cos\left(\dfrac{\pi}{2}-x\right)=\sin x$

19.
$$\tan^2 x - \cot^2 x = \left(\sec^2 x - 1\right) - \left(\csc^2 x - 1\right)$$
$$= \sec^2 x - \csc^2 x$$

21.
$$\frac{\sin\theta\cdot\cos\theta}{1-\sin^2\theta} = \frac{\sin\theta\cdot\cos\theta}{\cos^2\theta}$$
$$= \frac{\sin\theta}{\cos\theta} = \tan\theta$$

23.
$$\frac{1}{\sin x+1}+\frac{1}{1-\sin x}-2 = \frac{1-\sin x+1+\sin x}{1-\sin^2 x}-2$$
$$= \frac{2}{\cos^2 x}-2$$
$$= 2\left(\sec^2 x - 1\right)$$
$$= 2\tan^2 x$$

25.
$$\frac{\sin A\cdot\cos B+\cos A\cdot\sin B}{\cos A\cdot\cos B-\sin A\cdot\sin B}\cdot\frac{\dfrac{1}{\cos A\cdot\cos B}}{\dfrac{1}{\cos A\cdot\cos B}}$$

$$= \frac{\dfrac{\sin A}{\cos A}+\dfrac{\sin B}{\cos B}}{1-\dfrac{\sin A}{\cos A}\cdot\dfrac{\sin B}{\cos B}}$$

$$= \frac{\tan A+\tan B}{1-\tan A\cdot\tan B}$$

27. False

29. False

31. True

33. $\sin\left(x-\dfrac{\pi}{2}\right)=1 \;\Rightarrow\; x-\dfrac{\pi}{2}=\dfrac{\pi}{2} \;\Rightarrow\; x=\pi$

35. $\dfrac{e^x - e^{-x}}{e^x + e^{-x}}\cdot\dfrac{e^x}{e^x} = \dfrac{\left(e^x\right)^2-1}{\left(e^x\right)^2-1} = \dfrac{e^{2x}-1}{e^{2x}+1}$

6.2 Sum & Difference Identities

1. $\dfrac{\sqrt{2}+\sqrt{6}}{4}$

3. $\dfrac{\sqrt{2}-\sqrt{6}}{4}$

5. $\dfrac{\sqrt{2}-\sqrt{6}}{4}$

7. $-2\left(\sqrt{3}+2\right)$

9. $\dfrac{17}{5\cdot\sqrt{13}}$

11. $17/6$

13. $-5\cdot\sqrt{13}$

15. $\cos 120° = -\dfrac{1}{2}$

17. $\sin\left(\dfrac{12\pi}{35}\right)$

19.
$$\sin\left(\dfrac{3\pi}{2}+x\right)=\sin\left(\dfrac{3\pi}{2}\right)\cos x+\cos\left(\dfrac{3\pi}{2}\right)\sin x$$
$$= (-1)\cos x+(0)\sin x$$
$$= -\cos x$$

21. $63/65$

23. False

25. False

27. True

29. A.
$$\sin\left(A-B\right)=\sin A\cdot\cos B-\sin B\cdot\cos A$$
$$= \frac{5}{13}\cdot\frac{3}{5}-\frac{4}{5}\cdot\frac{12}{13} \;=\; -\frac{33}{65}$$
　　B.　$0° < A < B < 90°$
　　C.　Quadrant IV

31. Quadrant I

33. $x=\dfrac{11\pi}{6},\dfrac{\pi}{2}$

35. $\sin C$

6.3 Double Angle &Power-Reducing …

1. $-\dfrac{\sqrt{2+\sqrt{3}}}{2}$

3. $+\dfrac{2}{\sqrt{2-\sqrt{2}}}$

5. $-\dfrac{\sqrt{2-\sqrt{2}}}{2}$

7. $12/5$

9. $7/25$

11. $3/\sqrt{10}$

13. $-\sqrt{3}/2$

15. $\sin\left(\dfrac{2\pi}{9}\right)$

17. $\dfrac{1}{1-2a^2}$

19.
$$1-\cos(2x)\sec^2 x = 1-\left(2\cos^2 x-1\right)\cdot\sec^2 x$$
$$= 1-2+\sec^2 x$$
$$= \sec^2 x-1$$
$$= \tan^2 x$$

21.
$$\frac{\cos(2m)}{1-\sin(2m)} = \frac{\cos(2m)}{1-\sin(2m)}\cdot\frac{\left(1+\sin(2m)\right)}{\left(1+\sin(2m)\right)}$$
$$= \frac{\cos(2m)\cdot\left(1+\sin(2m)\right)}{\cos^2(2m)}$$
$$= \frac{1+\sin(2m)}{\cos(2m)}$$

23. $x\in(0,\pi) \Rightarrow \dfrac{x}{2}\in\left(0,\dfrac{\pi}{2}\right)$, so all

trigonometric functions have positive values.
$$\cot\left(\frac{x}{2}\right) = +\sqrt{\frac{1+\cos x}{1-\cos x}}$$
$$= \sqrt{\frac{1+\cos x}{1-\cos x}\cdot\frac{1+\cos x}{1+\cos x}}$$
$$= \sqrt{\frac{\left(1+\cos x\right)^2}{\sin^2 x}}$$
$$= \frac{\left(1+\cos x\right)}{\sin x}$$

25. $\dfrac{1}{2}(\sin x+\cos x)(2-\sin 2x) =$
$$= \frac{1}{2}(\sin x+\cos x)(2-2\sin x\cos x)$$
$$= (\sin x+\cos x)(1-\sin x\cos x)$$
$$= \sin x-\sin x\cdot\cos^2 x+\cos x-\sin^2 x\cdot\cos x$$
$$= \sin x\cdot\left(1-\cos^2 x\right)+\cos x\cdot\left(1-\sin^2 x\right)$$
$$= \sin x\cdot\sin^2 x+\cos x\cdot\cos^2 x$$
$$= \sin^3 x+\cos^3 x$$

27. $-\sqrt{\dfrac{3-\sqrt{5}}{6}}$

29. True

31. False

33. $+\sqrt{\dfrac{12\pm\sqrt{119}}{24}}$

35.
$$\cos(2\theta) = \cos(\theta+\theta)$$
$$= \cos\theta\cdot\cos\theta-\sin\theta\cdot\sin\theta$$
$$= \cos^2\theta-\sin^2\theta$$
$$\tan(2\theta) = \tan(\theta+\theta)$$
$$= \frac{\tan\theta+\tan\theta}{1-\tan\theta\cdot\tan\theta}$$
$$= \frac{2\tan\theta}{1-\tan^2\theta}$$

37. A. $\sqrt{\dfrac{1-\cos 120°}{\cos 120°+1}} = \sqrt{\dfrac{1-(-1/2)}{(-1/2)+1}} = \sqrt{\dfrac{3/2}{1/2}} = \sqrt{3}$

B. $\sqrt{\dfrac{1-\cos 120°}{\cos 120°+1}} = \tan\left(\dfrac{1}{2}\cdot 120°\right) = \tan 60° = \sqrt{3}$

39.
$$\cos(2x) = \cos x$$
$$2\cos^2 x-1 = \cos x$$
$$2\cos^2 x-\cos x-1 = 0$$
$$(2\cos x+1)(\cos x-1) = 0$$
$$\cos x = \{-1/2 \text{ or } 1\}$$
$$x = \begin{cases} 2\pi/3+2\pi k \\ 4\pi/3+2\pi k \\ 0+2\pi k \end{cases}, k\in\mathbb{Z}$$

41.
$$4\sin x\cdot\cos x = 1$$
$$\sin(2x) = 1/2$$
$$2x = \left\{\frac{\pi}{6},\frac{5\pi}{6},\frac{13\pi}{6},\frac{17\pi}{6}\right\}$$
$$x = \left\{\frac{\pi}{12},\frac{5\pi}{12},\frac{13\pi}{12},\frac{17\pi}{12}\right\}$$

43. Iman is correct. Because $n\in\mathbb{Z}$,
$$\left(\cos^2\left(\frac{\pi n}{2}\right)-\sin^2\left(\frac{\pi n}{2}\right)\right)^2 = \left(\cos\left(2\cdot\frac{\pi n}{2}\right)\right)^2$$
$$= \left(\cos(\pi n)\right)^2$$
$$= (\pm 1)^2$$
$$= 1$$

6.4 Many Familiar Formulae & …

1. It is a right triangle, so the given information uniquely defines a single triangle.

3. 21.965

5. $b = 7.151$, $c = 9.334$, $m\angle A = 40°$

7. $c = 187.344$, $m\angle A = 27.354°$, $m\angle B = 106.809°$

9. $m\angle B = 32.464°$, $m\angle C = 106.536°$, $c = 16.073$

11. $a = 9\sqrt{3}$, $m\angle A = 60°$, $m\angle C = 90°$

13. Same answers as #5.

15. Same answers as #7.
17. False
19. True
21. False
23. $8.608\,\text{cm}^2$
25. 143.005 ft
27. 61.432 km
29. Since the sine of an angle equals the sine of its supplement, it is questionable as to which of the two angles corresponds to the given sine.
31. It turns into the Pythagorean theorem since $\cos 90° = 1$.
33. A. Two angles are given, defining a unique triangle.
    B. It is a linear system with the variables being side lengths; there can only be one solution for sides.
35. A. It is a SSA situation.
    B. From the Law of Sines, $A = 53.604°$, whose supplement is too large to be an angle in this triangle.
    C. The Law of Cosines results in a quadratic in c with one positive and one negative solution; since c represents a side length and is therefore positive, there is only one solution.
    D. It is a system of linear equations and has one solution.

6.5 Sums & Differences of Sinusoids

1. $\sqrt{2}\sin\left(x + \dfrac{\pi}{4}\right)$

3. $4\sqrt{2}\sin\left(2x + \dfrac{5\pi}{12}\right)$

5. Envelopes: $y = \pm 2\cos\left(\dfrac{1}{2}x\right)$

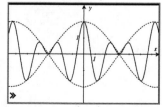

7. Envelopes: $y = \pm 2\sin\left(\dfrac{1}{2}x\right)$

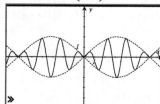

9. True
11. True
13. True

15. $\sqrt{21}\sin\left(2x + 0.333\right)$
17. Each addend function can be converted into the other by a simple horizontal translation.
19.
$$\sin A - \sin B = \sin\left(A\right) + \sin\left(-B\right)$$
$$= 2\sin\left(\dfrac{A + \left(-B\right)}{2}\right)\cos\left(\dfrac{A - \left(-B\right)}{2}\right)$$
$$= 2\cos\left(\dfrac{A + B}{2}\right)\sin\left(\dfrac{A - B}{2}\right)$$

21. Envelopes: $y = \pm 6\sin\left(\dfrac{3}{2}x\right)$

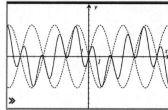

23. Envelopes: $y = \pm 10\cos\left(-\dfrac{1}{2}x - \dfrac{\pi}{4}\right)$

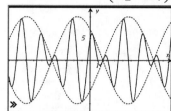

25. Envelopes: $y = \pm 10x\cos\left(-\dfrac{1}{2}x - \dfrac{\pi}{4}\right)$

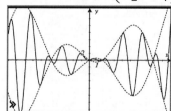

27. $\cos x + \cos\left(2x\right) = 2\cos\left(\dfrac{3x}{2}\right)\cos\left(\dfrac{x}{2}\right) = 0$

Because the outer envelope is $y = 2\cos\left(\dfrac{x}{2}\right)$, the *smallest* magnitude zeros must come from the "inner" function.
$$\cos\left(\dfrac{3x}{2}\right) = 0 \Rightarrow \dfrac{3x}{2} = \pm\dfrac{\pi}{2}$$
$$\Rightarrow x = \pm\dfrac{\pi}{3}$$

<u>Unit 7 – Parametric Functions & Vectors</u>
7.1 A Review of Parametric Functions

1. Parabola

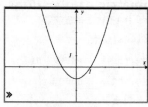

3. Line

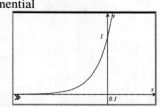

5. Circle

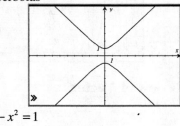

7. Exponential

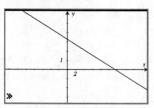

11. It relates only x and y; does not work for parametrically defined functions.

13. Example: $\begin{cases} x = t \\ y = \dfrac{42t - 9}{137} \end{cases}$

15. Example: $\begin{cases} x = 2\tan t \\ y = 2\sec t \end{cases}$

17. True

19. False

21. Hyperbolas

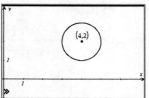

$y^2 - x^2 = 1$

23. Example: $\begin{cases} x = 3\cos t \\ y = 3\sin t \end{cases}$

25. $\begin{cases} x = t + 2 \\ y = \dfrac{t}{3} + 4 \end{cases}$, $t \in [-9, 0]$

27. Example: $\begin{cases} x = t^2 \\ y = t \end{cases}$

7.2 Remaining in Control

1.   A.   Starts at $(-1, 0)$, graphs clockwise

     B.   Starts at $(0, 1)$, graphs clockwise

     C.   Starts at $(0, -1)$, graphs counterclockwise.

3. π

5.   A.

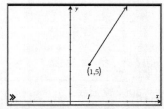

     B.   Oscillates infinitely from ∞ to $(1, 5)$ back to ∞.

7.   A.

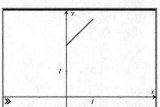

     B.   Oscillates along the line segment with endpoints $(0, 2)$ and $(1, 3)$.

9.

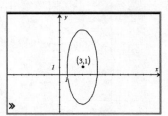

11.

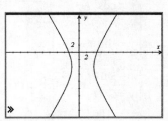

13. Example: $\begin{cases} x = -2 + 6t \\ y = 5 + 2t \end{cases}$, $t \in [0, 1]$

15. Example: $\begin{cases} x = 4 \\ y = t \end{cases}$

17. Example: $\begin{cases} x = \cos t + 1 \\ y = \sin t + 7 \end{cases}$

19. Example: $\begin{cases} x = 3\cos t \\ y = 8\sin t \end{cases}$

21.

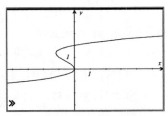

23.

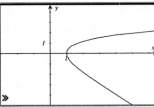

25. $\begin{cases} x = \cos 2\pi t \\ y = \sin 2\pi t \end{cases}$

27. False

29. False

31. A. $\begin{cases} x = -\sin t \\ y = \cos t \end{cases}$

    B. $\begin{cases} x = -\sin t \\ y = -\cos t \end{cases}$, $t \in [0, \pi]$ or

               $\begin{cases} x = -\sin\left(\pi \sin^2 t\right) \\ y = -\cos\left(\pi \sin^2 t\right) \end{cases}$

    C. $\begin{cases} x = \sin t \\ y = -\cos t \end{cases}$, $t \in \left[0, \dfrac{3\pi}{2}\right]$ or

               $\begin{cases} x = \sin\left(\dfrac{3\pi}{2}\sin^2 t\right) \\ y = -\cos\left(\dfrac{3\pi}{2}\sin^2 t\right) \end{cases}$

    D. $\begin{cases} x = \cos\left(t - \dfrac{2\pi}{3}\right) \\ y = \sin\left(t - \dfrac{2\pi}{3}\right) \end{cases}$

    E. $\begin{cases} x = \sin t \\ y = \cos t \end{cases}$, $t \in \left[0, \dfrac{\pi}{2}\right]$ or

               $\begin{cases} x = \sin\left(\dfrac{\pi}{2}\sin^2 t\right). \\ y = \cos\left(\dfrac{\pi}{2}\sin^2 t\right) \end{cases}$

33. $\begin{cases} x = 3 - 2t \\ y = 2 - 7t \end{cases}$, $t \in \left[-\dfrac{1}{7}, 1\right]$

7.3 Parametric Applications

1. $t = 0.113$, $t = 2.737$

3. $21\cos(0.91)t = 44 \Rightarrow t = 3.413\,\text{sec}$

   $y_{jill}(3.413) \approx 1.194$ m

5. No, she will need to move.

7. False

9. True

11. True

13. A. 1.851 sec

    B. 9.201 m

    C. $t = 1.040\,\text{sec}$, $t = 2.511\,\text{sec}$

    D. Yes, at $t = 1.286\,\text{sec}$ or $2.233\,\text{sec}$

15. A. 32.514 m

    B. $t = 0$

    C. 83.510 m

17. $y = 2\sin(20°)\left(\dfrac{x - 2\sin(20°)}{2\cos(20°)}\right) + 2\cos(20°)$

19. A. π

    B.

    $\begin{cases} x = -15\sin\left(\dfrac{\pi}{8}t\right) + 32\sin\left(\dfrac{\pi}{10}t\right) \\ y = -15\cos\left(\dfrac{\pi}{8}t\right) - 32\cos\left(\dfrac{\pi}{10}t\right) + 50 \end{cases}$

    C. 3 ft.

    D. 80 sec.

    E.

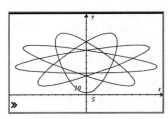

    F.

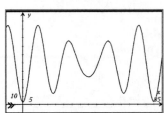

7.4 A Review of Vectors

1. $\langle 1, 7 \rangle$

3. $[12, \pi]$

5. $\left[\sqrt{13}, 0.588\right]$

7. $\pm\dfrac{1}{\sqrt{13}}\langle 2, 3 \rangle$

9. $\vec{v}-\vec{u}$

11. $\vec{v}-\dfrac{1}{2}\vec{u}$

13. $\langle x,y\rangle=\langle -1,-2\rangle+t\langle 5,9\rangle$

15. Example: For number 13, $\begin{cases} x=-1+5t \\ y=-2+9t \end{cases}$

17. True

19. True

21. False

23. A. $\langle t-4,7t-41\rangle$

    B. $\sqrt{(t-4)^2+(7t-41)^2}$

    C. $t=5.82$

    D. 1.838

    E. $\langle 1.82,-0,26\rangle$

25. 86.834 lbs at $-0.323°$

27. 11.775 mph; $4.872°$ East of South.

7.5 Dot Products & Work

1. 23

3. 0

5. 25

7. $25.56°$

9. $122.312°$

11. The pairs in B are normal

13. None

15. $\dfrac{17}{26}\langle 5,1\rangle$

17. 253.571 ft-lbs

19. 13

21. True

23. True

25. True

7.6 Cross Products

1.

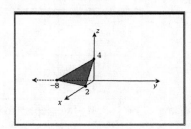

3.

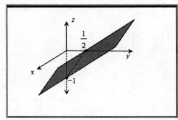

5. $\langle -23,7,6\rangle$

7. $\langle 5,-4,14\rangle$

11. $\dfrac{x}{3}+\dfrac{y}{5}-\dfrac{z}{9}=1$

13. $5x-y+4z=34$

15. True

17. True

19. False

21. $122.462°$

23. $a=3$, $b=2$

25. $21x-11(y-5)+-29(z-4)=0$

27. $\langle x-a,y-b,z-c\rangle\cdot\langle n_1,n_2,n_3\rangle=0$, where (a,b,c) is any point on the plane and $\langle n_1,n_2,n_3\rangle$ is any vector normal to the plane.

Unit 8 – Extensions of Trigonometry
8.1 A New Way to Define Functions

1. $\left[3\sqrt{2},\dfrac{3\pi}{4}\right]$, $\left[-3\sqrt{2},-\dfrac{\pi}{4}\right]$, $\left[3\sqrt{2},-\dfrac{5\pi}{4}\right]$

3. $\left[2,\dfrac{\pi}{3}\right]$

5. $\left[\sqrt{5},3.605\right]$

7. $(-0.347,1.969)$

9. $(-8,0)$

11. $r=\dfrac{3}{\cos\theta+\sin\theta}$

13. $r=3$

15. $r^2=\dfrac{36}{9\cos^2\theta+4\sin^2\theta}$

17. 3.718

19. $x^2+y^2-8y=0$; circle

21. $x^2+y^2-2x+4y=1$; circle

23. $x=5$; vertical line

25. $x^2-y^2=1$; hyperbola

27. False

29. True

31. A. $x^2+y^2-ax=0$

    B. $a^2/8$

33. Example: $r^2=\cos(2\theta)$

8.2 Polar Graphing I

1. $x^2+y^2-5y=0$

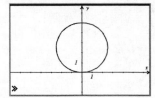

3. $2x - ey = 7$

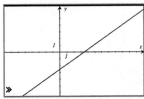

5. $x^2 + y^2 = 25$

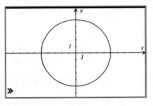

7. $x^2 + y^2 - 2x + 8y = 0$

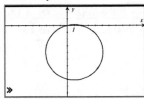

9. Note: This graph is for $\theta \geq 0$.

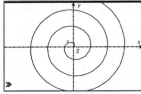

11. Counterclockwise rotation of $\dfrac{\pi}{3}$.

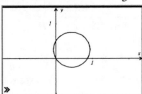

13. Reflection over the y-axis.

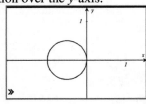

15. $r = 2$
17. $\theta = 2\pi / 3$
19. y-axis
21. None of them
23. False
25. False
27. True
29. $a^2 / \sqrt{2}$
31.

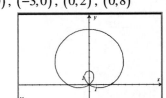

8.3 Polar Graphing II

1. Range: $r \in [-2, 8]$; Intercepts: the pole,
   $(3, 0)$, $(-3, 0)$, $(0, 2)$, $(0, 8)$

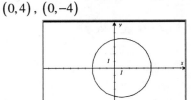

3. Range: $r \in [3, 5]$; Intercepts: $(5, 0)$, $(-3, 0)$,
   $(0, 4)$, $(0, -4)$

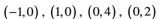

5. Range: $r \in [-4, 2]$; Intercepts: the pole,
   $(-1, 0)$, $(1, 0)$, $(0, 4)$, $(0, 2)$

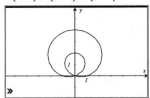

7. Range: $r \in [0, 4]$; Intercepts: the pole,
   $(-4, 0)$, $(0, -2)$, $(0, 2)$

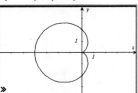

9. Range: $r \in [-3, 3]$; Intercepts: the pole,
   $(-3, 0)$, $(3, 0)$, $(0, 3)$, $(0, -3)$

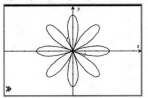

11. Range: $r \in [-4,4]$; Intercepts: the pole and
$(4,0)$

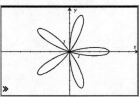

13. Range: $r \in [0,2]$; Intercepts: the pole,
$(0,-1)$, $(0,1)$, $(2,0)$

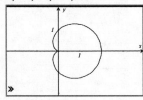

15. $r = 3\cos(4\theta)$

17. $r = 2 + \cos\theta$

19. $r = 3 - 3\cos\theta$

21. $r = 2 + \sin\left(\theta - \dfrac{\pi}{4}\right)$;   Axis: $y = -x$

23. $r = -\cos(3\theta)$

25. $\left(3, \pm\dfrac{\pi}{3} + 2\pi n\right)$, where $n \in \mathbb{Z}$.

27. There are four points of intersection:
$(\pm 0.296, -0.5)$, $(\pm 2.532, -0.5)$

29. False; it has 16 petals

31. True

33. $|A| > |B|$

35. $|B| - |A|$

37. $r = A + B\sin\theta$
$\Rightarrow r = A + B\sin 0 \ or \ r = A + B\sin\pi$
$\Rightarrow r = A \ \Rightarrow \ x = \pm A$

39. When D is odd, there are D distinct petals,
When D is even, there are $2D$ distinct petals.

41.

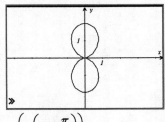

43. $r = 3\sin\left(4\left(\theta - \dfrac{\pi}{8}\right)\right)$

8.4 The Polar Form of Complex Numbers

1. $4\sqrt{2}\,\text{cis}\left(\dfrac{\pi}{4}\right)$

3. $2\,\text{cis}\left(\dfrac{5\pi}{6}\right)$

5. $2\,\text{cis}\,0$

7. $5\left(-\dfrac{1}{2} + \dfrac{\sqrt{3}}{2}i\right)$

9. $-2.053 + 2.378i$

11. 4

13. $4 - 4i\sqrt{3}$

15. $1/16$

17. $\left(8\,\text{cis}\left(\dfrac{\pi}{2}\right)\right) = 8i$

19. $5\,\text{cis}\left(\dfrac{\pi}{3}\right)$, $5\,\text{cis}(\pi)$, $5\,\text{cis}\left(\dfrac{5\pi}{3}\right)$

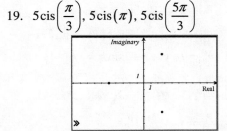

21. $z = 2^{1/6}\,\text{cis}\left(-\dfrac{\pi}{36} + \dfrac{\pi}{3}n\right)$, $n \in \mathbb{Z}$, so
$$z = \begin{cases} \sqrt[6]{2}\,\text{cis}(-\pi/36), \sqrt[6]{2}\,\text{cis}(11\pi/36), \\ \sqrt[6]{2}\,\text{cis}(23\pi/36), \sqrt[6]{2}\,\text{cis}(35\pi/36), \\ \sqrt[6]{2}\,\text{cis}(47\pi/36), \sqrt[6]{2}\,\text{cis}(59\pi/36) \end{cases}$$

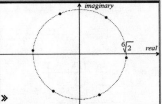

23. True

25. False

27. $\dfrac{z_1}{z_2} = \dfrac{r_1 \cdot \text{cis}\,\theta_1}{r_2 \cdot \text{cis}\,\theta_2} = \dfrac{r_1}{r_2} \cdot \dfrac{\cos\theta_1 + i\cdot\sin\theta_1}{\cos\theta_2 + i\cdot\sin\theta_2}$

$= \dfrac{r_1}{r_2} \cdot \dfrac{\cos\theta_1 + i\cdot\sin\theta_1}{\cos\theta_2 + i\cdot\sin\theta_2} \cdot \dfrac{\cos\theta_2 - i\cdot\sin\theta_2}{\cos\theta_2 - i\cdot\sin\theta_2}$

$= \dfrac{r_1}{r_2} \cdot \dfrac{\left[\begin{array}{c}\cos\theta_1\cos\theta_2 + \sin\theta_1\sin\theta_2 \\ + i(\sin\theta_1\cos\theta_2 - \cos\theta_1\sin\theta_2)\end{array}\right]}{\cos^2\theta_2 + \sin^2\theta_2}$

$= \dfrac{r_1}{r_2} \cdot \dfrac{\cos(\theta_1 - \theta_2) + i\cdot\sin(\theta_1 - \theta_2)}{1}$

$= \dfrac{r_1}{r_2} \cdot \text{cis}(\theta_1 - \theta_2)$

29. The roots of any complex number always lie on the tips of some rose curve.

8.5 A New Trigonometry

1. 0

3. $5/3$

5. $\dfrac{x^2-1}{x^2+1}$

7. True

9. False

11. True

13. $h(x)=e^x$

15. A. $\cos x+i\sin x=\dfrac{e^{ix}+e^{ix}}{2}+i\cdot\dfrac{e^{ix}-e^{ix}}{2i}=e^{ix}$

B.
$$e^{-ix}=e^{i\cdot(-x)}=\cos(-x)+i\sin(-x)$$
$$=\cos(x)-i\sin(x)$$

17. $1+e^{i\pi}=1+\cos\pi+i\sin\pi=1+(-1)=0$

19. $\cosh x$, $i\cdot\sinh x$

21. $\sinh(2x)=2(\sinh x)(\cosh x)$

23. $\tanh(2x)=\dfrac{2\tanh x}{1+\tanh^2 x}$

8.6 Graphs & Inverses of Hyperbolic Functions

1.

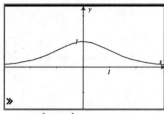

3. $\tanh^{-1}x=\dfrac{1}{2}\ln\left(\dfrac{x+1}{1-x}\right)$

5. True

7. True

9. True

11. $x=0$